Lecture Notes in Geoinformation and Cartography

Series editors

William Cartwright, Melbourne, Australia
Georg Gartner, Wien, Austria
Liqiu Meng, München, Germany
Michael P. Peterson, Omaha, USA

The Lecture Notes in Geoinformation and Cartography series provides a contemporary view of current research and development in Geoinformation and Cartography, including GIS and Geographic Information Science. Publications with associated electronic media examine areas of development and current technology. Editors from multiple continents, in association with national and international organizations and societies bring together the most comprehensive forum for Geoinformation and Cartography.

The scope of Lecture Notes in Geoinformation and Cartography spans the range of interdisciplinary topics in a variety of research and application fields. The type of material published traditionally includes:

- proceedings that are peer-reviewed and published in association with a conference;
- post-proceedings consisting of thoroughly revised final papers; and
- research monographs that may be based on individual research projects.

The Lecture Notes in Geoinformation and Cartography series also includes various other publications, including:

- tutorials or collections of lectures for advanced courses;
- contemporary surveys that offer an objective summary of a current topic of interest; and
- emerging areas of research directed at a broad community of practitioners.

More information about this series at http://www.springer.com/series/7418

Peter Kiefer · Haosheng Huang
Nico Van de Weghe · Martin Raubal
Editors

Progress in Location Based Services 2018

Springer

Editors
Peter Kiefer
Institute of Cartography
    and Geoinformation
ETH Zurich
Zürich
Switzerland

Nico Van de Weghe
Cartography & GIS,
    Department of Geography
Ghent University
Ghent
Belgium

Haosheng Huang
Department of Geography, Geographic
    Information Science
University of Zurich
Zürich
Switzerland

Martin Raubal
Institute of Cartography
    and Geoinformation
ETH Zurich
Zürich
Switzerland

ISSN 1863-2246   ISSN 1863-2351   (electronic)
Lecture Notes in Geoinformation and Cartography
ISBN 978-3-319-89076-0   ISBN 978-3-319-71470-7   (eBook)
https://doi.org/10.1007/978-3-319-71470-7

Printed on acid-free paper

This Springer imprint is published by Springer Nature
The registered company is Springer International Publishing AG
The registered company address is: Gewerbestrasse 11, 6330 Cham, Switzerland

Preface

Over the last twenty years, location based services (LBS) have become increasingly popular and have expanded into many areas of our daily lives. The success of LBS has been facilitated, driven, and accompanied by research activities of an active and growing community. Research on LBS still continues to improve and shape the future of LBS, driven by activities on topics including indoor and outdoor positioning, mapping, privacy, novel user interfaces, big data, smart environments, and citizen participation—just to name a few.

The LBS conference has become one of the main international research venues focusing on LBS. The 2018 edition of this conference is the first to be hosted at the Swiss Federal Institute of Technology (ETH Zurich, Switzerland), after 13 successful predecessor events in Vienna (2002, 2004, 2005), Hong Kong (2007), Salzburg (2008), Nottingham (2009), Guangzhou (2010), Vienna (2011), Munich (2012), Shanghai (2013), Vienna (2014), Augsburg (2015), and Vienna (2016).

This book contains sixteen full papers which have been accepted for LBS 2018 after a rigorous peer-reviewing process with a 42% acceptance rate. It is structured into four equal parts, covering a variety of ongoing and timely research topics in the fields: positioning, mapping, landmarks and mobility, location based social media, and citizen participation.

We would like to thank all authors for their excellent work and all reviewers for their critical and constructive comments. We hope you will find these papers interesting and relevant for your own work and look forward to your participation in one of the future LBS conferences.

Zürich, Switzerland   Peter Kiefer
Zürich, Switzerland   Haosheng Huang
Ghent, Belgium   Nico Van de Weghe
Zürich, Switzerland   Martin Raubal
November 2017

Reviewers

We would like to thank all the following experts who have helped to review the papers published in this book.

Gennady Andrienko, Fraunhofer Institute IAIS & City University London, Germany/UK
Masatoshi Arikawa, University of Tokyo, Japan
Thierry Badard, Laval University, Canada
Kate Beard-Tisdale, University of Maine, USA
Pia Bereuter, University of Applied Sciences and Arts Northwestern Switzerland FHNW, Switzerland
Susanne Bleisch, University of Applied Sciences and Arts Northwestern Switzerland FHNW, Switzerland
William Cartwright, RMIT University, Australia
Christophe Claramunt, Naval Academy Research Institute, France
Keith C. Clarke, University of California at Santa Barbara, USA
Weihua Dong, Beijing Normal University, China
Matt Duckham, RMIT University, Australia
Sara I. Fabrikant, University of Zurich, Switzerland
David Forrest, University of Glasgow, UK
Peter Fröhlich, AIT Austrian Institute of Technology, Austria
Georg Gartner, TU Vienna, Austria
Ioannis Giannopoulos, ETH Zurich, Switzerland
Gyözö Gidofalvi, KTH Royal Institute of Technology Stockholm, Sweden
Amy Griffin, UNSW Canberra, Australia
Muki Haklay, University College London, UK
Frédéric Hubert, Laval University, Canada
Mike Jackson, University of Nottingham, UK
Bin Jiang, University of Gävle, Sweden
David Jonietz, ETH Zurich, Switzerland
Hassan Karimi, University of Pittsburgh, USA
Farid Karimipour, University of Tehran, Iran

Contents

Part I
Positioning

Locations Selection for Periodic Radio Map Update in WiFi Fingerprinting

Germán M. Mendoza-Silva, Joaquín Torres-Sospedra and Joaquín Huerta

Abstract The construction and update of a radio map are usually referred as the main drawbacks of WiFi fingerprinting, a very popular method in indoor localization research. For radio map update, some studies suggest taking new measurements at some random locations, usually from the ones used in the radio map construction. In this paper, we argue that the locations should not be random, and propose how to determine them. Given the set locations where the measurements used for the initial radio map construction were taken, a subset of locations for the update measurements is chosen through optimization so that the remaining locations found in the initial measurements are best approximated through regression. The regression method is Support Vector Regression (SVR) and the optimization is achieved using a genetic algorithm approach. We tested our approach using a database of WiFi measurements collected at a relatively dense set of locations during ten months in a university library setting. The experiments results show that, if no dramatic event occurs (e.g., relevant WiFi networks are changed), our approach outperforms other strategies for determining the collection locations for periodic updates. We also present a clear guide on how to conduct the radio map updates.

Keywords Wifi fingerprinting · Radio map update · Regression · Optimization
Genetic algorithm

G. M. Mendoza-Silva (✉) · J. Torres-Sospedra · J. Huerta
Institute of New Imaging Technologies, Universitat Jaume I, Avda. Vicente Sos Baynat S/N,
Castellón, Spain
e-mail: gmendoza@uji.es

J. Torres-Sospedra
e-mail: jtorres@uji.es

J. Huerta
e-mail: huerta@uji.es

© Springer International Publishing AG 2018
P. Kiefer et al. (eds.), *Progress in Location Based Services 2018*, Lecture Notes
in Geoinformation and Cartography, https://doi.org/10.1007/978-3-319-71470-7_1

1 Introduction

As location-based services have grown in importance during recent years, the indoor positioning has increasingly drawn attention from the research community. The WiFi fingerprinting has been a very popular indoor positioning method for this community. Reasons for its popularity include a large number of WiFi access points (AP) already deployed in many environments, the generalized usage of WiFi-enabled smartphones, and a positioning accuracy that is acceptable for many applications (He and Chan 2016; Yiu et al. 2017). This method, however, has two known drawbacks: the WiFi measurements radio map construction and update.

The radio map construction and update for WiFi fingerprinting usually involve a person, or dedicated receiver, that collects WiFi measurements at some known locations. Thus, the collection process has a cost, either in the time that a paid person employs, or in the cost of deploying and maintaining receivers. The reduction of that cost is referred as mapping, calibration or radio map construction/update effort reduction.

It is acknowledged that, at least to some extent, the larger the number of measurement locations in the target area, the better the accuracy of the WiFi fingerprinting is Kanaris et al. (2016), Wang et al. (2016), Hernández et al. (2017), Yiu et al. (2017), but also the more costly the collection process is. To address this issue, methods that require only a few collection locations have been proposed (Alonazi et al. 2015). Such methods involve regression (interpolation/extrapolation) approaches or turning to collaborative or crowd-sourced approaches. If the collected data reliability is a hard concern, the one option is collecting measurements at all relevant locations. If data reliability is soft concern, another option is to collect measurements only at some location and then estimate measurement values for the remaining locations using a regression approach.

The studies proposing regression approaches generally show that the estimations made by their methods can be used instead of some of the actual measurements without significantly harming the localization accuracy provided by an Indoor Positioning System (IPS). These studies usually specify elimination procedures to drop some of the original locations in order to test their methods. However, those elimination procedures are not to be understood as suggested strategies for determining collection locations. The random locations distribution is a common approach (Ali et al. 2017), despite the locations distribution is very important (Li et al. 2014) for radio map construction. It is also acknowledged that the radio map needs periodic updates so that the positioning method can be robust to changes in the target environment and in the relevant APs (Wang et al. 2016; Hossain and Soh 2015).

The importance of the collection locations is intuitive and has been formally acknowledged in other subjects for other phenomena. Specifically, several papers have addressed the optimal (or quasi-optimal) placement of sensors that best measure a given phenomenon (Rowaihy et al. 2007; Joshi and Boyd 2009). A set of WiFi measurements collected at known locations by a person can be viewed as measurements of a set of sensors. Therefore, choosing the best locations for an individual to collect the WiFi measurements can be thought as optimizing the placement of a set of sensors.

This paper presents a novel approach for determining the collection locations for periodic WiFi radio map updates. The approach requires initial measurements, taken at a relatively dense set of known locations. The initial measurements are used to determine a set of locations that establishes a compromise between a small set's size and its goodness for estimating the Received Signal Strength (RSS) values at the remaining locations through a regression method. This paper suggests to find such set using a genetic algorithm optimization approach with a specific fitness function. The found set of locations, called the solution set, is proposed to be used as collection locations for the radio map periodic updates.

The proposed approach was tested using SVR as a regression method and a WiFi RSS database collected during ten months at one floor of a university library. The database contains measurements for training and test purposes. The training measurements for the first month were used to determine the solution set. The goodness of the solution set for selecting the measurement collection locations for the periodic radio map updates was explored across the following nine month in terms of: (1) RSS difference between the measurements and the RSS estimations provided by a regression fitted for the solution set, and (2) the effects of using the above RSS estimations for radio map update on the accuracy of a fingerprinting-based IPS, considering the test sets collected at each month. The experiments' results have shown the suitability of using our approach for determining the locations for periodic radio map updates in the tested environment.

In summary, in this paper we propose an alternative to common strategies for locations selection for WiFi radio map update and we experimentally show its benefits. While following those goals, we:

1. Present some drawbacks of the previous common strategies.
2. Describe how to determine a set of locations (solution set) where measurements should be taken in order to obtain fine RSS estimations for the remaining locations through regression.
3. Briefly describe how the proposal can be used to find challenging sets of locations to test regression approaches for WiFi fingerprinting.
4. Experimentally show how to use the estimations obtained from the solution set to update a WiFi radio map.

The remainder of the paper is organized as follows: Sect. 2 provides an overview of fingerprinting calibration efforts reduction, focusing mainly in regression-based approaches. Section 3 presents our proposal for measurement locations determination for WiFi radio map update. Section 4 provides the experimental testing of our proposals. Finally, Sect. 5 summarizes the ideas presented in this paper and proposes its continuation lines.

2 WiFi Radio Map Construction and Update

WiFi fingerprinting is performed in two phases: the offline training phase and the online (operational or query) localization phase (He and Chan 2016; Yiu et al. 2017). In the training phase, WiFi fingerprints are collected in the target area. A WiFi fingerprint is a vector of RSS values of the detected APs measured at a given time. Each training fingerprint is usually labeled with the location at which it was collected. The fingerprints are stored in a training database, which is also called radio map. In the localization phase, an IPS uses the training database to estimate location labels for new, unlabeled fingerprints.

Radio maps with measurements collected at relatively dense sets of locations provide higher positioning accuracies than those with measurement collected at sparse locations (Kanaris et al. 2016; Wang et al. 2016; Hernández et al. 2017; Yiu et al. 2017). Additionally, periodic radio map updates are needed because WiFi signals are prone to changes, due to either changes in the environment or in relevant APs (including reallocation, replacement and transmission power reconfiguration) (Hossain and Soh 2015; Wang et al. 2016). The effort reduction on radio map construction and the methods robustness to environment's changes has been targeted by WiFi fingerprinting researchers for over 10 years, with many of the attempts included in reviews like Hossain and Soh (2015), Pei et al. (2016), Wang et al. (2016). Some examples of the attempts are found in Yang et al. (2013), Alonazi et al. (2015), Majeed et al. (2016), Gu et al. (2016a). The study of Yang et al. (2013), instead of directly using the RSS values, used order relations between AP's RSS values. The authors in Alonazi et al. (2015) collected WiFi measurements at a few reference points (RPs) located at the ends of corridors and later enriched the radio maps with user-supplied new RSS values. In Majeed et al. (2016), the authors combined a small calibration set, the coordinates of all target locations and several simultaneous operational RSS measurements using semi-supervised alignment of manifolds to estimate the operational measurements' locations. Gu et al. (2016a) used the AP intensity order as similarity score to deal with the changes in relevant APs and mobile device diversity, and tested its approach with a database collected during 6 months.

The above solutions for effort reduction differentiates on whether the measurements are collected by (1) collaborative/crowd-sourced means or by (2) a dedicated collector. Each approach have its own benefits and drawbacks (Pei et al. 2016). In the first approach, the cost is almost negligible, but quality and completeness are concerns. In the second approach, the cost is reduced by making collection at only a few locations and then estimating (mainly performing a regression) the RSS values at the remaining locations.

The collaborative/crowd-sourced approaches include explicit or implicit user collaboration (He and Chan 2016; Wang et al. 2016; Hossain and Soh 2015). In the explicit case, the user is required to label all fingerprints, or at least a subset of them, with the location where they are taken. When there are unlabeled fingerprints, their labels are estimated using techniques that consider additional information, such as readings from other sensors (e.g., using pedestrian dead reckoning (PDR) Xiao et al.

2015) or environment knowledge. The environment knowledge may, for example, indicate the likely corresponding path segments or the intrinsic relations between neighboring fingerprints using models like Markov-chain (Lin et al. 2016). Also, floor plans and APs locations knowledge can be used to generate each AP radio map using propagation models (Ali et al. 2017). In the implicit case, location hints are opportunistically used to label WiFi measurements with the location without the user interaction. The location hints may come from other sensors, like a GPS sensor, or through estimations such as those used for unlabeled fingerprints in the case of the explicit user collaboration. The collaborative/crowd-sourced approaches are also used for radio map update. These approaches have a well known challenge: the labels quality (Wang et al. 2016).

The approaches that do not rely on collaborative/crowd-sourced contributions try to reduce the amount of locations required for constructing the initial radio map. Fingerprints are collected at a small amount of locations and the RSS values at the remaining target locations are estimated using regression (interpolation and extrapolation). The following subsection deepens on this subject.

2.1 Collection Effort Reduction for Fingerprinting Using Regression Approaches

Regression for RSS radio map enrichment is applied as follows. An initial, small set of locations with known coordinates $L_{n \times 2}$ is chosen for the target area. Then, if s measurements are made for each location and m wireless networks are detected in the whole campaign, the initial database is the set $D_{n \times m \times s} = \{r_{ijk}\}$, where r_{ijk} is the RSS value measured at the ith location, for the jth AP, and in the kth location sample. For each wireless network a, the regression method fit a function $f_a(L) = R_a$, with $R_a = \{r_{iak}\}$. Each function f_a is then used to predict RSS values for locations $\hat{L}$. If the points in $\hat{L}$ lie inside the convex hull of L, the estimation is usually called interpolation, and if they lie outside, it is called extrapolation. Extrapolation methods (extrapolation functions) are known to be less accurate, and thus more challenging and less used than the interpolation ones (Talvitie et al. 2015).

The regression methods has been used for reducing the calibration effort in fingerprinting for more than 10 years (Krumm and Platt 2003; Li et al. 2005). Among the methods found in literature are: linear interpolators (Talvitie et al. 2015), radial basis interpolators (Krumm and Platt 2003; Ezpeleta et al. 2015), Gaussian Process regression (Yiu et al. 2017), and Support Vector Regression (SVR) (Hernández et al. 2017). Some studies particularly focused on the spatial relations of measurements and the spatial characteristics of the environment for regression. They included methods widely used in spatial analysis like Inverse Distance Weighting (IDW) and Kriging (Li et al. 2005; Liu et al. 2015; Jan et al. 2015), Voronoi Tessellation (Lee and Han 2012), Sparsity Rank Singular Value Decomposition (SRSVD) (Gu et al. 2016b) and other particular heuristics (Bong and Kim 2012). Studies like Zhu et al. (2014) have also taken into account the time dimension for regression.

In the cited studies, the authors first collect a relatively dense dataset of RSS measurements, and, through elimination strategies, produce new datasets. Their regression methods are then applied to the new datasets in order to obtain estimated RSS measurements for the removed collection locations. The regression goodness is usually evaluated as (1) the difference in RSS values between measurements and estimations and (2) the difference in localization error of some IPS, between using dataset with a high percentage of removed points and the original dataset for training. The elimination strategy is an important factor in the results obtained in such evaluations (Talvitie et al. 2015). The regression performance found in literature varies significantly, from discrete but reasonably results of 50% location reduction (Ezpeleta et al. 2015) to astonishing results of 5% locations reduction (Gu et al. 2016b) with very little RSS or localization error difference.

Most of studies found in literature indicate the percentage of collection locations (with respect to all target locations) required for their regression methods to provide proper localization accuracies. However, they do not mention a methodology for determining the number of collection locations for a given environment (though it has been shown to be very important (Li et al. 2014)), or how to determine where those locations should be. An intuitive approach is to choose the amount of collection points as a function of the target area size and randomly determine their positions in that area. Some studies have used similar approaches.

In Kanaris et al. (2016), the authors proposed an algorithm that suggested a collection's sample size given a small preliminary set of measurements. They suggested the definition of a grid of locations in a target area and randomly choosing locations in the amount determined by the sample size calculation. Specifically, for the case of database update, collecting measurements at random locations in a target area is a common approach (Ali et al. 2017). Indeed, depending on the update frequency, the collaborative, crowd-sourced or opportunistic approaches can be also considered strategies of collecting update measurements at random locations.

The elimination strategies used for evaluating the goodness of regression methods found in the research literature have hinted on possible strategies for determining the locations for training set collection. The work of Krumm and Platt (2003) proposed an elimination strategy consisting in running a k-means clustering algorithm, and selecting only the k locations nearest the k cluster centroids. Other studies have resembled in their proposed elimination strategies the types of collection absences that may happen in regular collection processes, like random isolated absent points, zones with higher or lower percentage of elimination (Ezpeleta et al. 2015) or random blocks of absent points (Talvitie et al. 2015).

The following section presents the approach we propose to determine the set of locations where fingerprints for WiFi radio map update are to be taken.

3 Locations Set Determination for Radio Map Update

As seen in Sect. 2, studies found in literature have hinted possible approaches for choosing the collection locations. These approaches, however, have some drawbacks that are experimentally shown in Sect. 4. It is almost intuitive that neither the number of locations, nor their actual distribution, should be chosen randomly without any restriction. In addition, a uniformly spaced locations distribution may not take into account obstacles influencing the WiFi signals propagation. Therefore, a person with experience in WiFi-based indoor localization generally chooses the amount and distribution of the collection locations. Regardless of this person expertise, the previous task is not a trivial one.

This study harnesses the similitudes between (1) choosing a subset of measurement locations for estimating the values at remaining locations through regression and (2) choosing the placement of sensors for field estimation. The problem of sensor placement, related to sensor selection (activation), is a well-known problem that has long been addressed for wireless sensor networks. The sensor selection problem can be stated as choosing a set of k measurements from a set of m possible sensor measurements, which minimizes the error in some parameters estimation (Joshi and Boyd 2009). We suggest that the approaches for solving the previous problem can also be applied to finding the set of k locations from m possible ones, where the WiFi measurements will be collected so that the WiFi signal intensities for the remaining locations can be obtained through regression with a small error. What is more, we do not consider a fixed number of locations, but instead, obtain a compromise between the location set's size and the goodness of the regression.

The approaches to deal with the sensor placement/selection problem vary depending on the usage of the sensor measurements (Rowaihy et al. 2007). Specifically, some studies have proposed approaches for the case of using the sensor measurements for estimating a field of values (Joshi and Boyd 2009; Ranieri et al. 2014; Roy et al. 2016). The combinatorial nature of the problem (Joshi and Boyd 2009) makes it unfeasible to explore the whole solution space. If the total number of locations is very small, e.g., six, it is feasible to manually determine fine sets of locations where the measurements are to be taken. However, if a target environment has a (still small) set of 24 locations, and measurements are to be taken at 12 of those locations, the number of different possible sets of locations is $\binom{24}{12} = 2,704,156$. If the number of measurement locations is not already decided, the number of possible combinations rises to $2^{24} = 16,777,216$.

This paper determines the set of locations in a way simpler than those presented in Joshi and Boyd (2009), Ranieri et al. (2014), Roy et al. (2016) for sensor placement. Those studies have harnessed some property of the target problem or forced some form for the solution. We have used an optimization strategy based on genetic algorithms. Sensor placement optimization has already been addressed using genetic algorithms (Yao et al. 1993; Macho-Pedroso et al. 2016), even for indoor acoustic localization (Macho-Pedroso et al. 2016).

3.1 Genetic Programming for Locations Set Determination

The approach proposed in this study uses a genetic optimization algorithm to find a set of locations that includes only a small number of locations and the goodness of the regression obtained using these locations should be similar to the one obtained using the whole set of possible locations. The explanation presented here for genetic algorithms, as well as the library used in the experiments, are based on Mitchell (1998).

The genetic algorithms try to efficiently find solutions to problems that have huge spaces of candidate solutions. Each candidate solution for a problem is called an individual. Commonly, an individual is encoded as a bit string, where each bit represents the presence ('1') or absence ('0') of a trait. These algorithms start by considering a population of random individuals, and iteratively evolves it. The population of each iteration is called a generation. The following generation is the result of applying genetic operators on the current generation. The selection operator selects pairs of individuals whose traits are combined using a crossover operator to produce offspring. A fitness value is computed for every individual in a generation and those with higher fitness values are more likely to be chosen by the selection operator. A mutation operator is applied to the offspring to produce subtle changes in the resulting traits. Some of the new individuals can be randomly discarded, but the population size is maintained.

In this paper's proposal, the set of all locations $L = \{l_1, \ldots, l_n\}$ from the initial, dense collection represents the possible traits that each individual may have. The location set L have associated WiFi RSS measurements $D = \{r_{ijk}\}$. Assume a function $fmap(A, B) \rightarrow C$ so that A is a set of RSS values, B is a set of locations and C is the set of RSS values in A associated to locations in B. Then, $D_{l_p} = fmap(D, \{l_p\}) = \{r_{pjk}\}$ are the RSS measurements associated to location l_p. An individual represents a subset L_I of L. The size of the population, as well as the number of generations considered for population evolution are parameters of the algorithm that are presented in Sect. 4. We have designed the fitness value calculation of an individual so that larger subsets and differences between measured and estimated RSS values are penalized. Specifically, the fitness computation steps are:

1. Fit regressions f_a, for every detected access point a, using L_I and their associated measurements $fmap(D, L_I)$.
2. Use regressions f_a to estimate RSS values $E = \{\bar{r}_{ia}\}$ for locations of $\hat{L}_I = L - L_I$.
3. Compute the AP-wise and location-wise RSS absolute differences between E and $fmap(D, \hat{L}_I)$. Let MRD be the maximum value of those differences.
4. The individual's fitness is $(MRD + 2MRD\frac{ab}{tb})^{-1}$, where ab and tb are the number of '1' bits and the total number of bits, respectively. If for some reason the target number of locations is already predefined, say k, the individual's fitness can become $(MRD + 2MRD|ab - k|)^{-1}$.

After a given number of generations, the individual with higher fitness value, called the elite individual, could be chosen as the set of locations where WiFi

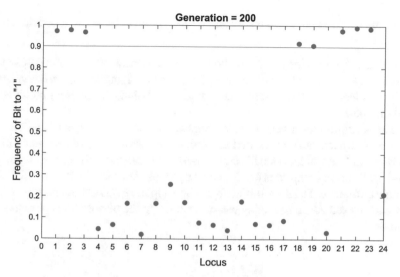

Fig. 1 Locations (bits) frequency. Blue dots represent how often a location has been included in individuals of generation 200. The blue line represents the frequency threshold

measurements are to be collected. The elite individual represents the set that has so far achieved the best compromise between a small number of locations and little degradation of the regression goodness. The genetic algorithm does not guarantee that the elite individual would be the optimal solution for a given problem, but is a fair alternative to an exhaustive search given the combinatorial nature of the problem.

This paper's main goal is not selecting the best locations for making a one-time regression. The main goal is determining the suitable locations for conducting periodic WiFi radio map updates so that the new RSS measurements help in estimating RSS values for remaining, target locations. The elite individual may represent a solution that is over-fitted for the initial measurements. Therefore, we propose to look at the occurrence frequency of each location in the final population and select only those with high frequency. We call this set of highest frequency locations the *solution set*. Figure 1 shows an example of the location's frequency for a population of (200) evaluated sets of locations after 200 iterations. The number of traits, i.e., the number of locations in the initial, dense collection is 24. The bit frequency represents how often a location is found in sets of locations. If we chose a high frequency threshold of 0.9, the solution set would be $\{1, 2, 3, 18, 19, 21, 22, 23\}$.

In summary, the steps needed for selecting the locations where the periodic update measurements are to be collected are:

1. Collect a relatively dense WiFi RSS training database.
2. Use a genetic algorithm, such as the one described in pages eight and nine of Mitchell (1998), using the fitness function described above in this section, to determine the locations' frequency in the population of sets of locations.

3. Choose as the solution set the locations with frequencies above a certain threshold.

Section 4 also provide a guide on how to use new measurements collected at the solution set for updating the radio map. We advise applying our approach independently for clearly unrelated zones, i.e., zones that belong to different buildings or different floors.

Besides suggesting a very good placement for the measurement locations, the proposed approach can be also used for testing the performance of regression methods. By computing an individual's fitness as $MRD + 2MRD\frac{ab}{tb}$, the genetic algorithm would determine a compromise between a large number of locations and a high RSS absolute difference. The set resulting from choosing the n highest fitness sets of locations can be used as a challenging test for evaluating the performance of regression methods.

4 Experiments

The approach proposed in Sect. 3 was tested using a WiFi RSS database collected in a university library during ten months (30 days of separation time, approximately). The database contains training and test sets for each month. Figures 2 and 3 show the collection locations for the training and test sets, respectively, using colored circles. The location label for each fingerprint is expressed in local coordinates in a 2D

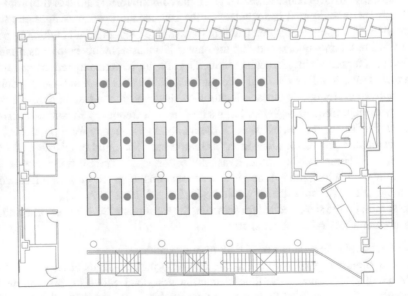

Fig. 2 Collection locations for the training sets. The colored rectangles represent the bookshelves

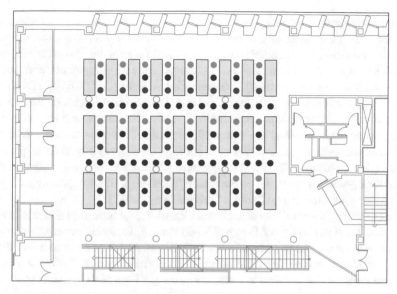

Fig. 3 Collection locations for the test sets. A circle's color identifies to group to which it belongs: red are groups 1 and 5; blue is group 2; green is group 3; and violet is group 4

Euclidean space. The collection locations are among bookshelves in the third floor of the library building. The database is part of a larger effort to gather data for studying short and long term RSS variations and for developing positioning method robust to those changes. Twelve fingerprints were collected at each location. The tests sets were divided into five groups for their collection. Each group had a particular location distribution and collection directions. Most of the experiments presented in this section used only the training sets. The test sets were only used for the evaluation using the kNN fingerprinting presented at the end of this section.

For the training sets and the groups 1, 2 and 3 of the test sets, the collector (a trained person) faced the "up" direction when collecting the first six fingerprints of each point, and the "down" direction when collecting the other six fingerprints. For groups 4 and 5 of the test sets, the faced directions were "right" and "left" instead of "up" and "down", respectively. For data dimensionality reduction, the APs detected in less than 5% of the fingerprints were removed. The device used for collection was a Samsung Galaxy S3 smartphone.

Section 3 defines the set determination without establishing any explicit restriction for the regression to use. However, an implicit restriction exists: The regression method should enable both interpolation and extrapolation, because there may be target measurement locations lying outside the convex hull of the locations in the solution set. This implicit restriction is also important because the extrapolation usage is mandatory for environments that, at collection time, contain areas where measurements cannot be taken (e.g., because of a meeting in an office).

As providing recommendations on regression methods for WiFi fingerprinting was not among the goals of our study, we tested only a few regression methods: IDW (interpolation and extrapolation), radial basis function interpolators like those in Ezpeleta et al. (2015), combinations of interpolation (linear, nearest, and natural) and extrapolation (linear, nearest) as provided by MathWorks® (2017a), and SVR as provided by MathWorks® (2017b). We chose SVR, using a Radial Basis Function (RBF) kernel and performing predictor data standardization, as regression method as it provided the best results regarding RSS absolute differences between RSS measurements and estimations and because it has been successfully used in previous studies (Hernández et al. 2017). We suggest to perform the regression method evaluation for a given environment before making a choice. Guides regarding interpolation and extrapolation can be found in Talvitie et al. (2015).

For evaluating the goodness of each set of locations, we have used a metric defined as the maximum value of the AP-wise RSS absolute differences between the original RSS measurements and the estimated ones. We have preferred the maximum difference over other measures (e.g., the mean) that may mask high RSS differences that are significant for distance-based techniques like kNN-based fingerprinting.

4.1 Evaluation for the Initial Month

Section 2 hinted on approaches for determining where to collect the RSS measurements to be used for fitting a regression. This subsection shows the evaluation results of three strategies for determining the collection locations. The strategies, which were applied to the training set corresponding to first month of our WiFi RSS database, are:

1. Random Sets of Points,
2. Manually-defined Sets of Points,
3. Optimized Set of Points.

The first approach considers differently sized sets of random locations. The second approach uses sets manually defined by an expert. The third approach finds a set of locations that establishes a compromise between the set's size and the regression goodness. The following subsections provide more details about each approach and its evaluation.

4.1.1 Random Sets of Points

This is an intuitive approach for the selection of the collection locations. The algorithm proposed in Kanaris et al. (2016) may allow determining the number of measurement locations. We instead decided to explore several numbers of locations, ranging from 6 to 18 points, which accounts for 25–75% of the 24 total locations in the target area, and they represent reasonable effort reductions. Table 1 presents

Table 1 Minimum and maximum values of RSS error metric for sets of randomly chosen locations

Set Size	Metric Min (dBm)	Metric Max (dBm)
6	19.13	44.74
8	15.71	43.01
10	12.12	41.35
12	11.18	42.10
14	11.99	35.58
16	10.52	35.22
18	9.56	28.59

the maximum and minimum of the RSS error metric previously defined. The experiment for each amount of points was repeated 200 times.

Table 1 shows two main facts. First, the more points are used for fitting the regression, the better the estimations are. Second, and more important, the RSS estimation quality heavily depends on the distribution of the randomly chosen locations, as absolute differences between the maximum and minimum metric values are up to 30.92 dBm.

4.1.2 Manually-Defined Sets of Points

As previously seen, selecting random points creates much uncertainty in the quality of the RSS estimations. A logical alternative is to manually define the set of locations. Better choices are done when the extent of the collection locations and the influence of the building layout and the furniture are taken into account. This subsection presents six alternative sets we considered that are likely choices and could provide fine RSS estimations through regression. The process of determining the tentative sets of locations is time-consuming, and it is especially cumbersome due to the large number of alternatives for each set's size. Table 2 presents the value of the RSS error metric for each alternative set. The ID of each set indicates its amount of locations. Figure 4 shows the location distribution of each set.

Table 2 Values for RSS error metric for manually defined sets of locations

Set ID	Metric (dBm)
6A	21.39
8A	23.45
8B	19.63
12A	23.13
12B	11.90
14A	12.20

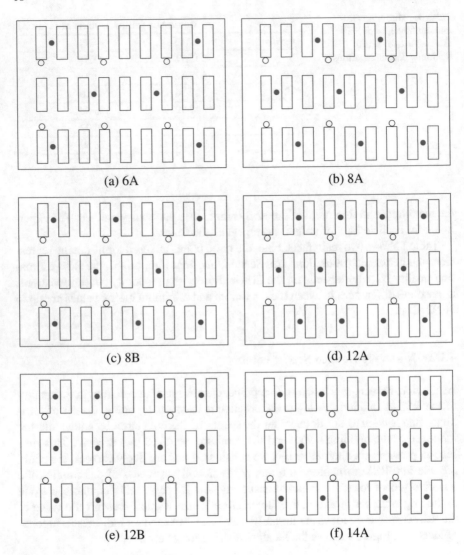

Fig. 4 Manually chosen sets of locations

The results presented in Table 2 reinforce the importance of the distribution of the collection locations. The estimation quality does not strictly decrease with the increase of the number of locations used for regression fit, as seen when comparing the set 6A with set 8A, set 8B with set 12A, and set 12B with set 14A. The locations distribution of each set, as shown in Fig. 4, sheds some light on the previous fact. The convex hulls of sets 6A and 8B include more of the target area than those of sets 8A and 12A, respectively. Nevertheless, the convex hulls of sets 12B and 14A are the same, and the set 12B provide better estimations than set 14A despite having a smaller number of locations. The above facts lead to conclude that even a well-designed set of locations may not be the best choice. Additionally, a set of locations that is optimal for a given environment, may not be optimal for another environment, a fact that we leave unproven because is beyond the focus of this paper.

4.1.3   Optimized Set of Points

As described in Sect. 3, with an optimization strategy based on a genetic algorithm it is possible to search for fine locations for fitting the regression. Specifically, the genetic algorithm implementation provided in Burjorjee (2009), which is in turn based on Mitchell (1998) was used for the experiments. We defined a population size of 200 individuals, used the fitness function proposed in Sect. 3, and run the algorithm for 100 generations. After testing several values, the numbers of 200 and 100 for population size and algorithm generations were the ones that provided higher stability (reproducibility) in the outputted solution. The obtained elite individual (best set of locations found for fitting a regression) and the solution set (described in Sect. 3) using a higher frequency threshold of 0.9, are depicted in Fig. 5. The value of the RSS error metric for the elite set (11 locations) was 8.8453 dBm, which is lower than any of the values obtained using the previous two strategies. The metric value for the solution set (eight locations) is 11.18 dBm, which is still lower than most of the values obtained using the previous two strategies.

Figure 5 shows two important facts. First, all target locations are contained in the convex hull of both optimized sets, which avoids the usage of extrapolation. Second, and more important, the location distributions of the optimized sets do not resemble those of the strategies explored in the Sect. 4.1.2, nor they are intuitive. Therefore, the locations chosen for fitting a regression should not be random, and determining a small set of locations that provides good estimations when used for fitting a regression, is not a trivial problem.

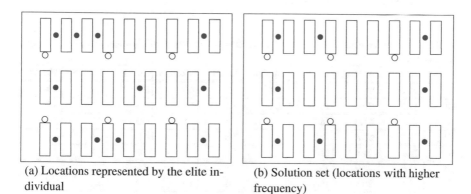

(a) Locations represented by the elite individual

(b) Solution set (locations with higher frequency)

Fig. 5 Set of locations obtained through optimization using a genetic algorithm

The following subsection explores the usage of the solution set for WiFi radio map update, following the procedure presented in Sect. 3.

4.2 Usage of the Solution Set for RSS Radio Map Update

The Sect. 4.1.3 presented the locations that our approach suggested for conducting the periodic updates to the WiFi radio map. This section presents the results of experiments that explored the goodness of those updates along 9 months (month 2–10) in terms of RSS difference between estimations and real measurements, and in terms of the accuracy of a fingerprinting-based IPS.

Regarding RSS differences between estimations and real measurements, the experiments tried three sets of locations and three RSS difference metrics. Table 3 shows the results of these experiments. Each table header indicates the usage of a particular set of locations and a specific RSS difference metric.

To explore the suitability of a set of locations for each month, a regression was fit using their associated measurements of the month training set. Besides the solution set (GA), sets 8A and 8B were also used for regression fitting. The sets 8A and 8B, previously introduced in Sect. 4.1.2, are now used for baseline comparisons. As RSS difference metrics, the experiments used:

1. MRD: The MRD value introduced in Sect. 3.1 for the fitness function definition.
2. Mean: Its value is computed in a way similar to MRD, but the mean value is used instead of the maximum. This metric is included because it is frequently used in the literature for evaluation of WiFi RSS regression methods.
3. MeanP: It is calculated as: Compute the AP-wise RSS absolute differences between RSS measurements and estimations. Compute per each location the mean of those differences. Take the maximum of those mean values. This metric is included because indicates how much the RSS difference may affect a RSS distance-based method like kNN.

Table 3 Values for RSS differences (dBm) according to metrics MRD, Mean and MeanP for sets 8A, 8B and GA

Month	MRD8A	MRD8B	MRDGA	Mean8A	Mean8B	MeanGA	MeanP8A	MeanP8B	MeanPGA
02	21.1	21.2	19.4	1.8	1.5	1.7	5.2	3.7	3.8
03	19.9	20.8	16.9	1.6	1.4	1.5	5.3	3.7	3.0
04	20.5	21.0	18.7	1.7	1.5	1.5	4.6	3.8	3.0
05	19.7	23.8	17.2	1.5	1.3	1.3	4.9	4.6	2.9
06	23.9	21.8	21.9	1.6	1.4	1.5	4.7	3.6	3.7
07	21.5	24.9	26.6	1.5	1.4	1.5	4.8	4.2	3.3
08	26.5	30.0	18.9	1.6	1.4	1.4	5.0	4.3	3.5
09	22.9	25.4	20.2	1.5	1.3	1.2	4.5	3.4	3.3
10	23.7	22.0	18.0	1.8	1.6	1.6	5.3	4.3	4.1

The results presented in Table 3 indicate that the solution set is a better choice than the other two sets as a set of collection locations for periodical updates. Regarding the MRD metric, the solution set provides the best result for most months. It is noticeable that for month number seven, the value for the solution set is 5.1 dBm worse than the one for the set 8A. Some insights on that behavior will be later provided when analyzing the set effects on fingerprinting-based IPS accuracy. Regarding the Mean metric, the solution set is consistently better than the set 8A, and slightly worse than the set 8B for some months. As for the MeanP metric, the solution set is much better than the 8A set. In comparison with the set 8B, the solution set is notably better for five of the months, and only slightly worse for two of them.

The experiments also explored how the localization accuracy of an IPS is affected by the usage of the proposed update approach, i.e., by taking the training RSS measurements of each month only at the solution set and using regression to estimate the RSS values for the remaining locations. As IPS, we tested a kNN fingerprinting approach. Given a training set of fingerprints with known location labels, a query fingerprint, and two parameters specified by the value of k and a distance metric on the fingerprint space, the kNN method finds the k fingerprints in the training set that are closest to the query fingerprint. The location label is estimated as the centroid of the location labels of the selected k closest fingerprints.

To measure the accuracy of an IPS, a test set of query fingerprints is usually used. The location labels are also known for the test set fingerprints, so that, for each fingerprint, the location estimation provided by the IPS and its original location label are used to compute a positioning error distance. In this paper, the positioning distance has been calculated using the Euclidean distance and the localization accuracy has been explored using the 75 percentile of the computed distances for test set.

The tested kNN used the RSS Euclidean distance as fingerprint distance metric. The k parameter value was experimentally determined using the training and test sets of the first month of the WiFi RSS database. Figure 6 shows the resulting localization accuracies. The value of k that provides the best metric value is nine, and it is the one used for kNN in the remaining experiments. This value may appear large, but it is a reasonable value given that 12 fingerprints were taken at each location and no aggregation operation was performed for fingerprints with the same location label.

For comparisons, the experiments included an evaluation of the radio map update at each month using all the training measurements collected at that month. Two updates strategies were tested: Replacement and addition. With the replacement strategy, all training fingerprints collected at one month replaced all fingerprints from the previous month in the WiFi radio map. With the addition strategy, the fingerprints of each month were added without any replacement or deletion from the previous months' fingerprints. The kNN method was used to estimate, for each month, the locations associated to the fingerprints of the test set of that month. Figure 7 shows the behavior of each update strategy along the time.

The strategy of addition provides values for the localization error metric that are smaller and smoother than those provided by the replacement strategy. The metric values for the strategy of addition ranges from 3.25 to 2.84 m. For the replacement strategy, however, the localization error metric ranges from 4.10 to 3.14 m. The

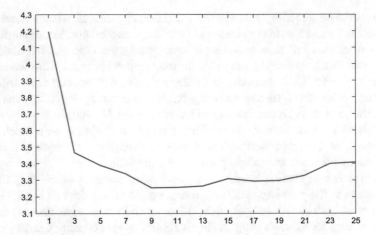

Fig. 6 75 percentile of positioning error using kNN for the first month

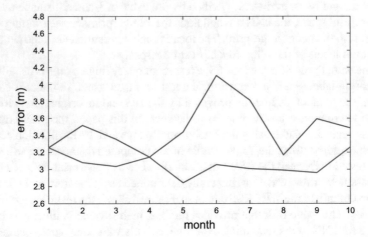

Fig. 7 Comparison of the strategies of replacement (red) and addition (blue). Measurements for all locations are available

months 6 and 7 have the highest metric values, which may indicate that the training values for those months were not as good (representative) as they were for other months.

The evaluation of using the solution set for radio map update was conducted as follows. For each month, the solution set was used to fit a regression, and the RSS values were estimated for the rest of locations. However, the estimation provides one fingerprint per location, while the training and test sets in the database contain 12 fingerprints per location. Additionally, the k value determined above for the kNN fingerprinting is the best under the assumption that there are 12 fingerprints per point. Therefore, we decided to create 12 fingerprints per location using the one fingerprint per location obtained through regression estimation and adding a random value.

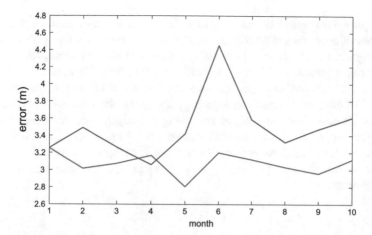

Fig. 8 Comparison of the strategies of replacement (red) and addition (blue). Measurements are available only for the solution set

In the training set from first month, the AP-wise standard deviations values were less than 6 dBm in 80% of cases. The added random value is then uniformly chosen in the interval [−6;6]. The random value addition is specific to the evaluation presented in this study and will not be needed for an IPS radio map update, for which it may be desirable to collect only one fingerprint per point. The fingerprints newly estimated for each month were considered for radio map update following the strategies of replacement and addition described above. Figure 8 presents the localization accuracy metric values for both strategies.

The results obtained using the strategy of addition and the RSS estimations from the solution set are very similar to those using that strategy and the measurements available for all locations. The localization error metric for the strategy of addition ranges from 3.25 to 2.81 m, which is the same interval obtained when using the RSS measurements for all locations. The strategy of replacement showed larger metric values, with higher variations, than the addition strategy. When compared to using the same strategy and the measurements from all locations, the usage of the estimations from the solution set caused larger variability, with the metric values ranging from 4.45 to 3.06 m, having a steeper variation for month 6.

The above results suggest that the usage of the solution set as collection locations for WiFi radio map update is a reasonable choice for the tested environment. The approach of determining the solution set is automatic, so that the specialized and cumbersome task of manually determining a proper set of collections locations is avoided. The MeanP value, i.e., the maximum of the location-wise mean RSS differences between estimations and measurements, was lower than the detected AP-wise standard deviation for all months. Additionally, the accuracy of the tested kNN fingerprinting had a similar behavior when using measurements from all location and estimations to when using the estimations obtained from the measurements of locations in the solution set.

The previous results have been obtained for an initial radio map collection month and across nine months of radio map updates. During the ten months period, no drastic changes in the presence or power configuration of the APs were observed, apart from wireless networks with very low presence in the data. Therefore, we advise the usage of our proposal for environments that allow an initial relatively dense collection and as long as no drastic change happens to the detected APs. Such changes could be detected, for example, by reviewing the appearance or disappearance of AP with strong RSS values in a significant number of fingerprints. We acknowledge that the required initial collection is costly, but if the cost is assumed, the knowledge of the locations for performing periodic updates could translate into a better IPS performance.

5 Conclusions

This paper has presented an approach for determining a subset of the target locations in a goal area where RSS measurements are suggested to be collected for periodic radio map updates. The measurements at the remaining locations are proposed to be obtained through regression. The subset, called the solution set, is determined from initial measurements taken at all target locations through an optimization approach based on a genetic algorithm. The proposed approach was tested using a database collected over ten months in a university library. The regression method tested in the experiments was SVR regression. The experiments' results support the suitability of using the estimation determined using the measurements of the solution set for periodic WiFi radio map updates. The suitability has been shown in terms of the RSS estimation accuracy and in terms of its effects on the localization accuracy of a fingerprint-based IPS.

We consider that the proposed approach may be of particular interest for future efforts devoted to automate the fingerprint collection process using dedicated devices or robotic agents. Future continuation lines of this study include (1) testing our approach in a larger and less densely collected environment, (2) explore the effects of drastic changes in the environment APs, and (3) explore the variant to our approach proposed in Sect. 2 for finding challenging sets for testing regression approaches used in WiFi fingerprinting.

Acknowledgements Germán M. Mendoza-Silva gratefully acknowledges funding from grant PRE-DOC/2016/55 by Universitat Jaume I.

References

Ali MU, Hur S, Park Y (2017) Locali: calibration-free systematic localization approach for indoor positioning. Sensors 17(6). https://doi.org/10.3390/s17061213

Alonazi A, Ma Y, Tafazolli R (2015) Less-calibration wi-fi-based indoor positioning. In: 2015 IEEE international conference on communications (ICC), pp 2733–2738. https://doi.org/10.1109/ICC.2015.7248739

Bong W, Kim YC (2012) Fingerprint wi-fi radio map interpolated by discontinuity preserving smoothing. In: International conference on hybrid information technology. Springer, pp 138–145

Burjorjee KM (2009) SpeedyGA: a fast simple genetic algorithm. https://es.mathworks.com/matlabcentral/fileexchange/15164-speedyga--a-fast-simple-genetic-algorithm

Ezpeleta S, Claver JM, Pérez-Solano JJ, Martí JV (2015) Rf-based location using interpolation functions to reduce fingerprint mapping. Sensors 15(10):27, 322–27, 340

Gu Y, Chen M, Ren F, Li J (2016a) HED: handling environmental dynamics in indoor WiFi fingerprint localization. In: 2016 IEEE wireless communications and networking conference, pp 1–6. https://doi.org/10.1109/WCNC.2016.7565019

Gu Z, Chen Z, Zhang Y, Zhu Y, Lu M, Chen A (2016b) Reducing fingerprint collection for indoor localization. Comput Commun 83:56–63. https://doi.org/10.1016/j.comcom.2015.09.022

He S, Chan SHG (2016) Wi-Fi fingerprint-based indoor positioning: recent advances and comparisons. IEEE Commun Surv Tutor 18(1):466–490. https://doi.org/10.1109/COMST.2015.2464084

Hernández N, Ocaña M, Alonso JM, Kim E (2017) Continuous space estimation: increasing wifi-based indoor localization resolution without increasing the site-survey effort. Sensors 17(1)

Hossain AKMM, Soh WS (2015) A survey of calibration-free indoor positioning systems. Comput Commun 66:1–13. https://doi.org/10.1016/j.comcom.2015.03.001

Jan SS, Yeh SJ, Liu YW (2015) Received signal strength database interpolation by kriging for a wi-fi indoor positioning system. Sensors 15(9):21, 377–21, 393. https://doi.org/10.3390/s150921377

Joshi S, Boyd S (2009) Sensor selection via convex optimization. IEEE Trans Signal Process 57(2):451–462

Kanaris L, Kokkinis A, Fortino G, Liotta A, Stavrou S (2016) Sample size determination algorithm for fingerprint-based indoor localization systems. Comput Netw 101:169–177. https://doi.org/10.1016/j.comnet.2015.12.015

Krumm J, Platt J (2003) Minimizing calibration effort for an indoor 802.11 device location measurement system. Microsoft Research, November

Lee M, Han D (2012) Voronoi tessellation based interpolation method for wi-fi radio map construction. IEEE Commun Lett 16(3):404–407. https://doi.org/10.1109/LCOMM.2012.020212.111992

Li B, Wang Y, Lee HK, Dempster A, Rizos C (2005) Method for yielding a database of location fingerprints in wlan. IEE Proc—Commun 152(5):580–586. https://doi.org/10.1049/ip-com:20050078

Li L, Shen J, Zhao C, Moscibroda T, Lin JH, Zhao F (2014) Experiencing and handling the diversity in data density and environmental locality in an indoor positioning service. ACM—Association for Computing Machinery

Lin K, Chen M, Deng J, Hassan MM, Fortino G (2016) Enhanced fingerprinting and trajectory prediction for iot localization in smart buildings. IEEE Trans Autom Sci Eng 13(3):1294–1307. https://doi.org/10.1109/TASE.2016.2543242

Liu C, Kiring A, Salman N, Mihaylova L, Esnaola I (2015) A kriging algorithm for location fingerprinting based on received signal strength. In: 2015 sensor data fusion: trends, solutions, applications (SDF), pp 1–6. https://doi.org/10.1109/SDF.2015.7347695

Macho-Pedroso R, Domingo-Perez F, Velasco J, Losada-Gutierrez C, Macias-Guarasa J (2016) Optimal microphone placement for indoor acoustic localization using evolutionary optimization.

In: 2016 international conference on indoor positioning and indoor navigation (IPIN), pp 1–8. https://doi.org/10.1109/IPIN.2016.7743609

Majeed K, Sorour S, Al-Naffouri TY, Valaee S (2016) Indoor localization and radio map estimation using unsupervised manifold alignment with geometry perturbation. IEEE Trans Mob Comput 15(11):2794–2808. https://doi.org/10.1109/TMC.2015.2510631

MathWorks® (2017a) Extrapolating scattered data, in MATLAB® R2017b. https://es.mathworks. com/help/matlab/math/scattered-data-extrapolation.html

MathWorks® (2017b) Support vector machine regression, in MATLAB® R2017b and statistics and machine learning toolbox™. https://es.mathworks.com/help/stats/support-vector-machine-regression.html

Mitchell M (1998) An introduction to genetic algorithms. MIT press

Pei L, Zhang M, Zou D, Chen R, Chen Y (2016) A survey of crowd sensing opportunistic signals for indoor localization. Mob Inf Syst 2016

Ranieri J, Chebira A, Vetterli M (2014) Near-optimal sensor placement for linear inverse problems. IEEE Trans Signal Process 62(5):1135–1146

Rowaihy H, Eswaran S, Johnson M, Verma D, Bar-Noy A, Brown T, La Porta T (2007) A survey of sensor selection schemes in wireless sensor networks. Proc SPIE 6562:A1–A13

Roy V, Simonetto A, Leus G (2016) Spatio-temporal sensor management for environmental field estimation. Signal Process 128:369–381

Talvitie J, Renfors M, Lohan ES (2015) Distance-based interpolation and extrapolation methods for rss-based localization with indoor wireless signals. IEEE Trans Veh Technol 64(4):1340–1353. https://doi.org/10.1109/TVT.2015.2397598

Wang B, Chen Q, Yang LT, Chao HC (2016) Indoor smartphone localization via fingerprint crowdsourcing: challenges and approaches. IEEE Wirel Commun 23(3):82–89. https://doi.org/10.1109/MWC.2016.7498078

Xiao Z, Wen H, Markham A, Trigoni N (2015) Robust indoor positioning with lifelong learning. IEEE J Select Areas Commun 33(11):2287–2301. https://doi.org/10.1109/JSAC.2015.2430514

Yang S, Dessai P, Verma M, Gerla M (2013) Freeloc: calibration-free crowdsourced indoor localization. In: 2013 proceedings IEEE INFOCOM, pp 2481–2489. https://doi.org/10.1109/INFOCOM.2013.6567054

Yao L, Sethares WA, Kammer DC (1993) Sensor placement for on-orbit modal identification via a genetic algorithm. AIAA J 31(10):1922–1928

Yiu S, Dashti M, Claussen H, Perez-Cruz F (2017) Wireless rssi fingerprinting localization. Signal Process 131:235–244

Zhu JY, Zheng AX, Xu J, Li VOK (2014) Spatio-temporal (s-t) similarity model for constructing wifi-based rssi fingerprinting map for indoor localization. In: 2014 international conference on Indoor positioning and indoor navigation (IPIN), pp 678–684. https://doi.org/10.1109/IPIN.2014.7275543

Task-Oriented Evaluation of Indoor Positioning Systems

Robert Jackermeier and Bernd Ludwig

Abstract The performance of indoor positioning systems is usually measured by their accuracy in meters. This facilitates the comparison of different systems, but does not necessarily give information about how well they perform in real-life scenarios, e. g. during indoor navigation of walking persons. In this paper, we present a task-oriented evaluation that adapts the idea of landmark navigation: Instead of specifying the error metrically, system performance is measured by the ability to determine the correct segment of an indoor route, which in turn enables the navigation system to give correct instructions. We introduce the area match metric in order to identify areas where positioning proves problematic. In order to evaluate the described metric, we use a pedestrian dead reckoning approach to compute indoor positions. Without any external correction, the correct segment of the test route is identified in 88.4% of all trials. Based on these results, we explore options how to identify and predict erroneous situations during the navigation process as well as beforehand.

1 Introduction and Motivation

Indoor positioning for pedestrian navigation is still an open problem as no positioning system that delivers absolute—such as GPS—coordinates is available. Many solutions providing precise (sub meter) localization require additional technical devices (Guo et al. 2015; Pham and Suh 2016; Romanovas et al. 2013). However, they are not at disposal in everyday life situations at which pedestrian navigation systems target.

R. Jackermeier (✉) · B. Ludwig
Chair for Information Science, University Regensburg, Universitätsstraße 31, D-93053,
Regensburg, Germany
e-mail: robert.jackermeier@ur.de

B. Ludwig
e-mail: bernd.ludwig@ur.de

© Springer International Publishing AG 2018
P. Kiefer et al. (eds.), *Progress in Location Based Services 2018*, Lecture Notes
in Geoinformation and Cartography, https://doi.org/10.1007/978-3-319-71470-7_2

Technically simpler solutions use sensors that come with every smartphone, such as accelerometers, gyroscopes, and step counters (Basso et al. 2015; Verma et al. 2016). Such approaches provide relative positioning data, and often suffer from cold start problems (Harle 2013). Solutions based on GSM, LTE, and WiFi receivers can compute absolute coordinates of an area in which a user is located, but fail in tracking a user's movement precisely (Waqar et al. 2016).

Independently from the used sensor technology, the approaches described above have in common that they do not take spatial knowledge into account which we understand to describe the ways persons can take in an indoor area: information about the user's current position limits the options for a position update. While — as Waqar et al. (2016) point out—such technology can advantageously be used to implement the described solutions quickly at any location (that eventually provides GSM, LTE or WiFi infrastructure), its major drawback is that these positioning algorithms do not provide effective means for error cancellation in terms of the navigation task a person is currently solving.

We address exactly this issue. We describe an approach that models spatial knowledge with an indoor navigation graph (see Fig. 1 as an example) that contains all routing decisions and path segments between any two arbitrary locations in an indoor/outdoor environment requesting a decision from the user how to execute a system's routing instruction. We apply our indoor navigation graphs to cancel the positioning error of relative positioning algorithms (e.g. step counters) that accumulates while a person is navigated. A similar approach is taken by Link et al. (2013). In contrast to our approach the authors rely on OpenStreetMap data that are much more complex to handle and therefore less flexible for adapting a system to a new environment where there doesn't already exist a corresponding model. Further, the authors use the OpenStreetMap data for a metric evaluation of the approach, while we relate success to navigation instructions: we consider positioning to be successful if it allows a navigation system to give the correct next instruction relative to the users position on the route, or, more generally, if it allows a location based service relying on it to function properly.

For rendering this definition operational, we introduce the **area match score** as a new metric for indoor positioning. We analyze the performance of an algorithm we implemented on the basis of this definition. From results of the analysis we derive effective criteria for an automatic decision when the computed position estimate data is no longer reliable. In such cases, dead reckoning approaches cannot recover the error. We conclude that effective positioning algorithms for pedestrian navigation must be able to apply different techniques for sensing the user's environment in order to provide optimal position estimates.

In this paper, we first report the relevant state of the art, then we explain how we relate landmark-based navigation and indoor positioning and develop our mathematical model for the posed problem. Next, we present an empirical evaluation for a navigation task that required test persons to continuously walk on an indoor route on one of the floors of a complex and therefore cognitively demanding building on the

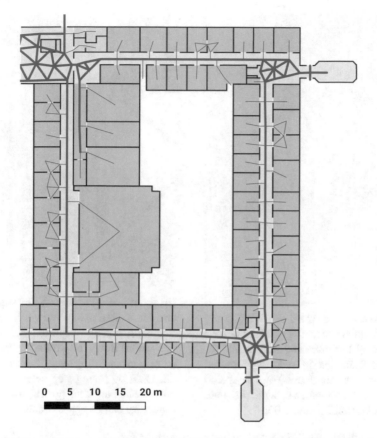

Fig. 1 Indoor navigation graph in the test area. Main edges used for localization are shown as thick lines. Notice the mesh-like topology in larger open areas

campus of our university. Finally, we discuss the obtained results in the light of our task-based performance metric. We analyze limitations of the approach and derive relevant issues of future work from the insights obtained from the evaluation.

2   State of the Art in Indoor Positioning

As already stated in the introduction, many localization techniques based on different types of sensors have been proposed for pedestrian indoor navigation systems and indoor positioning in a broader sense. Despite all these research efforts, there is still no technology established as a widely accepted state of the art similarly to GPS for outdoor areas.

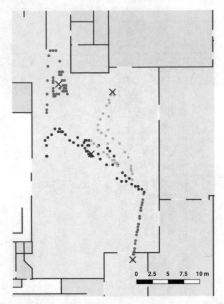

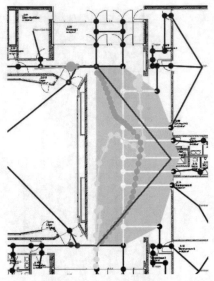

(a) Result of a previous WiFi study in an indoor area using Fraunhofer's awiloc. While standing still at the positions indicated in red, the reported locations (colored dots) are scattered around the area with a root mean square error between 2.1 and 5.0 m.

(b) Result of a previous WiFi study in an indoor area using Fraunhofer's awiloc. The position reported by awiloc (yellow) follows the ground truth (red, from bottom to top) very loosely, if at all. Only by fusing step detection data with a Kalman filter (green) the actual trajectory can be approximated.

Fig. 2 Performance of WiFi signals in indoor positioning tasks

As our work is focused on positioning algorithms that do not need sensors beyond those available in a smartphone anyway, the following review of the state of the art leaves aside more exotic approaches that require special sensors or hardware.

WiFi-based indoor localization can be widely deployed in modern buildings where a sufficient WiFi infrastructure is usually available, but suffers from multiple problems, as Davidson and Piché (2016) point out: Creation and maintenance of radio maps is time consuming and therefore expensive. A low scan rate on current smartphones leads to disjointed position estimates. Furthermore, device heterogeneity, the smartphones orientation, and the attenuation of signals by humans are identified as disadvantages. Due to these issues, WiFi-based systems generally achieve an accuracy of at most a few meters and are suited to determine the approximate position, but not for continuous tracking.

Our own findings confirm these claims: Fig. 2a shows some of the results from an earlier study, where the location reported by Fraunhofer's WiFi-based awiloc system[1] wanders around in an indoor area even if the test person is not moving.

---

[1]https://www.iis.fraunhofer.de/en/ff/lv/lok/tech/feldstaerke/rssi/tl.html.

Nevertheless, the RMSE of the position reaches up to 5 m. Even more problems arise when the test person is moving (an example can be seen in Fig. 2b), where the location updates usually are lagging behind and do not match the path that was actually taken.

On the other hand, Bluetooth Low Energy (BLE) beacons as another wireless localization technique are designed to be more accurate, but are far less widespread and therefore more expensive to deploy. In particular, as they are mounted at fixed positions and send signals over a small distance only, persons could be tracked continuously only if beacons were mounted along all paths persons can walk on.

A recent development for getting an rough estimate of the user's position bases on sending a sound signal via the smartphone's loudspeaker and recording it immediately with the microphone. Rooms have particular acoustic characteristics that can be recognized to identify in which room out of a set of trained rooms the smartphone is currently located (Rossi et al. 2013).

In summary, several approaches exist that provide a rough estimate of the user's current position, but not of the user's movement, and therefore can be applied in a hybrid approach to reinitialize a dead reckoning algorithm after it has failed to determine a reliable position estimate.

For tracking movements, pedestrian dead reckoning (PDR) is widely used. Several solutions do not need any external infrastructure, but rely solely on inertial sensors available in smartphones.

True inertial navigation by the double integration of acceleration values is not feasible since the sensor measurements are much too noisy. Instead, PDR is mostly accomplished by a variant of so called Step and Heading Systems (SHS), that detect the user's steps and try to estimate their length and direction (Harle 2013). Step detection on smartphones is historically achieved through the accelerometer using various techniques (see Muro-de-la Herran et al. 2014; Sprager and Juric 2015; Susi et al. 2013). Lately, dedicated step detector sensors are available in more and more devices. The heading can be inferred from a combination of magnetic compass and gyroscope, while step length can be either assumed as fixed or dynamic, e.g. based on the frequency (Harle 2013).

The main disadvantage of any dead reckoning solution is the need for an initial position from which the relative positioning can start as SHS by their nature cannot compute absolute positions. Furthermore, the positioning error increases over time due to noisy sensor data. Given both of these problems either error correction through external sensors or an algorithm that matches the sensor data to a final position estimate are necessary to employ dead reckoning for more complex tasks such a navigating a user or other location based services.

As in our work external sensors should be avoided, matching algorithms are the only option for solving the indoor positioning problem. Maps are often represented as discrete graphs (see e.g. Thrun et al. 2005) and have been used successfully to locate robots in complex environments. For pedestrian indoor localization, graph models of the environment were first introduced by Liao et al. (2003). They introduced the particle filtering method on a Voronoi graph in order to make the position estimation more robust and efficient. Since then, other researchers have adapted and improved this approach (e.g. by adding multiple sensor modalities): The system recently presented by Hilsenbeck et al. (2014) is operating on a graph generated from a 3D model of the environment. Herrera et al. (2014) use existing material from OpenStreetMap and enrich it with information about the indoor areas of a building. Ebner et al. (2015) generate a densely connected graph from the floor plan of a building. All these approaches have in common that creating a map is either time consuming or expensive (due to the need for special hardware) or relies on existing data. Furthermore, normally the resulting graphs do not contain any information besides the geometry of the building, making them unsuitable to use as data source for the path planner of a navigation system.

As far as the evaluation of indoor positioning is concerned, the state of the art can be surveyed best by looking at recent competitions that aim to compare the performance of indoor positioning systems. Held regularly, the provide an opportunity to gain insights into established evaluation methods. Potortì et al. (2015) report the results of the EvAAL-ETRI competition held in conjunction with the IPIN 2015 conference. To assess the error of the participating systems, they add a penalty for wrongly detected floors or buildings to the actual positioning error. The final ranking is determined by the 75% quantile of the resulting errors. In their evaluation of the 2015 EvAAL-ETRI WiFi fingerprinting competition, Torres-Sospedra et al. (2017) use the mean error as the metric to rank the competitors: The mean error of the tested system is not lower than 6 m. Lymberopoulos et al. (2015) again use the mean positioning error to rank the systems participating in the 2014 Microsoft Indoor Localization Challenge. Interestingly however, they remark that the mean error or other commonly used metrics do not represent the performance of a system in its entirety. We follow this assessment and argue for a task-oriented view on the performance of a positioning approach that we introduce below.

3   Data Model for Landmark-Based Navigation

While the mean positioning error is definitely of interest for building autonomous systems that can navigate in indoor environments (e.g. robots for ambient assisted living), for the implementation of many location based services it is an inappropriate performance metric. In our view, this is due to the fact that users of location based services experience the environment from a cognitive perspective that assigns meaning to perceivable objects. E.g. a pedestrian can walk to a distant object without continuous technical assistance while a robot cannot. Therefore, in applications involving humans it may often be sufficient to know that the user is close to a semantically meaningful object (e.g. a door at the end of a corridor or a certain cloth shop in a shopping mall). The main consequence of this hypothesis is that the precision of indoor positioning has to be measured in terms of the user's relative position to objects relevant for his current task instead of meters in a coordinate system that the user cannot even perceive.

For an implementation of this idea, we rely on graphs to formally represent a map of the environment and define and locate relevant objects. With such a representation of the environment, also the task of navigating a user can be based on relevant objects which are commonly known as landmarks in the GIS literature (see e.g. Ohm et al. 2015).

At this point, we can state the major contribution of the present paper. We propose an approach to combine a graph-like representation of an environment with the minimally necessary metric information to correctly align the data computed by a SHS in the graph in order to assign the user's position to more easily perceivable objects in the environment which we call **areas** or **landmarks** depending whether we refer to a part of a path the user should walk on or a relevant object the user can perceive in the environment. It has to be noted that we do not assume a particular sensor technology or SHS algorithm. We only assume to receive vectors that quantify the step length and direction of a pedestrian's movement.

This approach for an indoor navigation system shares similarities with other work (in particular Link et al. 2013). As a new contribution, we introduce the concepts of **areas** and the **area match score** that link indoor positioning based on SHS with landmark based navigation (Sect. 4). By doing this, we relax performance requirements for positioning algorithms as we no longer need to optimize the metric errors at any time of the navigation process. Instead, it is sufficient to identify the correct area a user is currently walking on: given the current area, the system can generate a navigation instruction that incorporates a landmark easily perceivable from the estimated position of the user. Under normal conditions, users are able to walk towards the indicated landmark without further assistance. Then, in order to continue the navigation process the positioning algorithm has to determine whether the user is *close* to the landmark and can switch to the next instruction. Mostly, this task is much easier than continuously determining the exact position.

To introduce our approach, we describe the concepts we build our knowledge graphs on, and the algorithm we rely on to compute routes that we then split into path segments.

Our data model is based on the graph representation described in Ohm et al. (2015). We adapted their concept for environment models in order to generate indoor navigation graphs such as the one in Fig. 1. The graphs are created manually in a web tool by drawing on top of floor plans of a building. Edges represent paths users can walk on. Edges can connect multiple floors or buildings, allowing for the modeling of arbitrary building geometries. Nodes connected by edges are used to formalize decision points where users eventually have to change their direction. Landmarks can be integrated seamlessly in the graph structure in the form of special nodes that can optionally be enriched with images.

In summary, the system relies on a single data structure for routing and localization, minimizing the effort needed for map creation and maintenance.

Fig. 3 User interface of the
data collection app. The
current area is highlighted in
light red

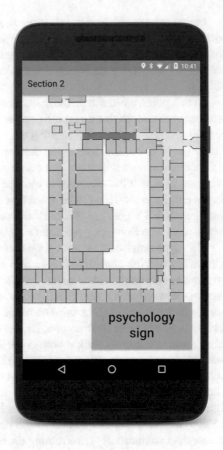

In order to transform such a graph into an indoor navigation graph, adjacent edges
are combined into **areas** if the part of the environment captured by an area is perceivable as a unique object with salient landmarks (e.g. a corridor, a foyer, a staircase).
An example can be seen in Fig. 3, which shows part of a corridor as an area consisting
of a several adjacent edges. In Fig. 4 a typical landmark is displayed: the billboard
shown on the map is also referred to in the navigation instruction.

For landmark based navigation, it is crucial to give navigation instructions at the
right time in order not to confuse users and to guarantee good usability as well as
reaching the destination. A navigation instruction is given at the right time if it does
not refer to any landmark that is not yet visible from the user's current position or
that the user has already passed before.

Fig. 4 Example of a
landmark-based navigation
instruction

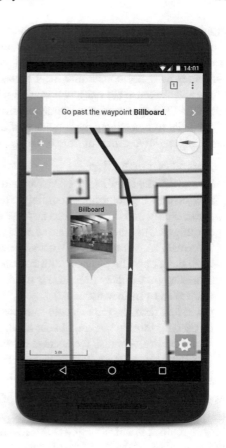

4 Graph-Based Localization

Consequently, in order to give the right instructions at the right time, the user's relative position towards landmarks referred to in navigation instructions needs to be known. Our indoor positioning algorithm computes this position by mapping sensor data to areas in the indoor navigation graph.

For this mapping, we implemented a recursive stochastic filter that after each measurement assigns a probability to each area proportional to the likelihood of the user to currently walk on a certain area.

The filter is implemented as a particle filter (see Thrun et al. 2005). This family of algorithms represents a probability distribution by means of a representative sample, a set of so called particles. Around the expected position the number of particles is high while elsewhere it is low according to the small probability mass. Using a sampling and resampling strategy the set of particles is updated after each measurement in order incorporate the new information (see Thrun et al. 2005 for details): Far away from the expected position the particles diminish while new ones are generated for positions with high probability mass. Our implementation can also incorporate input from multiple sensors such as detected steps or WiFi signals in order to implement the advocated hybrid approach for indoor positioning. Furthermore, information contained in the indoor navigation graph stabilizes and corrects the position estimates as many constraints for the user's current location can be derived from the graph structure (in particular invalid positions receive probability zero while in standard SHS approaches the same locations are possible positions).

In order to investigate the influence of the precision of the SHS on our approach, we compared two different state of the art algorithms:

- motionDNA by Navisens
  The motionDNA SDK by Navisens is a well-known commercial state-of-the-art motion tracking solution. According to the company's website,[2] it relies on inertial sensors only and does not need any external infrastructure to operate. The sensor readings are updated with a rate of 24 Hz on our test device and include a variety of information such as the user's activity and the device orientation and position. For this study, only the position information (relative to the initial position, measured in meters in X and Y direction) is used.
- Android's built-in sensors
  On recent devices, the Android framework gives access to many sensors that can be used for motion tracking. In our case, the step detection sensor tells us whenever a step occurs, whereas the average orientation during the step as provided by the rotation vector sensor is used as step direction. We assume that the user orients the smartphone in his walking direction and use a fixed step length.

In Fig. 5 the data computed by each of both algorithms for a single walk on the test route is plotted into the map of the building. Many position estimates are far off the route. This observation illustrates that information about the environment is indispensable for the position estimates to be used in an indoor navigation system.

In order to map SHS estimates to the indoor navigation graph, we apply the described particle filter. Initially the probability is distributed uniformly over all edges.

Whenever a step is detected, a Gaussian naive Bayesian classifier updates the probability distribution for the edges starting in the current node. The update takes the motion model for the user (i.e. the distance and direction of the detected step) and the orientation of the considered edges into account. The probability of the user to walk on an edge increases if this edge is parallel to the direction detected by the SHS. The increment for an edge not parallel to the detected direction decreases proportionally to the angle between the direction and the orientation of the edge.

Unlike other approaches, the algorithm does not immediately select the edge with the highest probability as the current position estimate. Instead, it updates the set of particles each of which represents a different hypothesis for the user's current position. A similar approach has been successfully applied to localization in robotics (see Thrun et al. 2005) and allows to

---

[2]http://navisens.com.

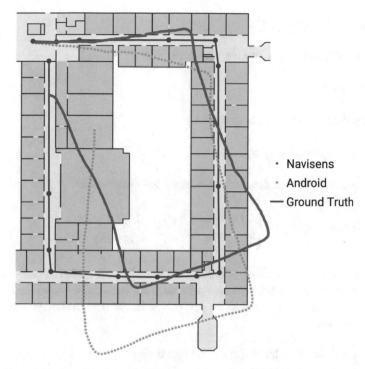

Fig. 5 The trajectories of both motionDNA (blue) and raw step data (green) of a typical walk along the test route. Ground truth (starting top left, then clockwise) and the boundaries of the areas defined for the evaluation experiment are drawn in red

- account for noise in the SHS data, which may stem from the rotation vector sensor (or rather the underlying magnetic compass) or the way the device is held in the hand,
- account for differences in step length while a person is walking, and
- account for different step lengths of different users.

More formally, each particle's state is defined by the vector $\{n_t, d_t, e_t\}$, where n_t denotes the starting node at time t, d_t the distance walked since leaving the node, and e_t a discrete probability distribution for the edges adjacent to the node. On every step, the state is updated according to

$$\{n_t, d_t, e_t\} \sim p(n_t, d_t, e_t | n_{t-1}, d_{t-1}, e_{t-1}, z_{\theta,t}, z_{l,t}, G), \tag{1}$$

where $z_{\theta,t}$ and $z_{l,t}$ are the measured step direction and length, and G is the graph of the building. Applying the procedure detailed in Hilsenbeck et al. (2014), the update rule can be decomposed to its independent parts. The noisy step length measurement with the empirically determined variance σ_l^2 is modeled by

$$l_t \sim p(l_t|z_{l,t}) \sim \mathcal{N}(z_{l,t}, \sigma_l^2),\tag{2}$$

leading to the updated cumulative step distance of

$$d_t \sim d_{t-1} + l_t.\tag{3}$$

Similarly, the step direction is updated by

$$\theta_t \sim p(\theta_t|z_{\theta,t}) \sim \mathcal{N}(z_{\theta,t}, \sigma_\theta^2)\tag{4}$$

and subsequently used to determine the new edge distribution:

$$e_t^i \sim p(e_t^i|e_{t-1}^i, \theta_t, G) \sim \begin{cases} \mathcal{N}(\Delta\theta_t^i, \sigma_e^2) * e_{t-1}^i & \text{if } |\Delta\theta_t^i| \leq 100 \\ 0 & \text{otherwise} \end{cases}\tag{5}$$

Here, e_t^i denotes the probability of the user to currently walk on the i-th edge adjacent to the current node and $\Delta\theta_t^i$ the angle difference between the step and the i-th edge. Finally, the decision whether the user has completed an edge and moved to the next is formalized as:

$$n_t \sim \begin{cases} \text{no} & \text{if } d_t \leq \text{length}(e) \wedge e = \text{argmax}_i(e_t) \\ \text{yes} & \text{if } d_t > \text{length}(e) \wedge e = \text{argmax}_i(e_t), \end{cases}\tag{6}$$

i. e. whenever d_t exceeds the length of the currently most probable edge e. In this case the starting node has to be updated: n_t is set to the sink node of the previous edge and d_t is reset to zero. Since the walked distance usually does not align exactly with the edge length, the difference is added to the position estimate and the step bias is reinitialized to $\mathcal{N}(z_{l,t}, \sigma_l^2)$ as the prior distribution for the new current edge e_t.

After the update step, the particle importance weights are distributed according to the non-normalized probability of the most probable of all adjacent edges:

$$\omega_t = \omega_{t-1} * p(z_{l,t}, z_{\theta,t}|n_t) \sim \omega_{t-1} * \max_i(e_t)\tag{7}$$

Finally, stochastic universal sampling is performed, which guarantees low variance and a representation of the samples in the new particle distribution that is proportional to their importance weights (see Thrun et al. 2005 for further details).

In order to estimate the user's position, the expected value of the particle distribution is calculated. From there, the closest point that is located on either an edge or a node of the graph is computed as the final position estimate. This snap to the indoor navigation graph ensures that the position estimate is a location that is accessible to the user and—differently to the pure SHS algorithms—prevents the positioning algorithm to assume impossible movements, e.g. through walls.

5 Task-Oriented Evaluation

The purpose of this study is to test how well the correct area on a route can be determined by the localization system, which—as noted above—is a requirement for correct navigation instructions and for successful navigation in general. Based on those findings, it is our goal to identify patterns that point to problematic areas and to propose ways to mitigate the issue.

5.1 Experimental Setup

As detailed above, the indoor navigation graph is a simplified model of the environment. Perfectly accurate position tracking is feasible only if the user actually walks on the edges of the graph. In reality however, the user's movement is not constrained to the graph structure, and edges and nodes do not necessarily have a perceptible counterpart that can ease the user's orientation. As an example we consider the foyers at the end of the long corridors in Fig. 1. They are modeled as a dense sub graph in order to approximately represent different paths a user may take through the foyer. The user however perceives the foyer as a single object that can be traversed arbitrarily instead of a discrete graph.

In order to take the user's moving and orienteering behavior into account, we introduce the **area match score** as a performance metric that enables us to investigate the problem in a task-oriented manner. Each area corresponds to a section of the route, which in turn is comprised of multiple edges in the indoor navigation graph. Strictly speaking, an area is matched if the position estimate is located in the same area the user is currently walking on.

However, the introduction of artificial segments inevitably leads to matching errors at the boundaries of two adjacent areas as the user's position cannot estimated without error. Therefore, areas may be mismatched.

In order to account for this problem, we relax the strict definition of a match by adding the mean position error to the boundaries of each area. The final area match score for the whole route is then defined as the percentage of position updates that match the correct area according to the relaxed definition above.

To collect data for an evaluation of the implemented indoor positioning algorithm, we conducted an empirical study in the ground floor in an university building. There, we defined a test route spanning 182 m. The route leads through 4 corridors in a rectangular shape. Three of the corners are modeled as small foyers (see Fig. 1). The only obstacles on the route are several glass doors that had to be passed in order to reach the destination of the route.

The route was segmented into areas. Their boundaries were set at positions where semantically relevant objects—i.e. salient landmarks—are located. For determining salient landmarks along the route, we followed the approach described in Kattenbeck (2016): 19 persons rated 32 objects in the test area regarding different aspects of their salience. We selected the objects with the highest predicted overall salience as landmarks for the navigation instructions in our experiment. These landmarks included e. g. a glass cabinet, a wall painting, a bench and a sign for the department of psychology. Additionally, architectural features such as the aforementioned glass doors or the beginning and end of foyers were used to segment the route into areas. For each area, we formulated a navigation instruction that should explain to the test persons how to proceed the route. Finally, the route consisted of fourteen sections of varying size (see Fig. 5). The main factor that influences the size of the sections is the visibility of the landmark at their end: some can be referenced unambiguously from further away, while for others one has to be closer, thus causing smaller sections.

Acquisition of Positioning Data

Starting from a defined position, 7 different persons who were familiar with the area and the landmarks performed a total of 15 walks along the test route. Data collection took place over the course of several days, with an LG Nexus 5X running Android 7.1.2 as the test device. Before each test run, the compass was calibrated and its proper functionality was verified. During the experiment, the phone was held in the hand in front of the body, pointing in the direction the person was heading toward.

For data collection, a custom Android application was developed. It is able to capture data from various sensors of the device:

- Steps detected by the built-in Android step detection sensor.
- Orientation data from Android's orientation vector sensor, which in turn fuses magnetometer and gyroscope readings.
- Data from Navisens' motionDNA SDK. First and foremost, this includes the relative position, but heading direction, orientation of the device, as well as detected user activity are also logged.
- The signal strength of WiFi access points in the area (not used in this study).
- A video recording of the device's back-facing camera, capturing the test person's feet and the area immediately in front of them.

The app's user interface consists of a map of the test area and a single button that allows the user to start the test run. After the localization on the starting node and sensor systems were initialized, the first area to walk through was highlighted on the map and the button text changed to the first instruction. When a test person reached the landmark, he or she pressed the button in order to set the ground truth for the transition between two adjacent areas, and the interface was updated with information for the recently entered area.

Validation of the Collected Data

With this experimental setup, we collected sensor data for the test route and a ground truth labeled by experts in a single run of the experiment. We avoided to make use of other, technically very complex methods to label the logged sensor data with the correct area.

In order to verify whether the collected samples were representative for average persons walking straight ahead, several gait characteristics were calculated.

- The mean gait speed during a walk can easily be determined by the quotient of route length and the time needed to complete the route, measured by the difference of timestamps between last and first step. The result is a mean speed of 1.30 m/s $(SD = 0.14$ m/s), which is well within the margin reported by Bohannon and Williams Andrews (2011).
- In order to calculate the step length, the steps are counted manually for each walk by means of the recorded video, revealing that Android's step detector misses about 5.8% of steps on average.
- The mean step length amounts to 0.73 m $(SD = 0.077$ m), which is classified as fast gait according to the study from Oberg et al. (1993). This can be explained by the fact that the test persons knew the area and the route very well.

In summary, the collected data is representative for "average" persons who currently perform a similar navigation task.

Analysis of the Collected Data

The analysis of the raw data shows—quite expectedly—that the error quickly accumulates, leading to a high mean location error of 11.5 m (Android sensors) respectively 12.0 m (motionDNA). Figure 5 shows the trajectories of a typical walk. Navisens' motionDNA often struggles with substantial drift towards the left early on, but otherwise manages to track the overall shape quite well. The version relying on the Android step counter usually shows drifts in different directions throughout the walk due to the lack of correction. Additionally, the reported distances differ between the tracking methods: motionDNA's paths are usually shorter $(M = 174.1$ m, $SD = 15.37$ m), Android's longer $(M = 189.4$ m, $SD = 16.49$ m) than the ground truth of 182.0 m.

Before the motionDNA data could be used as input to the particle filter, some preprocessing was inevitable: Since the update frequency of about 20 Hz was rather high (about an order of magnitude higher than the step frequency), the data was split in batches of ten measurements that were treated as a single step. In two of the 15 recorded walks, the relative location reported by motionDNA unexpectedly was set back to the starting point of the route. Therefore, the area in which the reset occurred was eliminated from the data set.

5.2 Localization Results

Since it was not feasible to run both Navisens' and our indoor localization implementation at the same time on one device, we processed the collected data in an offline simulation of our indoor positioning algorithm.

In order to extend the data set, from each actual walk a stochastic motion model was learned and used to generate 20 additional walks proportional to the learned model. 10 of them were generated using motionDNA for simulating steps of the user and 10 others using the Android step counter.

Fig. 6 Empirical
cumulative distribution
function showing the
accuracy with the two
different motion tracking
methods

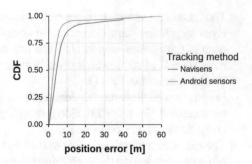

The extended data set was used for the evaluation of the implemented algorithm. In the remainder of this section, we present our evaluation results and discuss their impact on the appropriateness of the proposed area match score for localizing users during indoor navigation.

Accuracy Metrics for the Sensor Data

Figure 6 shows a comparison of the two motion tracking solutions regarding their positioning accuracy, i.e. the distance from estimated position to ground truth, after their raw data has been processed by the particle filter.

The mean and median error of motionDNA amount to 7.02 and 4.28 m respectively, while the Android sensors lead to an accuracy of 4.39 (mean) and 2.60 (median) meters. This performance gap is likely caused by two factors:

- The drift at the beginning that motionDNA often suffers from is propagated throughout the whole walk, causing a mismatch between step directions and the graph edges.
- Open spaces at the ends of the corridors allow for some overshooting, which benefits the approach using Android sensors and its slightly longer steps. The too short distance reported by motionDNA however can often not be compensated by the particle filter.

Analysis of the Area Match Score

On average, 60.6% (motionDNA) resp. 74.7% (Android sensors) of position updates match their area. Figure 7 visualizes the area match score for each area on the route. Obviously, the choice of the SHS influences the overall performance of our positioning algorithm. It cannot repair arbitrary errors of the SHS as positions too far away from any edge and directions very different from the orientation of the edges nearby the user's current position decrease the probability of the particles for these edges significantly (see Eq. 5).

As Fig. 8 illustrates, the area match score and the positioning error are inversely correlated ($r(58) = -0.87$, $p < 0.05$). From these observations we conclude that in order to support indoor navigation effectively any indoor positioning needs to be able to reliably estimate a user's relative movements. While in this study we only analyzed walking, this observation in a more general setting equally applies to other kinds of movement (e.g. climbing stairs, taking an elevator, etc.).

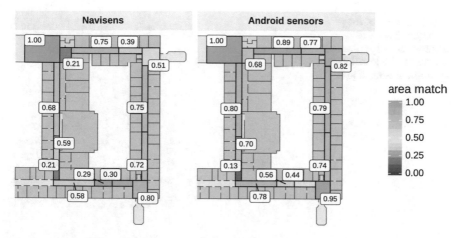

Fig. 7 Area match scores for the two motion tracking methods

Fig. 8 The area match score correlates inversely with the median positioning error. Each point represents the mean of 10 simulated runs for one actual walk

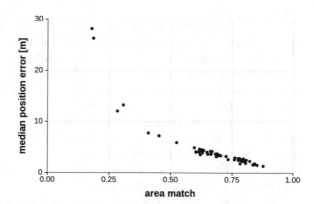

Influence of the Navigation Graph on the Area Match Score

While from the preceding analysis we learn the lesson that the area match score's precision depends on the quality of the step detection, in the following we identify other sources for area match errors.

The first source is the indoor navigation graph. Its usage introduces artifacts for the actual movement of a person as it always has to be snapped on one of the edges— sometimes a very crude discretization of the actually available degrees of freedom how to move.

While in corridors no problems may arise, Fig. 7 illustrates that in junctions and foyers, the area match score tends to decrease. In such a situation, there is only a single correct edge that can be hypothesized as the current position. However, the particle resampling may fail when the SHS misses the user's turn or at least recognizes it too late. In this case only few or even no particles are generated for the current edge while the majority of the particles hypothesizes the user to continue to

Fig. 9 Closeup of the part
of the graph that was
changed. The edges drawn in
red were added to stabilize
the position estimation at
this junction

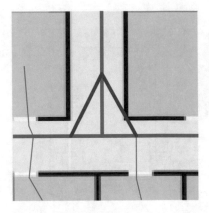

walk straight ahead. This phenomenon is particularly obvious for the junction on the bottom left in Fig. 7.

Contrarily, the three other larger foyers are represented by a densely connected net that enables the system to track almost arbitrary paths within these areas (see Fig. 1). In these areas, the area match score remains high.

This circumstance teaches us that indoor navigation graphs should not only model accessibility relations between locations in the modeled environment, but also approximate the geometry of the locations.

We tested this hypothesis by connecting the nodes adjacent to junctions with additional slanted edges as depicted in Fig. 9, the benefits of which are twofold: Firstly, it models more natural paths where the test person cuts the corner slightly; secondly, it allows for the compensation of step length differences since now multiple paths lead into the corridor that is branching off.

Using the new graph structure, we repeated the computation of the area match score. The result was not only an improvement in the area after the junction, but in all subsequent sections as well. In the small but critical area immediately after the junction, the area match score was improved by 38% (from 0.21 to 0.29) for motionDNA, and almost tripled (from 0.13 to 0.36) for the Android sensors (Fig. 10). The improvement is even statistically significant for the remainder of the route after the change: a Wilcoxon rank sum test indicates that the area match score is greater for the graph model with additional edges ($Mdn = 0.78$) than for the original version ($Mdn = 0.73$), $W = 33942$, $n_1 = n_2 = 300$, $p < 0.05$.

For reference, the median position error when calculated for the whole route also decreased from 2.60 to 2.52 m for the Android sensors, and from 4.28 to 4.11 m for the version running with motionDNA.

We conclude that by applying a systematic methodology to design an indoor navigation graph, we can almost completely eliminate the negative influence of the discretization of the physical environment that is inevitable to construct the representation of the environment.

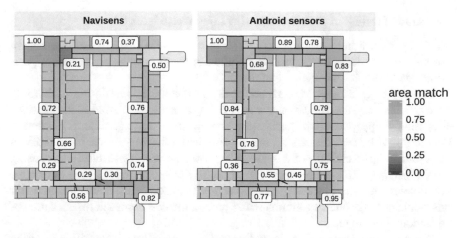

Fig. 10 Area match scores for the two motion tracking methods (with additional edges in the lower left corner)

In the remainder of this analysis, we only discuss results obtained by applying best practices learned so far: we use the internal Android SHS on the indoor navigation graph with additional edges for junctions (as shown in Fig. 9).

Influence of Area Transitions on the Area Match Score

Dividing a route into areas as explained above introduces another artifact at the boundaries of adjacent areas. It may prove problematic that boundaries are strict while SHS is noisy. Therefore, measurements taken around boundaries may be randomly assigned to one of the areas and increase the area match error.

In particular, the smaller an area is, the higher the precision of the SHS has to be for the measurement to be matched to the correct area. Therefore, in order to eliminate the influence of this artifact on the area match score, it seems justified to smooth the boundaries, allowing positions up to 2.5 m (i. e. the median position error) away from the exact boundary still to count as a match.

By loosening the definition of an area match in this way, the score increases from 0.77 to 0.88 on average, almost cutting the remaining error in half. Considering only the middle part of each area, defined as those positions that are further than 2.5 m away from each of the area's boundaries, the area match score amounts to 0.87 (strict) respectively 0.91 (approximate). On the other hand, when looking at the boundaries themselves (i.e. the interval of ±2.5 m around the boundary), the scores amount to 0.73 for the parts immediately after a segment change and 0.84 for the part at the end of each segment.

In summary, we conclude that the SHS position estimate tends to lag behind more often than it precedes the actual position.

Automatic Prediction of Locations with Low Area Match Score

Manually identifying problematic areas with a controlled experiment and known ground truth as discussed above is no solution of the productive use of indoor positioning in a complex navigation system.

Instead, the goal has to be to automatically predict these areas during positioning a user. In this way, the navigation system can be empowered to avoid such situations altogether or apply situation-specific positioning strategies (e.g. mounting beacons, interacting with users, classifying raw sensors with other algorithms than SHS). In our opinion, it is a challenge for research in indoor positioning to substitute sensing the GPS signal with a hybrid approach to classify raw data into absolute or relative coordinates. One step towards this research goal is to compute confidence scores for positioning data in order to automatically predict the best approach for each area of an (indoor) environment.

We investigated the particle set continuously computed by our algorithm and tried to detect a predictor for a confidence score of the estimated current position: During navigation, divergence monitoring can be used to identify situations where the particle distribution deviates too far from the true posterior.

The extreme case where the sum of the non-normalized particle importance weights is close to zero is already handled by the system. This generally happens when the step direction does not even approximately coincide with any adjacent edge. If such a situation is detected the particle filter is re-initialized with a spread out normal distribution around the last known location, in order to allow the position tracking to pick up again.

Another promising way to predict erroneous situations is to find a correlation between the spread of the particles and the accuracy of the positioning, the assumption being that a more scattered set of particles leads to a wrong position more often. To measure the spread of the particles, we use the root mean square error with respect to the mean of the particles. It provides meaningful values even when one dimension of the coordinates is equal for all the particles. Nevertheless, interpreting the spread of the particles proved difficult since there are multiple factors at play: At the beginning, the particles are somewhat spread out due to the particle filter's initial normal distribution. Since they are forced to move along the graph structure, variance decreases whenever there is only one possible edge in the heading direction, i. e. in corridors. Meanwhile, variance increases due to the different step lengths as long as there is no turn in the route to filter out the wrong step length hypotheses. As a result, the data can only be interpreted properly after a few turns in the route, when the influence of the particle filter initialization has decreased.

And indeed, if the test route is considered as a whole, no correlation can be found between positioning accuracy and particle dispersion. However, taking only the second half of the route into account, area match and RMSE do correlate negatively, $r(10) = -0.83, p < 0.01$. As can be seen from Fig. 11, there are matching local maxima and minima in areas 8 and 11, corresponding to areas immediately after a turn in the test route. With this knowledge, problematic areas can be identified heuristically based on the graph structure and the calculated route before the actual navigation takes place.

Fig. 11 Development of median particle RMSE and area match score over the course of the test route

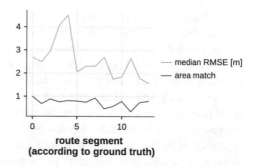

6 Conclusions

The main objective of our work was to introduce and validate the area match score as an approximate indoor positioning metric tailored to the needs of generating landmark based navigation instructions for pedestrians. An empirical evaluation of the score in a typical indoor environment displays promising results: The best value for a single run is 0.918, with a median position error of very accurate 1.07 m. The average score for the best configuration amounts to 0.770 with a median position error of 2.52 m (for a PDR system with map-matching). While these figures cannot be compared directly to those reported for the EvAAL-ETRI competition (Potortì et al. 2015), e.g. due to a different indoor layout, they show that the approach certainly can compete with the state of the art. Nevertheless, we identified room for improvements:

- **Environment Model**: In the present study, the turns in the route proved most problematic, causing a sharp decrease of accuracy in the areas after junctions. In our work, we developed a methodology how to overcome his issue. However, in almost any building, there are also other layouts such as foyers or other large open spaces. Therefore, one aspect for our future research is to generalize our findings for junctions to areas with different geometric characteristics.
- **Motion Model**: We assumed the user to steadily walk ahead on the same floor in order to be able to reliably analyze the SHS data. However, analogous models have to be developed for other ways to move (in areas of other environmental type), such as taking stairwells, elevators, or escalators. According to the most recent position estimate an indoor positioning algorithm will have to decide at runtime which of the models to be used for the analysis of raw sensor data.

The comparison of the two motion tracking solutions showed that the supposedly more sophisticated one does not outperform the built-in step counter when embedded in a more complex, non-metric approach to indoor positioning. Both suffer from a cold start problem and produce wrong estimates when indoor positioning starts. To Navisens' credit, we only used a small portion of motionDNA's capabilities and designed the experiment in a way that the Android sensors would have a reasonable chance at competing, e. g. by restricting the device location and only using a single floor for the test route.

As far as the practical purpose of implementing pedestrian navigation systems is concerned that can reliably generate instructions during the navigation phase our research and the high precision values achieved in the experiments point out a promising way to integrate metric sensor data, graph-like environment models for route calculation and landmarks and landmark based navigation strategies.

Even if only few test persons took part in our study, we argue that our experiment provides reliable results: the gait parameters of our test persons are within the margin reported by medical surveys. Furthermore, the sensor data from each walk is processed by the localization algorithm multiple times in order to eliminate noise stemming from the non-deterministic nature of the particle filter. All in all, there are 150 runs with tens of thousands of individual position updates for each configuration, which is enough data for robust results.

In our future work, we will evaluate hybrid system architectures with additional positioning data e.g. from WiFi or BLE in order to analyze their performance to reinitialize PDR after a complete failure. Furthermore, we will evaluate our approach to predict critical PDR errors in an online setting and investigate whether it allows to reduce PDR failures.

Finally, our experiments indicate that by machine learning techniques areas can be identified that lead to high errors in the area match score. Such an analysis enables us to systematically expand the indoor navigation graph in these areas in order to reduce the error rate. A second option is to prepare the environment in these critical areas e.g. by mounting BLE beacons. We will investigate whether this is a practicable strategy to further reduce the error rate or whether other hybrid strategies for indoor positioning have to be applied.

References

Basso S, Frigo G, Giorgi G (2015) A smartphone-based indoor localization system for visually impaired people. In: 2015 IEEE international symposium on medical measurements and applications (MeMeA) proceedings, pp 543–548

Bohannon RW, Williams Andrews A (2011) Normal walking speed: a descriptive meta-analysis. Physiotherapy 97(3):182–189

Davidson P, Piché R (2016) A survey of selected indoor positioning methods for smartphones. IEEE Commun Surv Tutor

Ebner F, Fetzer T, Deinzer F, Köping L, Grzegorzek M (2015) Multi sensor 3D indoor localisation. In: 2015 international conference on indoor positioning and indoor navigation (IPIN)

Guo H, Uradzinski M, Yin H, Yu M (2015) Indoor positioning based on foot-mounted imu. Bull Pol Acad Sci Tech Sci 63(629–634):3

Harle R (2013) A survey of indoor inertial positioning systems for pedestrians. IEEE Commun Surv Tutor 15(3):1281–1293

Herrera JCA, Plger PG, Hinkenjann A, Maiero J, Flores M, Ramos A (2014) Pedestrian indoor positioning using smartphone multi-sensing, radio beacons, user positions probability map and indoor-osm floor plan representation. In: 2014 international conference on indoor positioning and indoor navigation (IPIN), pp 636–645

Hilsenbeck S, Bobkov D, Schroth G, Huitl R, Steinbach E (2014) Graph-based data fusion of pedometer and wifi measurements for mobile indoor positioning. In: Proceedings of the 2014 ACM international joint conference on pervasive and ubiquitous computing

Kattenbeck M (2016) Empirically measuring salience of objects for use in pedestrian navigation. Dissertation, University Regensburg

Liao L, Fox D, Hightower J, Kautz H, Schulz D (2003) Voronoi tracking: location estimation using sparse and noisy sensor data. In: Proceedings 2003 IEEE/RSJ international conference on intelligent robots and systems (IROS 2003), vol 1, pp 723–728

Link JB, Smith P, Viol N, Wehrle K (2013) Accurate map-based indoor navigation on the mobile. J Locat Based Serv 7(1):23–43

Lymberopoulos D, Liu J, Yang X, Choudhury RR, Handziski V, Sen S (2015) A realistic evaluation and comparison of indoor location technologies: experiences and lessons learned. In: Proceedings of the 14th international conference on information processing in sensor networks, ACM, pp 178–189

Muro-de-la Herran A, Garcia-Zapirain B, Mendez-Zorrilla A (2014) Gait analysis methods: An overview of wearable and non-wearable systems, highlighting clinical applications. Sensors (Basel, Switzerland) 14(2):3362–3394

Oberg T, Karsznia A, Oberg K (1993) Basic gait parameters: reference data for normal subjects, 10–79 years of age. J Rehabil Res Dev 30(2):210–23

Ohm C, Müller M, Ludwig B (2015) Displaying landmarks and the user's surroundings in indoor pedestrian navigation systems. J Ambient Intell Smart Environ 7(5):635–657

Pham DD, Suh YS (2016) Pedestrian navigation using foot-mounted inertial sensor and lidar. Sensors 16(1)

Potortì F, Barsocchi P, Girolami M, Torres-Sospedra J, Montoliu R (2015) Evaluating indoor localization solutions in large environments through competitive benchmarking: the EvAAL-ETRI competition. In: 2015 international conference on indoor positioning and indoor navigation (IPIN). IEEE, pp 1–10

Romanovas M, Goridko V, Klingbeil L, Bourouah M, Al-Jawad A, Traechtler M, Manoli Y (2013) Pedestrian indoor localization using foot mounted inertial sensors in combination with a magnetometer, a barometer and RFID. Springer, Berlin, Heidelberg, pp 151–172

Rossi M, Seiter J, Amft O, Buchmeier S, Tröster G (2013) Roomsense: an indoor positioning system for smartphones using active sound probing. In: Proceedings of the 4th augmented human international conference, ACM, New York, NY, USA, AH '13, pp 89–95. https://doi.org/10.1145/2459236.2459252, http://doi.acm.org/10.1145/2459236.2459252

Sprager S, Juric MB (2015) Inertial sensor-based gait recognition: a review. Sensors 15(9):22, 089–22, 127

Susi M, Renaudin V, Lachapelle G (2013) Motion mode recognition and step detection algorithms for mobile phone users. Sensors (Basel, Switzerland) 13(2):1539–1562

Thrun S, Burgard W, Fox D (2005) Probabilistic robotics (Intelligent robotics and autonomous agents). The MIT Press

Torres-Sospedra J, Moreira A, Knauth S, Berkvens R, Montoliu R, Belmonte O, Trilles S, João Nicolau M, Meneses F, Costa A et al (2017) A realistic evaluation of indoor positioning systems based on wi-fi fingerprinting: the 2015 EvAAL-ETRI competition. J Ambient Intell Smart Environ 9(2):263–279

Verma S, Omanwar R, Sreejith V, Meera GS (2016) A smartphone based indoor navigation system. In: 2016 28th international conference on microelectronics (ICM), pp 345–348

Waqar W, Chen Y, Vardy A (2016) Smartphone positioning in sparse wi-fi environments. Comput Commun 73:108–117

An Original Approach to Positioning with Cellular Fingerprints Based on Decision Tree Ensembles

Andrea Viel, Andrea Brunello, Angelo Montanari and Federico Pittino

Abstract In addition to being a fundamental infrastructure for communication, cellular networks are employed for positioning through signal fingerprinting. In this respect, the choice of the specific strategy used to obtain a position estimation from fingerprints plays a major role in determining the overall accuracy. In this paper, a new machine learning approach, based on decision tree ensembles, is outlined and evaluated against a set of well-known, state-of-the-art fingerprint comparison functions from the literature. Tests are carried out with different tracking devices and environmental settings. It turns out that the proposed approach provides consistently better estimations than the other considered functions.

Keywords Positioning · Fingerprinting · Cellular · Machine learning Random forest

1 Introduction

Due to the large variety of its application fields, which range from location-based services to asset tracking and fleet management, location estimation is a quite active research area.

At the present days, the Global Positioning System (GPS) is the standard de facto Global Navigation Satellite System (GNSS) for positioning. However, despite of its widespread usage, GPS has some shortcomings: (i) its availability and performance

A. Viel (✉) · A. Brunello · A. Montanari
University of Udine, Udine, Italy
e-mail: viel.andrea@spes.uniud.it

A. Brunello
e-mail: andrea.brunello@uniud.it

A. Montanari
e-mail: angelo.montanari@uniud.it

F. Pittino
u-blox Italia SpA, Trieste, Italy
e-mail: fedepittino@gmail.com

© Springer International Publishing AG 2018   49
P. Kiefer et al. (eds.), *Progress in Location Based Services 2018*, Lecture Notes
in Geoinformation and Cartography, https://doi.org/10.1007/978-3-319-71470-7_3

are reduced in certain environments, such as indoor areas and urban canyons, (ii) sometimes, a long time is required to obtain a position fix, and (iii) it has a high energy consumption, which can be a problem for battery-powered devices (Li et al. 2014; Paek et al. 2011; Zhuang et al. 2010).

To overcome these weaknesses, GPS is often paired with localization methods that exploit cellular networks. Given the ubiquitous use of cellular modules as a means of communication, this combination turns out to be a cost-effective solution. In particular, an approach which can be thought of as complementary or alternative to GPS is provided by signal fingerprinting, that is, the localization of a device by means of the fingerprint of the received signals. Signal fingerprinting makes it possible to estimate the current position of a device by correlating the radio signals it detects (its fingerprint) with the set of past observations, each one tagged with the respective GPS position and stored in a database (Chen et al. 2006). Fingerprinting gives a quick response and has limited power consumption; its precision depends on the size of the cells and on the spatial distribution of the stored observations.

Machine learning techniques have already been successfully applied for the purpose of numeric prediction, such as scoring, in a variety of domains, ranging from education (Romero and Ventura 2010) to health-care (Tomar and Agarwal 2013) and economy (Ngai et al. 2011). Among such techniques, decision trees are a popular method for many predictive tasks. In the following, we show that they can be profitably employed also in fingerprint positioning systems, with the goal of evaluating the distance between fingerprints. One of the crucial features of a fingerprint positioning system is the method it uses to obtain a position estimation by comparing and matching the fingerprint submitted by a device with those stored in the database. In this paper, we demonstrate that decision trees can be a valid alternative to existing solutions for such a task.

We start with a short account of the most commonly used fingerprint comparison functions among those proposed in the literature, which takes into consideration a wide range of different applicative environments. Then, a novel approach based on machine learning is outlined, which is shown to produce position estimations with higher accuracy than previous methods. Unlike the other comparison functions analysed in this work, such a solution also provides a measure of the uncertainty of the position estimation. As a matter of fact, some machine learning approaches to position estimation have already been proposed in the literature. The one described in this paper differs from them in various respects; in particular, it is not limited by the number of received signals.

The rest of the paper is organized as follows. In Sect. 2, we provide some background knowledge on fingerprint positioning systems, with a special attention to the metric they adopt to measure the distance between fingerprints. Next, in Sect. 3, we introduce some basic notions about decision tree ensembles. In Sect. 4, we describe the datasets used in this work as well as the new fingerprint comparison method. Then, in Sect. 5, we report the outcomes of an extensive experimental evaluation of the described techniques on various, heterogeneous datasets. Finally, in Sect. 6, we briefly analyse related work. Section 7 summarizes the achieved results and outlines possible directions for future work.

2 Fingerprint Positioning Systems

In this section, we introduce the basic features of fingerprint positioning systems and the main metrics that they exploit. Fingerprint positioning systems, also known as database correlation methods (DCM), are a class of positioning systems that make use of the variance of signals received by a device in different positions. While they can be applied in various contexts and with several wireless technologies, in this paper we focus on cellular network signals obtained in an outdoor setting.

One of the main advantages of fingerprint position systems is that, unlike other cellular-based positioning systems, they do not require any assistance from the network operator and can be implemented using only parameters available on client side. The latter is an advantage also from a privacy point of view, as the process of location estimation can be entirely done on the device itself.

In cellular systems, fingerprints commonly include the received signal strength of the cells observed by the device, and they allow one to distinguish between the *serving cell*, which is the one the device is currently connected to, and the other cells, which are referred to as *neighbours*. Serving cells in GSM and LTE networks have an additional parameter, called Timing Advance (TA), which can be exploited to enrich the fingerprint with a discrete measure of the distance between the base station of the cell and the device. Structurally, a fingerprint can be viewed an array of arrays, where the outer one represents the list of observed cells (the serving cell and its neighbours), and each inner array provides detailed information about a cell like the network operator, the Cell-ID, and the signal strength.

Fingerprint positioning methods typically consist of two distinct phases, commonly referred to as the *off-line* (or training) and the *on-line* (or positioning) phases. In the off-line phase, fingerprints are collected at various locations by surveys (or are generated by simulation models) to form a fingerprint map (aka radio map), stored in a database. This is one of the common parts of every fingerprint positioning system. During the on-line phase, when a device asks for an estimation, its current signal fingerprint is compared to the fingerprints stored in the database to find the best match. In the Nearest Neighbour (NN) method, the estimated position corresponds to the one in which the most similar fingerprint was collected (according to a suitable distance function), while, in the k-Nearest Neighbours (k-NN) method, the k-closest neighbours are averaged, possibly employing distance-based weighting (Weighted k-Nearest Neighbours).

In the literature, several functions for assessing the similarity between fingerprints have been defined. In this work, we make an evaluation of the performance of the metrics proposed in the literature on both Wi-Fi and cellular fingerprinting, namely, Euclidean distance, Spearman correlation, hyperbolic fingerprinting, and relative RSS-based fingerprinting.

Euclidean Distance. Euclidean Distance is the earliest defined and most commonly used metric in fingerprint positioning systems (Bahl and Padmanabhan 2000). Formally, it measures the Euclidean distance between the fingerprints in a n-dimensional space, where n is the number of different beacons, such as cells or access points.

The coordinates of the points correspond to the signal strengths. Conceptually, the computed value evaluates how similar signal measurements are to each other: the smaller the Euclidean distance between two measurements is, the more similar they are. The formula is shown in Eq. 1, where $ss_{k,i}$ is the signal strength of the k-th beacon in the i-th fingerprint f_i:

$$d_{eu}(f_1, f_2) = \sqrt{\sum_{i=1}^{n} (ss_{i,1} - ss_{i,2})^2} \tag{1}$$

The formula can be applied straightforwardly on the cells appearing in both fingerprints, while for the others a penalty term is usually added. In this work, cells which are not in common between the two fingerprints are considered to have signal strength equal to zero, as if their signals were too weak to be received.

Spearman Correlation. Spearman Correlation (Zekavat and Buehrer 2011) measures the similarity between two ordered sets of values by looking at their rank, rather than their absolute values (as it happens with the more common Pearson correlation). Such an approach can be understood as follows. Consider two devices placed at the same position. For some reason, such as, for instance, the different gain of their antennas, they might record slightly different signal measurements originating from the same beacons. Still, such beacons should be similarly ranked by the two devices. The formula is shown in Eq. 2, where rg_{f_i} are the signal strengths of the i-th fingerprint converted to their rank:

$$\rho(f_1, f_2) = \frac{cov(rg_{f_1}, rg_{f_2})}{\sigma_{rg_{f_1}} \sigma_{rg_{f_2}}} \tag{2}$$

Hyperbolic Fingerprinting. Hyperbolic Fingerprinting measures the differences in signal strength ratios between pairs of beacons. Such a measure can be interpreted as follows: even if the signal itself tends to fluctuate, and different devices typically have different receiving capabilities, the ratios between the received signals should be stable. The method was proposed in Kjrgaard and Munk (2008) for Wi-Fi fingerprinting, where the problem is exacerbated by the lack of a standard metric for reporting the strength of the signal. After computing all the ratios in each fingerprint, the difference between them is evaluated as in the Euclidean distance formula (Eqs. 3 and 4):

$$r(x, y) = log(\frac{x}{y}) \tag{3}$$

$$d_{hyp}(f_1, f_2) = \sqrt{\sum_{i=1}^{n} \sum_{j=i+1}^{n} (r(ss_{i,1}, ss_{j,1}) - r(ss_{i,2}, ss_{j,2}))^2} \tag{4}$$

Relative RSS-based Fingeprinting. Relative RSS-based Fingerprinting (Meniem et al. 2013) can be viewed as a simplified version of Hyperbolic fingerprinting: instead of looking at signal strengths ratios, only the relative order (greater, less, equal) between the signal strength pairs is checked. Then, the similarity between two fingerprints is given by the ratio of the number of matching pairs over the total number of pairs. It is obvious that, since the number of observable cells at a given time instant is limited (usually up to seven cells), in practice there is just a fixed number of possible ratios.

The above metrics were chosen since all of them, with the exception of the first one, try to cope with two basic problems: (i) the natural fluctuation of the signal strengths and (ii) the different receiving capabilities of the devices. As for the former, it is well known that, even if placed at the same position, a device may receive different signals over time, as they are affected by various environmental factors. The latter refers to the fact that the types of mobile devices that are used for creating radio maps in the training phase may be different from the ones that are used in the positioning phase.

3 Decision Tree Ensembles

In this section, we introduce the machine learning tool we are going to use in the following, namely, decision tree ensembles.

Data mining can be defined as the process of analysing huge quantities of data in order to extract meaningful patterns, which were previously unknown, or only merely presumed. Such regularities can then be used to increase one's knowledge about the specific domain, or may be exploited to derive rules, for the purpose of automatic classification or prediction (Witten et al. 2011). Patterns are typically captured by models, which are inferred from the data by means of a suitable *machine learning algorithm*.

In this paper, we focus on the problem of *regression*, which is a form of *supervised learning*. The algorithm is given a set of training examples, each one characterized by a set of *predictor* attributes and a numerical *label*. The resulting model encodes a mapping between the predictors and the label, and can be used to assign a value to instances for which the value of the label is unknown.

Decision trees are a popular method for many supervised machine learning tasks, owing their success mainly to their efficiency, during both the learning and the prediction phases, as well as to their intuitive interpretability. However, a drawback of decision trees is that they are usually less accurate in their predictions than other methodologies, and have a tendency to overfit training data, that is, they have low bias, but very high variance (Hastie et al. 2009). A possible way to improve their accuracy, at the expense of a loss in the interpretability of the model and of a higher complexity in the training and prediction phases, is to build a set of different trees (ensemble), and then combine the single predictions in order to output the final result. Various methodologies can be used to build such an ensemble; one of the most famous is *bootstrap aggregating*, or *bagging*.

Consider a labelled dataset D, made by n instances, and a decision tree learning algorithm L. As noted before, decision tree learning algorithms have the characteristic of being *unstable*, meaning that little changes in the input may produce very different trees as output. In order to train an ensemble of k tree models, bagging exploits such an instability in the following way: for $i = 1, \ldots, k$, a dataset D^i is generated by randomly drawing $|D|$ instances from D with replacement (that is, the same instance may occur multiple times in D^i). Then, algorithm L is applied on each of the datasets, with the result of obtaining k different trees. In the prediction phase, the single tree outcomes are simply combined by voting (in a classification setting) or averaging (in case of regression).

Let k be the number of trees to generate in the ensemble. In the learning phase, the RF algorithm operates as follows: as in bagging, the dataset is repeatedly sampled with replacement for k times, obtaining k different datasets having the same cardinality as the original one. Then, a so-called *random tree* is built from each dataset, by selecting as the split criterion at each node the best attribute (according to a predefined measure) from a randomly determined set of predictors. The reason for considering only a subset of all attributes at each split is the correlation of the trees in an ordinary bagging approach: if one or a few attributes are very strong predictors for the response variable, these features will be selected in many of the trees in the ensemble, thus preventing one from achieving a high degree of variability among the models. Empirically, the RF algorithm has proven to be capable of obtaining very good performances in terms of prediction accuracy in many application domains (Hall et al. 2009).

4   Experimental Setup

In this section, we introduce the framework within which the various metrics for fingerprint positioning systems have been experimented and compared. In particular, we describe the employed dataset and tools.

4.1   Datasets

The performances of the various methods for comparing fingerprints are evaluated using four real-world datasets of GSM cellular fingerprints paired with a GPS position. Each dataset was obtained in a different setting, and it is identified by a letter from A to D.

Datasets A, B and C are provided by an external party and contain fingerprints coming from several devices with different characteristics. In particular, fingerprints in these datasets have been sparsely collected over large areas of different locations.

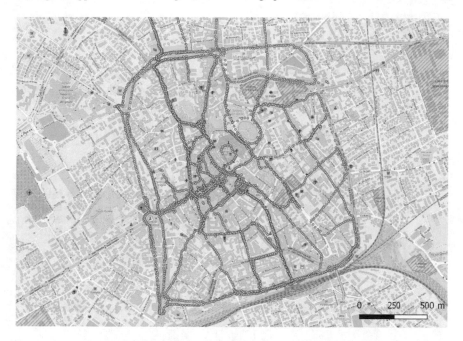

Fig. 1 Dataset D observations over the historical centre of Udine, Italy. Map Data © OpenStreetMap contributors, CC BY-SA

Table 1 Characteristics of the datasets used in the experiment

Dataset	A	B	C	D
Location	South Africa	Poland	Taiwan	Udine (Italy)
Kind of environment	Mixed	Mixed	Mixed	Urban
Number of fingerprints	4,330,382	265,276	324,755	4,802

The dataset D was collected by one of the authors by wardriving on foot and by bike over several days in the historical centre of the city of Udine, Italy. The main characteristics of the datasets are summarized in Table 1.

A graphical representation of the observations contained in Dataset D, plotted over the test area, is shown in Fig. 1. They were collected by a Sony Xperia Z3 Compact phone using an Android application developed for the purpose. Note that, because of API limitations, the observations in dataset D, differently from the others, do not contain the TA attribute.

The density of the collected fingerprints varies across the different datasets, and measuring its exact value is not trivial since it can considerably change also moving from one area to another one of the same dataset. As for the process of position estimation, instead of directly measuring such a density by the number of fingerprints

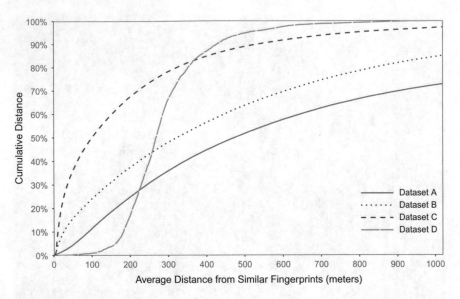

Fig. 2 Cumulative plot of the average distance between a fingerprint and its similar neighbours

in a certain area, it turns out to be more significant to consider the distribution of similar fingerprints.

A fingerprint is considered similar enough to the reference fingerprint if it has at most p different cells with respect to it (we use $p = 2$). This rough filtering reduces the number of candidate fingerprints for every position estimation, and it has been applied also in the rest of the paper. Figure 2 shows the cumulative plot of the average distance from a fingerprint and the surrounding ones which are similar enough to be used for the position estimation. It represents a measure of the density of the fingerprints from the point of view of their usefulness for position estimation. A high value means that similar fingerprints are close to each other, and this makes it easier to discriminate between different locations and thus to obtain good position estimations. Figure 3 reports information about the standard deviation, showing the regularity of the previous value across the entire dataset.

Putting together information about the density of the fingerprints, as emerging from Fig. 2, and information about their consistency, as represented in Fig. 3, it is possible to characterize the four datasets in the following way: datasets A and B have a low density of fingerprints, typically spread over a large area, while datasets C and D have a considerably higher concentration of observations, and this is especially true for dataset D, which was collected in a small urban area.

Since the training phase of the Random Forest model is memory intensive, two random subsets are extracted from each dataset by varying the initial seed. The first one has been used for training, while the latter has been kept aside for the evaluation process. In both subsets, each fingerprint is paired with others as explained in next section. The same test subset is also used for the evaluation of positioning methods

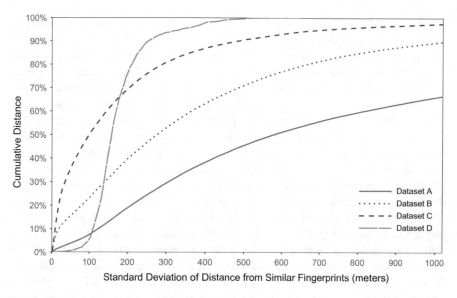

Fig. 3 Cumulative plot of the standard deviation of the distance between a fingerprint and its similar neighbours

Table 2 Size of training and evaluation subsets

Dataset	A	B	C	D
Training (%)	0.05	4	4	30
Evaluation (%)	0.05	4	4	30
Training pairs	146,308	257,500	313,279	96,302
Evaluation pairs	2,101,762	553,567	1,449,187	222,402

that do not require training, like the Euclidean one. The percentage of data assigned to each dataset is shown in Table 2. Since for the dataset D we use a large percentage of the whole dataset, we make sure that the same pair of fingerprints does not appear in both the training and the evaluation subset.

In the following, we illustrate the distinctive features of the fingerprint comparison method based on decision tree ensembles that we propose. In next section, we describe the process of model training; then, in the subsequent section, we present the process of position estimation.

4.2 A Novel Fingerprint Comparison Method: Model Training

A Random Forest model needs to be trained before its actual use. In previous contributions (Bozkurt et al. 2015; Jedari et al. 2015; Sánchez et al. 2012), machine learning models have been trained using the signal strength of each beacon as predictors, while the output of the model is the estimated position, usually one of the training points. Such an approach may be appropriate for indoor environments, where the number of beacons (usually Wi-Fi access points) is low and mostly fixed, but it does not scale in an outdoor setting, where there can be thousands of cells.

Here we follow a different approach. Models are trained on a set of instances, each obtained from pairs of fingerprints. The features describe the similarity between the two considered fingerprints, such as the number of common cells and their average signal strength difference. The output of the model is the estimated distance between the locations in which the two fingerprints were collected. It is worth pointing out that having a numeric distance, instead of a position, as output has the advantage of providing also a measure of the accuracy of position estimation.

For each dataset, the training instances are selected by taking a subset of the fingerprints, and then pairing them with all the other fingerprints (also outside the generated subset) having at most two different cells. Such a process is similar to the process of the position estimation, in which an input fingerprint is compared to all the similar fingerprints stored in the database to find the best match. In order to limit the size of the training set, and thus the computational burden of the learning phase, each starting fingerprint is matched only with the first hundred ones which are the closest geographically.

Such an approach was satisfactory for all datasets but dataset C. In dataset C, indeed, fingerprints are very dense, and the proposed approach was not sufficient as it left out fingerprints taken at larger distances. Thus, for this dataset, the training instances were chosen in such a way that they included also a certain number of farther observations. The rationale is that, although the classifier should be trained to discriminate especially between closely collected fingerprints (as this is the most difficult situation), the training set must include also fingerprints which are barely related in order to learn how to discern them.

Compared to the classical distance metrics, the machine learning model allows one to consider features which are not necessarily derived from the signal strength, like, for instance, the Timing Advance, or that describe the characteristics of the device. This is particularly convenient, because fingerprints collected from the same device model, or even from the same device, are more correlated to each other.

For each pair of fingerprints in the training set, a number of predictor attributes are generated to describe their differences:

- same device, indicating whether the devices that collected the two fingerprints are exactly the same or not;
- same device model, indicating whether the devices that collected the two fingerprints belong to the same model or not;
- same serving cell, indicating whether the devices that collected the two fingerprints had the same serving cell or not;
- serving cell signal strength difference, which takes a proper value if the devices had the same serving cell;

- Timing Advance difference, which takes a proper value only for the GSM case, and only if the devices had the same serving cell;
- number of common cells between the two fingerprints, expressed as an absolute value;
- number of common cells between the two fingerprints, expressed as a fraction of the total number of cells;
- number of cells that are not in common between the two fingerprints, expressed as an absolute value;
- number of cells that are not in common between the two fingerprints, expressed as a fraction of the total number of cells;
- average difference in signal strength over the common cells;
- average difference in signal strength over the cells that are not in common;
- average signal strength of the serving cells, if they are not the same but are still observed in the other fingerprint among the neighbour cells;
- average signal strength of the serving cells, if they are not the same and each one is not observed in the other fingerprint;
- Euclidean Distance value, as calculated between the two fingerprints;
- Spearman Correlation value, as calculated between the two fingerprints;
- Hyperbolic Fingerprinting value, as calculated between the two fingerprints;
- Relative RSS-based Fingerprinting value, as calculated between the two fingerprints.

The numerical label for the regression model is given by the actual distance between the points where the two fingerprints were collected.

Once the training set generation had been completed, a wrapper-based *Attribute Selection* has been carried out by making use of the methods available in Hall et al. (2009), and exploiting RF as the base classifier. As a matter of fact, after the application of the selection phase on data taken from the different datasets, it turned out that no predictors have been eliminated via this process, meaning that, potentially, all of them might influence the final prediction. Indeed, even the removal of a single attribute reduced the prediction performance of the ensemble.

4.3   A Novel Fingerprint Comparison Method: Position Estimation

The evaluation was carried out by randomly selecting a subset of fingerprints from each dataset, and by comparing them to the remaining ones by means of the different metrics. This is the same process as the one used in a fingerprint positioning system to estimate the position of a device given an input fingerprint.

The position estimation was done by the (single) Nearest Neighbour method, as we empirically observed that taking more than one candidate did not improve the positioning performance. In the presence of multiple fingerprints with the same scoring, their estimation was averaged. The most likely reason is that in sparse datasets the neighbours could be far from each other.

As far as Euclidean and Hyperbolic distances are concerned, the best fingerprint is the one with the smallest score, while for Spearman and Relative scoring the best fingerprint is the one with the highest value. Machine learning models have been trained to return a numeric distance, and thus the best fingerprint corresponds to the one with the smallest predicted value.

The Weka toolset provided the implementations of the machine learning algorithms used in this work. For the case of Random Forest, most of the parameters were left at their default settings, except for *numIterations*, which was set to 60, and *breakTiesRandomly*, which was set to True.

The non-machine learning methods, described in Sect. 2, were implemented in PL/pgSQL and executed on the PostgreSQL (PostgreSQL Global Development Group 2008) database where the fingerprint datasets reside.

5 Results

This section presents the results of the comparison of the different methods for computing position estimations across several datasets.

5.1 Positioning Performance

We evaluated the positioning accuracy of the different methods across the different datasets. Table 3 reports the average and median error for each of them. It is worth noticing that the average is always much higher than the median, meaning that there is a certain number of observations with a high positioning error. In addition, the error has a high variance between the different datasets. Such a phenomenon is probably due to the different characteristics of the datasets, especially for what concerns the fingerprint density. Nevertheless, independently from the considered dataset, the

Table 3 Average (and median) positioning error

Dataset	A	B	C	D
Euclidean	523m (53m)	256m (39m)	147m (31m)	33m (15m)
Spearman	637m (96m)	288m (49m)	153m (34m)	45m (23m)
Relative	772m (173m)	365m (95m)	145m (37m)	126m (92m)
Hyperbolic	627m (69m)	341m (54m)	165m (35m)	47m (17m)
Random forest	312m (22m)	219m (31m)	136m (30m)	22m (13m)

RF-based method provided a better position estimation than the one given by the classical approaches.

An interesting outcome of the experimentation is that in most datasets the Euclidean distance method performed better than the other methods discussed in Sect. 2. This comes as a surprise since the latter ones were designed to surpass it, especially in situations where the collection of the fingerprints was done by different devices, as in most of the considered datasets.

Relative scoring performed poorly in almost all datasets, most probably because, as pointed out in Sect. 2, the number of possible values is limited, and thus it does not provide a good way to discriminate between fingerprints.

In Fig. 4, Fig. 5, and Fig. 7 we show the cumulative plot of the positioning error across the dataset A, B, and D, respectively. It is easy to observe that there is a noticeable difference among the various methods and Random Forest is significantly better than all the others. Euclidean distance evaluation strategy tends to follow as the second best, as it performs the best among the classical ones.

As for dataset C, the improvement given by the Random Forest is less evident. As shown in Fig. 6, the differences among the various methods is modest and it becomes more difficult to discriminate the lines in the plot. Since the dataset C is the only one exhibiting this behaviour, the most probable reason for such an outcome is in the environment, e.g., differences in the deployment of the cellular network. The high density of the fingerprints can be thought of a further explanation of the phenomenon; however, it must be observed that dataset D has an even higher density, but it behaves in a quite different way.

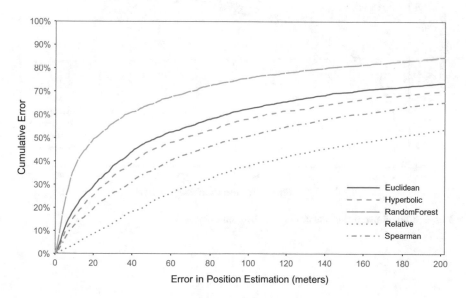

Fig. 4 Cumulative plot of the positioning error on Dataset A

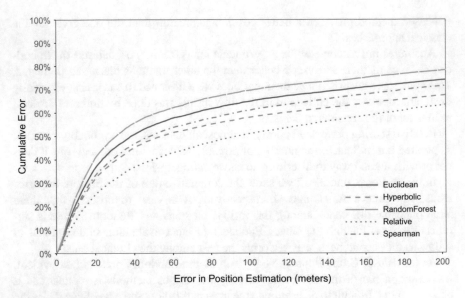

Fig. 5 Cumulative plot of the positioning error on Dataset B

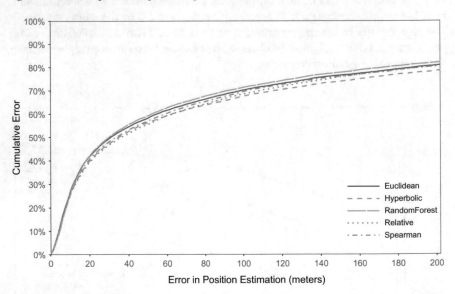

Fig. 6 Cumulative plot of the positioning error on Dataset C

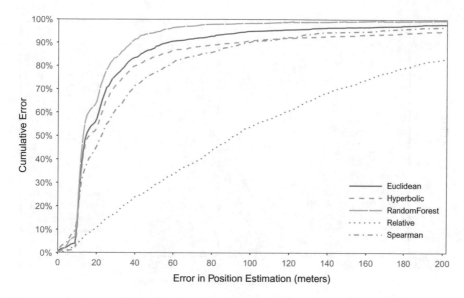

Fig. 7 Cumulative plot of the positioning error on Dataset D

Dataset D is the one with the lowest positioning error. This is due to the fact that such dataset consists of highly dense observations, collected in a single city centre. Moreover, all the fingerprints were collected by the same device. As a result, there is a strong correlation between the signals and the positions.

Unlike the other test cases, the observations in this dataset do not include the TA attribute, which, in principle, could have further improved the estimation. However, in this dataset the Random Forest model is already approaching the error of a GPS receiver, and thus it is difficult to expect any large improvement.

In the cumulative plot shown in Fig. 7, it can be easily noticed that the curves raise sharply around the values of 10 and 20 m. The reason is that the program that collected the fingerprints on the phone while war-driving was set to record them at approximately 10 m of distance each other.

5.2 Uncertainty Measure

As we already observed, the Random Forest models trained in this work are not used to directly obtain a position estimation, but to provide a measure of the distance from the estimation itself. Such a distance gives a direct measure of the predicted uncertainty around the position estimation. Ideally, this value should be as close to the actual one as possible. However, in order for this measure to be useful in practice, it is sufficient for it to be able to provide an upper bound on the error in the position estimation.

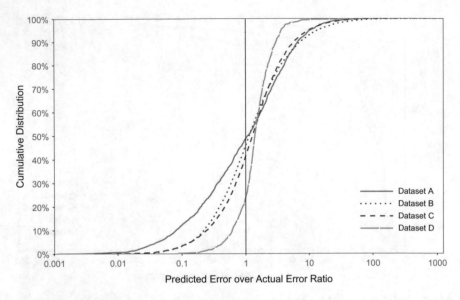

Fig. 8 Cumulative plot of the ratio between predicted and actual positioning errors

Figure 8 shows the plot of the ratio between the predicted error from the model and the actual error obtained from the distance between the position estimation and the GPS fix. The vertical line intersecting the horizontal axis at 1 divides the plot in two sides: on the left there are the estimations in which the error was underestimated, and on the right those in which the error was overestimated. In all of the four datasets, it is possible to observe that the models, most of the time, provide an overestimation of the error, since the curves are skewed on the right. The best case corresponds to dataset D, in which the trained model gives a correct estimation about 80% of the time, while the worst is given by dataset A, in which the corresponding model reports a correct error estimation just over 50% of the time.

Though the ratio gives us an intuition about the behaviour of the models, it is also interesting to assess the extent of the actual error. Figure 9 and Fig. 10 show the distribution of the difference between the predicted and actual error, respectively, when the latter is underestimated and overestimated.

Looking at Fig. 9, we may observe that datasets A, B, and C exhibit a similar behaviour, that is, roughly 20% of the time the error in the uncertainty is over 100 m. The model trained on dataset D performs better, with the error in the uncertainty rarely surpassing 10 m. This is somehow expected, since the overall positioning errors in dataset D are lower than in the other datasets, as we have already shown.

Even though an overestimated uncertainty is less serious than an underestimated one, Fig. 10 shows that the value rarely differs significantly from the actual positioning error and this is especially true for datasets A and D.

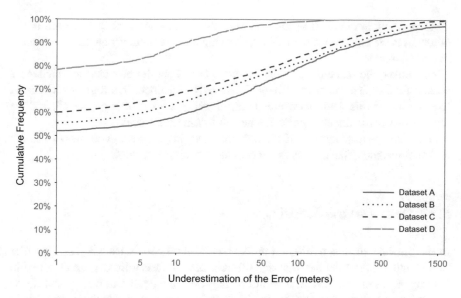

Fig. 9 Cumulative plot of the difference between predicted and actual positioning errors (when the error is underestimated)

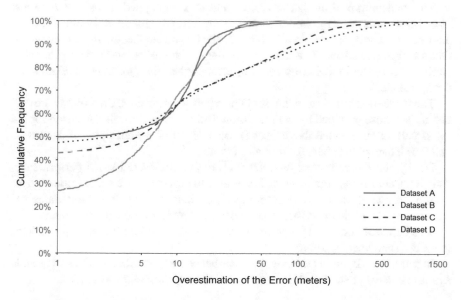

Fig. 10 Cumulative plot of the difference between predicted and actual positioning errors (when the error is overestimated)

Overall, more than half of the accuracy measures were correct, and even in the other cases the error was often limited, showing that the accuracy measure is reliable in most cases.

Moreover, the machine learning methods can generally be tuned to privilege a certain kind of error over the others. As already explained, in a regression setting, like the one in this work, an overestimation of the error is to be preferred, and we plan to systematically explore such a possibility in future work.

This is a clear advantage of the machine learning approach with respect to the classical methods, that cannot easily provide a similar measure.

5.3   Cross-Dataset Model

In the previous results, a different model has been trained on each dataset, in order to take into account the fact that the different sets of observations were obtained in heterogeneous environments. It would be interesting, however, to understand if it is possible to apply the same model to different datasets, possibly with a small loss in accuracy.

The models trained on datasets B, C, and D were applied on dataset A to test how the positioning error would change compared with the model trained on that dataset. The results are shown in Table 4. It can be noticed that the model trained on dataset B behaves on dataset A in a very similar way as the "native" model, producing similar average and median positioning errors. Also, the plot lines in Fig. 11 are nearly identical.

The model originated from dataset D provides worse results, even if they are still better than those provided by the Euclidean distance. The most likely reason for the lower performance is that the dataset D has a subset of the attributes of dataset A, and thus it cannot take benefit from all of them.

Finally, the model trained on dataset C showed the overall worst performance. This is not entirely surprising since it was able to provide only a small improvement even on its own dataset. In any case, this strengthens our hypothesis that dataset C has some intrinsic characteristics that make it different from the other datasets.

Together with the results reported in the previous section, it is now possible to identify a fundamental pattern.

In datasets A, B, and D there was a clear benefit in using Random Forest models against classical methods, while in dataset C the improvement was modest.

Table 4 Average (and median) cross-dataset positioning error

Dataset for training	A	B	C	D
Positioning error	321m (22m)	280m (23m)	584m (71m)	522m (41m)

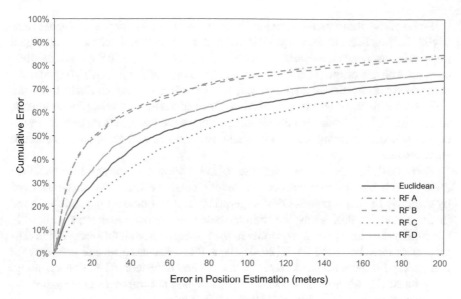

Fig. 11 Cumulative plot of positioning error when applying the different trained models on Dataset A

Moreover, the model trained on dataset B can work almost equally well on dataset A, and this is true up to a certain point also for the model trained on dataset D.

These findings suggest that there are characteristics in the datasets A and B that make them similar to each other, while dataset C is significantly different, thus most probably requiring a different approach to obtain similar results.

In any case, the exact way to recognize behaviourally similar datasets is a matter for future study.

6 Related Work

Various localization methods exploiting cellular networks have been proposed in the literature (Deblauwe 2008). Some of them make use of the observation of network signals, like Cell-ID or signal fingerprinting, others execute specific measurements of network parameters. Among the latter ones, we mention those based on Time of Arrival (Caffery and Stuber 1998) and those using Time Difference of Arrival (Spirito and Mattioli 1998), which make accurate time measurements in order to determine the distance from the device to the base station. To work properly, both these methods need to know the positions of the base stations with a high precision. Another solution is offered by a technique based on the Angle of Arrival, which requires antenna arrays at the base stations to determine the direction of the signal (Caffery and Stuber 1998; Deligiannis et al. 2007; Qi et al. 2006).

Some of the above methods require network operator assistance or impose non-trivial modifications to the network infrastructure. This is not the case with signal fingerprinting, which correlates the Received Signal Strength (RSS) from multiple beacons with the position of the device. An advantage of the fingerprinting method over the above-mentioned ones is that it is more tolerant to signal fluctuations and multi-path effects, which is a particularly helpful feature in urban environments where also GPS is mostly at a disadvantage. Such a method has been used in a variety of contexts, including Wi-Fi and cellular networks as well as indoor and outdoor environments.

Early work in this area includes the RADAR system (Bahl and Padmanabhan 2000), which demonstrates that accurate indoor location estimations can be achieved by using Wi-Fi access points. GSM fingerprinting in an outdoor context is analysed in Chen et al. (2006), where the authors show that the outcomes are significantly affected by the quality and the quantity of the fingerprints available for position estimation. In Retscher and Joksch (2016), the authors discussed several measures for comparing distances between vectors in the context of indoor Wi-Fi fingerprinting. They found that all measures do not provide significant improvements against the commonly employed Euclidean distance.

Data mining techniques, such as decision trees and Random Forest, have been already used in positioning-related applications in the past. In Jedari et al. (2015) and Sánchez et al. (2012), Random Tree and Random Forest classifiers were used for positioning using Wi-Fi fingerprinting in indoor contexts. In Bozkurt et al. (2015), several machine learning approaches for indoor fingerprinting localization were compared, and the authors concluded that decision trees were among the best available methods.

In all these contributions, however, the input of the machine learning models was made by the received signal strength of the Wi-Fi access points. This is one of the main differences with our work, where we use attributes that describe the differences between fingerprints instead of the raw signal strengths. This approach is scalable and independent from the number of beacons.

An approach more similar to the one proposed in this paper, which makes use of features based on difference between fingerprints, was proposed in Sohn et al. (2006). However, there are significant differences in the kind of data used, since they consider GSM traces and, in the scope of the work, logistic regression models were used for mobility detection and not for positioning.

7 Conclusions and Future Work

In this paper, we proposed a machine learning method, based on Random Forest, to significantly enhance the performance of fingerprint positioning systems in outdoor contexts using cellular signals. A novel approach for generating the attributes used by the machine learning algorithm was also devised. The developed solution is scalable, unlike those commonly used in Wi-Fi environments that assume a limited and/or fixed amount of beacons. The novel approach has been compared to various other methods already present in the literature, focusing on the ones that should have improved the position estimation, especially in certain problematic settings. The comparison was done using several datasets collected in different environments, both rural and urban, by devices with different characteristics.

As for future work, we are thinking of evaluating other machine learning models for positioning, like Support Vector Machines or Neural Networks, using the newly proposed approach for the generation of the attributes. In addition to its basic features, the presented strategy provides a measure of the uncertainty of the position estimation. This comes as an advantage over classic methods for position estimation, which are not able to provide a measure of their accuracy. A final subject for future investigation concerns the composition of the dataset to use for the training of the machine learning models. While the adoption of a specific model for each dataset provided the best accuracy, the results showed that it is possible to use a model trained on a different dataset than the one in which the model is then going to be applied. If the datasets are similar enough, this cross-dataset model application can be done without losing much of the accuracy. The exact way to distinguish, and possibly classify, the datasets for detecting a priori which of them might work well with the same models is, however, a matter for further investigation.

References

Bahl P, Padmanabhan VN (2000) Radar: an in-building RF-based user location and tracking system. In: Proceedings of of the 19th INFOCOM, vol 2. IEEE, pp 775–784

Bozkurt S, Elibol G, Gunal S, Yayan U (2015) A comparative study on machine learning algorithms for indoor positioning. In: International symposium on innovations in intelligent SysTems and applications (INISTA), pp 1–8. https://doi.org/10.1109/INISTA.2015.7276725

Caffery J, Stuber GL (1998) Subscriber location in CDMA cellular networks. IEEE Trans Veh Technol 47(2):406–416

Chen MY, Sohn T, Chmelev D, Haehnel D, Hightower J, Hughes J, LaMarca A, Potter F, Smith I, Varshavsky A (2006) Practical metropolitan-scale positioning for GSM phones. In: Proceedings of the 8th UbiComp. Springer, pp 225–242

Deblauwe N (2008) GSM-based positioning: techniques and applications. ASP

Deligiannis N, Louvros S, Kotsopoulos S (2007) Mobile positioning based on existing signalling messages in GSM networks. In: Proceedings of the 3rd MOBIMEDIA

Hall M, Frank E, Holmes G, Pfahringer B, Reutemann P, Witten IH (2009) The WEKA data mining software: an update. SIGKDD Explor Newsl 11(1):10–18. https://doi.org/10.1145/1656274.1656278

Hastie T, Tibshirani R, Friedman J (2009) The elements of statistical learning: data mining, inference and prediction, 2nd edn. Springer

Jedari E, Wu Z, Rashidzadeh R, Saif M (2015) Wi-fi based indoor location positioning employing random forest classifier. In: 2015 international conference on indoor positioning and indoor navigation (IPIN), pp 1–5. https://doi.org/10.1109/IPIN.2015.7346754

Kjrgaard MB, Munk CV (2008) Hyperbolic location fingerprinting: a calibration-free solution for handling differences in signal strength (concise contribution). In: 2008 Sixth annual IEEE international conference on pervasive computing and communications (PerCom), pp 110–116. https://doi.org/10.1109/PERCOM.2008.75

Li X, Zhang X, Chen K, Feng S (2014) Measurement and analysis of energy consumption on android smartphones. In: Proceedings of the 4th ICIST. IEEE, pp 242–245. https://doi.org/10.1109/ICIST.2014.6920375

Meniem MHA, Hamad AM, Shaaban E (2013) Relative RSS-based GSM localization technique. In: IEEE international conference on electro-information technology, EIT 2013, pp 1–6. https://doi.org/10.1109/EIT.2013.6632643

Ngai EWT, Hu Y, Wong YH, Chen Y, Sun X (2011) The application of data mining techniques in financial fraud detection: a classification framework and an academic review of literature. Decis Support Syst 50(3):559–569. https://doi.org/10.1016/j.dss.2010.08.006

Paek J, Kim KH, Singh JP, Govindan R (2011) Energy-efficient positioning for smartphones using cell-id sequence matching. In: Proceedings of the 9th MobiSys. ACM, pp 293–306

PostgreSQL Global Development Group (2008) PostgreSQL. http://www.postgresql.org

Qi Y, Kobayashi H, Suda H (2006) Analysis of wireless geolocation in a non-line-of-sight environment. IEEE Trans Wirel Commun 5(3):672–681

Retscher G, Joksch J (2016) Comparison of different vector distance measure calculation variants for indoor location fingerprinting. In: Proceedings of the 13th international conference on location-based services, ICA, pp 53–76

Romero C, Ventura S (2010) Educational data mining: a review of the state of the art. IEEE Trans Syst Man Cybern Part C 40(6):601–618. https://doi.org/10.1109/TSMCC.2010.2053532

Sánchez D, Quinteiro JM, Hernández-Morera P, Martel-Jordán E (2012) Using data mining and fingerprinting extension with device orientation information for WLAN efficient indoor location estimation. In: 2012 IEEE 8th international conference on wireless and mobile computing, networking and communications (WiMob). IEEE, pp 77–83

Sohn T, Varshavsky A, LaMarca A, Chen MY, Choudhury T, Smith I, Consolvo S, Hightower J, Griswold WG, de Lara E (2006) Mobility detection using everyday GSM traces. Springer, Berlin, Heidelberg, pp 212–224. https://doi.org/10.1007/11853565_13

Spirito MA, Mattioli AG (1998) On the hyperbolic positioning of GSM mobile stations. In: Proceedings of ISSSE '98. IEEE, pp 173–177

Tomar D, Agarwal S (2013) A survey on data mining approaches for healthcare. Int J Bio-Sci Bio-Technol 5(5):241–266

Witten IH, Frank E, Hall MA (2011) Data mining: practical machine learning tools and techniques, 3rd edn. Morgan Kaufmann Publishers Inc., San Francisco, CA, USA

Zekavat R, Buehrer RM (2011) Handbook of position location: theory, practice and advances, 1st edn. Wiley-IEEE Press

Zhuang Z, Kim KH, Singh JP (2010) Improving energy efficiency of location sensing on smartphones. In: Proceedings of the 8th MobiSys. ACM, pp 315–330

Jaccard Analysis and LASSO-Based Feature Selection for Location Fingerprinting with Limited Computational Complexity

Caifa Zhou and Andreas Wieser

Abstract We propose an approach to reduce both computational complexity and data storage requirements for the online positioning stage of a fingerprinting-based indoor positioning system (FIPS) by introducing segmentation of the region of interest (RoI) into sub-regions, sub-region selection using a modified Jaccard index, and feature selection based on randomized least absolute shrinkage and selection operator (LASSO). We implement these steps into a Bayesian framework of position estimation using the maximum a posteriori (MAP) principle. An additional benefit of these steps is that the time for estimating the position, and the required data storage are virtually independent of the size of the RoI and of the total number of available features within the RoI. Thus the proposed steps facilitate application of FIPS to large areas. Results of an experimental analysis using real data collected in an office building using a Nexus 6P smart phone as user device and a total station for providing position ground truth corroborate the expected performance of the proposed approach. The positioning accuracy obtained by only processing 10 automatically identified features instead of all available ones and limiting position estimation to 10 automatically identified sub-regions instead of the entire RoI is equivalent to processing all available data. In the chosen example, 50% of the errors are less than 1.8 m and 90% are less than 5 m. However, the computation time using the automatically identified subset of data is only about 1% of that required for processing the entire data set.

C. Zhou (✉) · A. Wieser
Institute of Geodesy & Photogrammetry, ETH Zürich, Stefano-Franscini-Platz 5,
8093 Zürich, Switzerland
e-mail: caifa.zhou@geod.baug.ethz.ch

A. Wieser
e-mail: andreas.wieser@geod.baug.ethz.ch

© Springer International Publishing AG 2018   71
P. Kiefer et al. (eds.), *Progress in Location Based Services 2018*, Lecture Notes
in Geoinformation and Cartography, https://doi.org/10.1007/978-3-319-71470-7_4

1   Introduction

Fingerprinting-based indoor positioning systems (FIPs) are attractive for providing location of users or mobile assets because they can exploit signals of opportunity and infrastructure already existing for other purposes. They require no or little extra hardware, (He and Chan 2016), and differ in that respect from many other approaches to indoor positioning e.g., the ones using infrared beacons (Lee et al. 2004), ultrasonic signals (Hazas and Hopper 2006), radio frequency identification (RFID) tags (Bekkali et al. 2007), ultra wideband (UWB) signals (Ingram et al. 2004), or foot-mounted inertial measurement units (IMUs) (Gu et al. 2017). FIPS benefit from the spatial variability of a wide variety of observable features or signals like received sigal strength (RSS) from wireless local area network (WLAN) access point (APs), magnetic field strengths, or ambient noise levels. FIPS are therefore also called feature-based indoor positioning systems (Kasprzak et al. 2013). The attainable quality of the position estimation using FIPS mainly depends on the spatial gradient of the features and on their stability or predictability over time (Niedermayr et al. 2014).

Key challenges of FIPS are discussed e.g., in Kushki et al. (2007) and more recently in He and Chan (2016). The former publication focuses on four challenges of FIPS utilizing vectors of RSS from WLAN AP as fingerprints. In particular, the paper addresses (i) the generation of a fingerprint database to provide a reference fingerprint map(RFM) for positioning, (ii) pre-processing of fingerprints for reducing computational complexity and enhancing accuracy, (iii) selection of APs for positioning, and (iv) estimation of the distance between a fingerprint measured by the user and the fingerprints represented within in the reference database. Extensions to large indoor regions and handling of variations of observable features caused by the changes of indoor environments or signal sources of the features (e.g., replacement of broken APs) are addressed in He and Chan (2016).

Two widely used fingerprinting-based location methods, which we also employ herein are, k-nearest neighbors (kNN) (Padmanabhan and Bahl 2000) and maximum a posteriori (MAP) (Youssef and Agrawala 2008). The time and storage computational complexity of both methods is proportional to the number of reference locations in the RFM (i.e. the area of the RoI) and the number of observable features. This means that these approaches become computationally expensive in large RoIs with many APs.

The goal of this paper is to propose three steps in order to facilitate accurate and flexible indoor positioning in a large region of interest (RoI) with potentially very high numbers of available features and feature availability varying across the RoI. For flexibility e.g., with respect to including different types of features, performing quality prediction of estimated locations, and performing quality control of measured and modeled feature values, we choose a Bayesian approach to position estimation using the maximum a posteriori (MAP) principle. The three steps proposed herein are intended to reduce the computational complexity in terms of processing time and storage requirements, in particular during the position estimation stage which may either have to be carried out for a single user on the mobile device or for a potentially large number of users concurrently on a server. Furthermore, the steps help to make the computational effort for position estimation almost independent of the size of the RoI and of the total number of observable features within the RoI, which are important aspects for application of FIPS in large areas.

The first step is the segmentation of the entire RoI into non-overlapping sub-regions. The next step is the identification of approximations of the user location with a granularity corresponding to the size of the sub-regions such that the actual position estimation can be restricted to a search or optimization within a few candidate sub-regions. We apply a modified Jaccard index, (Park et al. 2010; Jani et al. 2015), within this step, see Sect. 3.2. The final step is the identification of relevant features within each sub-region and the subsequent selection of a small number of relevant features available both in the measured fingerprint and in the RFM for the actual position estimation. We base this feature selection on a randomized least absolute shrinkage and selection operator (LASSO) approach, (Tibshirani 1996), see Sect. 3.3.

2 Related Work

2.1 Sub-region Selection

There are mainly two types of approaches for sub-region selection[1]: approaches based on clustering and approaches based on similarity metrics. Feng et al. (2012) and Chen et al. (2006) applied affinity propagation and a k-means algorithm, respectively, to divide the RoI into a given number of sub-regions according to the features collected within the RoI. Both papers present clustering-based sub-region selection and require prior definition of the desired number of sub-regions and knowledge of all features observable within the entire RoI. These clustering-based approaches take the fingerprint measured by the user into account during the clustering process which may thus have to be repeated with each new user fingerprint obtained.

Similarity metric-based sub-region selection instead identifies the sub-region whose fingerprints contained in the RFM are most similar to the fingerprint observed by the user. They differ depending on the chosen similarity metric. E.g., Kushki et al. (2007) use the Hamming distance for this purpose, measuring only the difference in terms of observability of the features, not their actual values. Still, these approaches typically need prior information on all observable features within the entire RoI when associating a user observed fingerprint with a sub-region. This may be a severe limitation in case of a large RoI or changes of availability of the features. Modified Jaccard

---

[1] In other publications, sub-region selection is called spatial filtering (Kushki et al. 2007), location-clustering (Youssef et al. 2003), or coarse localization (Feng et al. 2012).

index-based sub-region selection as used in this paper belongs to the latter category. However, the approach proposed herein requires only the prior knowledge of the features observable within each sub-region when computing the similarity metric between the observations in the RFM and in the observed user fingerprint.

2.2 Selection of Relevant Features

Approaches to selection of features actually used for positioning differ with respect to several perspectives. We focus on three: (i) whether they take the relationship between positioning accuracy and selected features into account, (ii) whether they help to reduce the computational complexity of position estimation, and (iii) whether they are applicable to a variety of features or only features of a certain type. The chosen features for positioning should be the ones allowing to achieve the best positioning accuracy using the specific fingerprinting-based positioning method or achieving a useful compromise between accuracy and reduced computational burden.

Previous publications focused on feature selection for FIPS using RSS from WLAN APs and consequently addressed the specific problem of AP selection rather than the more general feature selection. Chen et al. (2006) and Feng et al. (2012) proposed using the subsets of APs whose RSS readings are the strongest assuming that the strongest signals provide the highest probability of coverage over time and the highest accuracy. Kushki et al. (2007) and Chen et al. (2006) applied a divergence metric (Bhattacharyya distance and information gain, respectively) to minimize the redundancy and maximize the information gained from the selected APs. The limitations of these approaches are: (i) they are only applicable to the FIPS based on RSS from WLAN APs, and (ii) they only take the values of the features into account as selection criteria instead of the actual positioning accuracy. Kushki et al. (2010) proposed an AP selection strategy able to choose APs ensuring a certain positioning accuracy using a nonparametric information filter. However, this approach uses continuously measured fingerprints to select the subset of APs maximizing the discriminative ability with respect to localization. This method therefore needs several online observations for estimating one current position.

In this paper, we propose an approach based on randomized LASSO to choose the most relevant features for fingerprinting-based positioning. This method differs from previous ones in three ways: (i) it takes the positioning error into account, (ii) the feature selection can be pre-computed and thus allows reducing the computational complexity of position estimation, and (iii) it is a general feature selection method applicable also to fingerprints containing different types of features simultaneously.

3 The Proposed Approach

In this section, we briefly summarize the fundamentals of fingerprinting-based positioning and present the main contributions of this paper, contributing to reduced computational complexity independent of the size of the RoI. In particular we present (i) candidate sub-region selection using a modified Jaccard index, (ii) selection of relevant features using randomized LASSO, and (iii) MAP-based positioning benefiting from the previous two steps. Finally we briefly discuss the computational complexity of the proposed method.

3.1 Problem Formulation

Generally, an FIPS is realized using two stages: offline and online stage (Fig. 1). The result of the former is the reference fingerprint map (RFM), i.e., a model representing the relation between the observable features and location. At the online stage, the user's location is estimated by matching the currently measured fingerprint to the RFM using a fingerprinting-based positioning method (e.g., maximum a posteriori (MAP)-based positioning).

We chose a representation herein where the RFM is a discrete set of fingerprints associated with chosen reference positions throughout the region of interest (RoI). Each fingerprint is an associative array consisting of a collection of (key, value) pairs.

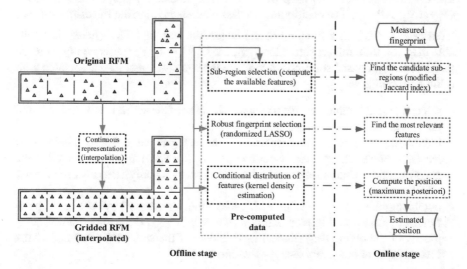

Fig. 1 The proposed framework. In order to make the number of reference points within all sub-regions equal we interpolate the reference data in the original reference fingerprint map (RFM) to provide a denser regular grid of reference points. This interpolated RFM is used to calculate the pre-computed data for online positioning

The key is a unique identifier of the respective feature, e.g., in case of WiFi-based fingerprinting an integer obtained by hashing the media access control (MAC) address of an AP. In case of a measured fingerprint the value is the measurement of the corresponding feature. In case of a fingerprint within the RFM the value is the expected value of the feature at the corresponding reference position.

Herein we assume that the measurements for establishing the RFM have been made at arbitrary locations across the RoI and the RFM is obtained by spatial interpolation using nearest-neighbor (Watson and Philip 1984) to obtain the fingerprints at a regular gird of reference points. Further details are given in Sect. 4.

For reducing the computational effort during the online stage, we propose to divide the RoI into N non-overlapping grid cells (sub-regions) g^i such that RoI $= \{g^1, g^2, \cdots, g^N\}$. In the 2D case each sub-region can simply be a rectangle or square in the coordinate space.[2] Let there be M^i fingerprints $\mathbf{O}^{ij} \in \mathbb{R}^{L^{ij} \times 2}$ within the ith grid cell of the RFM where $j \in \{1, 2, \ldots, M^i\}$ and L^{ij} is the number of features observed or observable at the corresponding location. Each of these fingerprints is associated with a position $\mathbf{l}^{ij} \in \mathbb{R}^D$ where D is the dimension of the coordinate space. Later on, we use the symbols $\mathbf{o}^{ij}_{keys}$ and $\mathbf{o}^{ij}_{values}$ to represent the vectors of keys and values separately such that $\mathbf{O}^{ij} = (\mathbf{o}^{ij}_{keys}, \mathbf{o}^{ij}_{values})$. For arguments where the sequence of the elements is irrelevant we will later use the same symbols to also indicate the sets of keys and values with $|\mathbf{o}^{ij}_{keys}| = |\mathbf{o}^{ij}_{values}| = L^{ij}$, where $|\cdot|$ denotes the cardinality of a set.

At the online stage a newly measured fingerprint $\mathbf{O}^u \in \mathbb{R}^{L^u \times 2}$ becomes available at the unknown user location $\mathbf{l}^u$ where L^u denotes the number of observed features at this location. This fingerprint is also represented by keys and values, i.e. $\mathbf{O}^u = (\mathbf{o}^u_{keys}, \mathbf{o}^u_{values})$. The positioning process consists in inferring the estimated user location $\hat{\mathbf{l}}^u = f(\mathbf{O}^u)$ as a function of the fingerprint and the RFM where f is a suitable mapping from fingerprint to location, i.e. $f : \mathbf{O} \mapsto \mathbf{l}$. We subsequently focus on the following proposed solutions to mitigate the computational load associated with offline and online stage:

- selecting the sub-region as a coarse approximation of the actual user location based on a modified Jaccard index;
- identifying the most relevant features within each grid cell using the randomized least absolute shrinkage and selection operator (LASSO) algorithm such that the actual location calculation can later be carried out using only those instead of using all features;
- combining the above two steps with a maximum a posteriori (MAP)-based positioning approach and implementing it in a way to keep the computational complexity of the online stage almost independent of the size of the RoI and of the total number of observable features within the RoI.

---

[2] An analysis of strategies for optimum definition of the sub-regions in terms of size and shape is left for future work.

3.2 Sub-region Selection Using Modified Jaccard Index

By analyzing the similarity between the keys of the measured fingerprints and the keys associated with the individual sub-regions in the RFM we can identify candidates of sub-regions most likely containing the actual user location. The subsequent estimation of the user location can then be limited to the selected candidate regions thus reducing the computational load assuming that the index calculation is computationally less expensive than the position estimation. We use a modified Jaccard index $c(g^i, \mathbf{O}^u) \in [0, 1]$, (Jani et al. 2015), as the similarity metric. It is calculated for the observed fingerprint $\mathbf{O}^u$ and each sub-region by:

$$c(g^i, \mathbf{O}^u) = \frac{1}{2}\left(\frac{|\mathbf{o}^i_{keys} \cap \mathbf{o}^u_{keys}|}{|\mathbf{o}^i_{keys} \cup \mathbf{o}^u_{keys}|} + \frac{|\mathbf{o}^i_{keys} \cap \mathbf{o}^u_{keys}|}{|\mathbf{o}^u_{keys}|}\right) \tag{1}$$

Here, $\mathbf{o}^i_{keys} \in \mathbb{R}^{L^i}$ is the set of unique keys representing the observable features within the ith grid cell, i.e., the union of the keys ($\mathbf{o}^{ij}_{keys}$ now considered as sets) of all fingerprints within this cell:

$$\mathbf{o}^i_{keys} = \cup_j \mathbf{o}^{ij}_{keys}, j = 1, 2, \cdots, M^i. \tag{2}$$

The first term in (1) is the Jaccard index (Park et al. 2010), which indicates the fraction of features common to the currently measured fingerprint and to the sub-region. The maximum value of 1 is obtained for this term (and the entire expression) if the features in the fingerprint are exactly all the features available within the sub-region. A lower value indicates that there are features which are missing either in the current fingerprint or in the RFM of the sub-region. The second term in (1) is a modifier causing the index to favor sub-regions containing all or most features observed by the user over sub-regions lacking some of these features. The underlying assumption is that the user may not be able to observe all actually available features while the RFM is nearly complete and it is therefore unlikely to observe features missing in the RFM.

The k sub-regions with the highest values of the modified Jaccard index are selected as candidate sub-regions for the subsequent positioning. Their cell indices are collected in the vector $s^u_k \in \mathbb{N}^k$ for further processing. If the sub-regions are non-overlapping, as introduced above, k needs to be large enough to accommodate situations where the actual user location is close to the border between certain sub-regions and small enough to reduce the computational burden of the subsequent user location estimation. We will further discuss this in Sect. 4.

3.3   Feature Selection Using Randomized LASSO

In a real-world environment there may be a large number of features available for positioning, e.g., hundreds of APs may be visible to the mobile user device in certain locations. Not all of them will be necessary to estimate the user location. In fact using only a well selected subset of the available signals instead of all may provide a more accurate estimate and will reduce the computational burden. Furthermore the number of observable features typically varies across the RoI e.g., due to Wifi AP antenna gain patterns, structure and furniture within a building. However, it is preferable to use the same number of features throughout the candidate sub-regions for assessing the similarity between the measured fingerprint and the ones extracted from the RFM during the online phase.

We therefore recommend selecting a fixed number h of features per candidate sub-region for the final position estimation. To facilitate this selection during the online phase, the relevant features within each sub-region are already identified beforehand once the RFM is available. We use an approach based on randomized LASSO, an L_1-regularized linear regression model (Tibshirani 1996), for this step. Each feature within the sub-region is associated with an estimated coefficient by this approach. If the coefficient is sufficiently different from zero the corresponding feature is identified as relevant. During the online phase h features (possibly different for each sub-region) are selected among the identified ones such that they are available both within the RFM and the user fingerprint.

The total number of observable features within the ith subregion is $|\mathbf{o}^i_{\text{keys}}| = L^i$. To represent all fingerprints of this sub-region in the RFM by vectors of the same dimension we replace each $\mathbf{o}^{ij}_{\text{keys}}$ by $\mathbf{o}^i_{\text{keys}}$, see (2), and the corresponding vector of values $\mathbf{o}^{ij}_{\text{values}}$ by a vector $\mathbf{f}^{ij} \in \mathbb{R}^{L^i}$ which contains just the corresponding element from $\mathbf{o}^{ij}_{\text{values}}$ for each feature whose key is in $\mathbf{o}^{ij}_{\text{keys}} \cap \mathbf{o}^i_{\text{keys}}$. For all other features it contains a value indicating that the feature is missing (e.g., a value lower than the minimum observable value of the corresponding feature).

Feature selection using LASSO is based on estimating the coefficients $\mathbf{P}^i \in \mathbb{R}^{L^i \times D}$ of a linear regression of position onto features according to:

$$\hat{\mathbf{P}}^i = \arg\min_{\mathbf{P}^i} \frac{1}{M^i} \sum_{j=1}^{M^i} \|\mathbf{P}^{i^{\mathrm{T}}} \mathbf{f}^{ij} - \mathbf{l}^{ij}\|_2^2 + \lambda \|\mathbf{P}^i\|_1 \qquad (3)$$

where λ is a hyperparameter which needs to be set appropriately, and the L_1-norm term on the right hand side is used for regularization. Any zero element within $\mathbf{P}^i$ indicates that the corresponding features does not contribute to the position estimation. Therefore, we identify the rows of $\hat{\mathbf{P}}^i$ whose absolute values exceed a given threshold (e.g., 10^{-4}) and consider the corresponding features relevant. Their keys are collected in the vector $\mathbf{q}^i$.

Table 1 Pseudocode of randomized LASSO

Algorithm: randomized LASSO
Input: **Data** = $\{\mathbf{f}^{ij}, \mathbf{l}^{ij}\}$ $j = 1, 2, \cdots, M^i$; sampling ratio $\epsilon \in (0, 1)$;
1:      number of randomizations $T \in \mathbb{N}$; threshold $h \in \mathbb{N}$
2: Output: relevant features $\bar{\mathbf{q}}^i$ of ith grid
3: for $t = 1, 2, \cdots, T$:
4:      $\widetilde{\mathbf{Data}}$ = sampling with replacement from **Data** with ratio ϵ
5:      $\mathbf{q}_t^i$ = LASSO-based fingerprint selection using $\widetilde{\mathbf{Data}}$
6: end for
7: computing the frequency of selection of each feature according to $\mathbf{q}_t^i, t = 1, 2, \cdots, T$
8: return $\bar{\mathbf{q}}^i$: set of features selected most frequently

However, the results are affected by the choice of λ and the optimum choice depends on the data. So, feature selection based on LASSO with any fixed priorly chosen value λ is unstable (Fastrich et al. 2015). In order to get an appropriate fixed value of λ, we use cross validation. However, the stability of LASSO-based feature selection can be improved by repeating the above process several times (e.g., 200 times) using a randomly sampled subset of fingerprints from the respective sub-region each time and finally taking the features most frequently contained in $\mathbf{q}^i$ as the actually most relevant ones. This approach is called randomized LASSO (Meinshausen and Bühlmann 2010; Wang et al. 2011). Although the computational cost of this process increases with increasing of size of the dataset (number of RPs and features, thus size of the RoI), it needs to be carried out only once at the offline stage. If need be, it can be implemented on a powerful computer and using parallel programming.[3] Table 1 displays its realization for the present application in terms of pseudocode, where $\bar{\mathbf{q}}^i$ represents the finally chosen vector of relevant keys of the ith sub-region.

3.4 MAP-based Positioning

Given an RFM, i.e. a database of fingerprints and associated reference positions, the aim of positioning is to infer the most probable location $\hat{\mathbf{l}}^u$ of the user according to the fingerprint $\mathbf{O}^u$ observed at the actual but unknown location $\mathbf{l}^u$. We use a variety of discrete candidate locations $\mathbf{l}$ and apply Bayes' rule to compute for each of them the degree of belief in the assumption that the current location of the user is $\mathbf{l}$ given the available RFM and the currently observed fingerprint. This is an MAP-based positioning method as proposed, e.g., by Park et al. (2010); Madigan et al. (2005).

---

[3]For the dataset used in Sect. 4, it took about 64 mins for one randomization on a Windows 10 PC with 6 cores Intel Xeon CPU, 32G RAM.

The posterior probability $\mathrm{Prob}(\mathbf{l}|\mathbf{O}^u)$ of being at location $\mathbf{l}$ is computed by:

$$\mathrm{Prob}(\mathbf{l}|\mathbf{O}^u) = \frac{\mathrm{Prob}(\mathbf{O}^u|\mathbf{l})\mathrm{Prob}(\mathbf{l})}{\mathrm{Prob}(\mathbf{O}^u)} \tag{4}$$

where $\mathrm{Prob}(\mathbf{O}^u|\mathbf{l})$ is the conditional probability of the fingerprint given the assumed location $\mathbf{l}$, and $\mathrm{Prob}(\mathbf{l})$ and $\mathrm{Prob}(\mathbf{O}^u)$ are the prior probabilities of location and fingerprint respectively. Since the prior probability of the fingerprint is independent of the candidate location the MAP estimate can be obtained from:

$$\hat{\mathbf{l}}^u = \arg\max_{\mathbf{l}} \left[\mathrm{Prob}(\mathbf{O}^u|\mathbf{l})\mathrm{Prob}(\mathbf{l})\right] \tag{5}$$

Assuming that the observable features are conditionally independent of each other (5) can be represented by the naïve Bayes model:

$$\hat{\mathbf{l}}^u = \arg\max_{\mathbf{l}} \left[\prod_{j=1}^{L^u} \mathrm{Prob}(\mathbf{O}_j^u|\mathbf{l})\mathrm{Prob}(\mathbf{l})\right] \tag{6}$$

where $\mathbf{O}_j^u$ denotes the jth observed feature at the current location. In this paper, we introduce sub-region and relevant feature selection into MAP-based positioning. Therefore, we only take candidate locations in the chosen sub-regions and calculate the posterior using only the previously selected most relevant features.

Thus, (6) is modified to be:

$$\hat{\mathbf{l}}^u = \arg\max_{\mathbf{l} \in g^i} \left[\prod_{j=1}^{|\bar{\mathbf{q}}^i|} \mathrm{Prob}(\mathbf{O}_j^u|\mathbf{l})\mathrm{Prob}(\mathbf{l})\right], \forall i \in \mathbf{s}_k^u \tag{7}$$

where $\mathbf{s}_k^u$ is the vector denoting the indices of the candidate sub-regions (see Sect. 3.2) and $\bar{\mathbf{q}}^i$ is the set of selected relevant features of the ith sub-region (see Sect. 3.3). The conditional probability $\mathrm{Prob}(\mathbf{O}_j^u|\mathbf{l})$, which models the density of the jth feature for a given location $\mathbf{l}$, is estimated using kernel density estimation with a Gaussian kernel from the observations stored in the RFM, see details e.g., in Scott (2015); Kushki et al. (2007).

Prior knowledge of the user location, e.g. derived from previous estimates of user locations and a motion model, could be used to represent the prior probability $\mathrm{Prob}(\mathbf{l})$ of the locations. However, as in Sect. 3.2, we assume also now that no such prior information is available and can hence use equal probability of $\mathbf{l}$ across all candidate sub-regions such that also $\mathrm{Prob}(\mathbf{l})$ can be dropped from (7).

3.5 Computational Complexity of Online Positioning

In this part, we analyze the computational complexity of the proposed approach and compare it to MAP-based positioning without sub-region and feature selection. For the latter, the computational complexity of estimating one position is $\mathcal{O}(\alpha N(|\mathbf{o}^{\mathrm{RoI}}_{\mathrm{keys}}| + 1))$, where α is the number of candidate locations in each of the N sub-regions, and $|\mathbf{o}^{\mathrm{RoI}}_{\mathrm{keys}}|$ denotes the total number of observable features in the entire RoI:

$$\mathbf{o}^{\mathrm{RoI}}_{\mathrm{keys}} = \bigcup_i \mathbf{o}^i_{\mathrm{keys}}, \ i = 1, 2, \cdots, N. \tag{8}$$

The computational complexity of the proposed method is $\mathcal{O}(\alpha k(\max_{i \in s^u_k}\{|\bar{\mathbf{q}}^i|\} + 1))$, where k is the number of selected sub-regions and $|\bar{\mathbf{q}}^i|$ is the number of selected features. So, clearly the computational complexity of the proposed approach is significantly less than for the MAP-based approach without sub-region and feature selection. Furthermore, it is independent of the size of the RoI and of the total number of available features within the RoI. The proposed approach is to constrain and limit the search to a set of candidate reference locations and selected features for the online positioning. Though we only give the analytical formula of the computational complexity of MAP, other fingerprinting-based location methods will also benefit from the proposed approach, because the computational complexity of fingerprinting-based positioning is proportional to the size of the search space.

The computational complexity of the proposed approach can also be kept low by an appropriate implementation strategy. Besides the RFM further data required during the online positioning stage can be precomputed already during the offline stage (Fig. 1). This holds in particular for:

- the set $\mathbf{o}^i_{\mathrm{keys}}$ of available feature keys of each sub-region required for calculating the modified Jaccard index at the online stage,
- the set $\bar{\mathbf{q}}^i$ of relevant features of each sub-region calculated using randomized LASSO,
- and the conditional distribution $(\mathrm{Prob}(\mathbf{O}_j|\mathbf{l}))$ of the selected relevant features within each sub-region obtained from kernel density estimation.

At the online stage these pre-computed data are cached to the user device to achieve location estimation while realizing mobile positioning. The proposed pre-processing steps also reduce the required storage space for saving the cached pre-computed data because these data only need to cover the selected relevant features instead of all the features observable within the RoI.

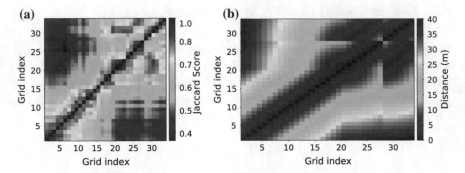

Fig. 2 Analysis of modified Jaccard index. **a**: Spatial distribution of the modified Jaccard index. **b**: Euclidean distance between centroids of each pair of sub-regions

4 Experimental Results and Discussion

In this section we analyze data obtained from real measurements collected using a Nexus 6P smart phone (for WLAN RSS) and a Leica MS50 total station (for position ground truth) within an L-shaped RoI of about 150 m^2 in an office building for fingerprinting-based WLAN indoor positioning, in which there are 399 observable access points. The shape of the actual floorplan corresponds to the shape given in Fig. 1.

We use a kinematically mapped RFM from about 2000 reference fingerprints obtained by recording data approximately every 1.5 seconds while a user walked through the RoI. The total station tracked a prism attached to the Nexus smart phone with an accuracy of about 5 mm. This approach is a compromise between the high accuracy attainable by stop and go measurements at carefully selected and previously marked reference positions and the low extra effort of crowd-sourced RFM data collection as outlined e.g., in Radu and Marina (2013). In order to evaluate the performance of the proposed approach independently an additional test data set was collected comprising fingerprints at approximately 500 test positions (TPs) located throughout the RoI.

The coordinates of the TPs as measured by the total station were later used as ground truth for calculating the positioning error in terms of mean squared error (MSE) of the Euclidean distance between estimated and true coordinates. Data processing according to the proposed algorithms, as outlined in Fig. 1, was implemented in Python using the scikit-learn package (Pedregosa et al. 2011).

We divided the RoI into 34 square grid cells (sub-regions) of approximately $2 \times 2\, m^2$ and densified the original RFM to a regular grid of about 100 reference points per m^2 (i.e. spacing about $0.2 \times 0.2\, m^2$) by interpolation. The resulting gridded RFM was used for all further processing steps.

Fig. 3 Empirical cumulative positioning accuracy for different choices of parameters. The positioning error herein is the Euclidean distance between the estimated and true coordinates of the TPs

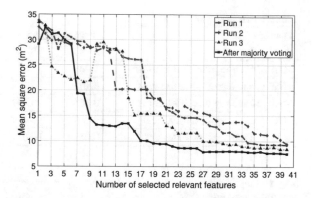

Figure 2a shows the modified Jaccard index for all pairs of sub-regions indicating that the index is related to the Euclidean distance (Fig. 2b). This corresponds to the expectation that the APs available in nearby sub-regions are similar while different APs are observed in sub-regions far from each other.

We have then investigated the number of relevant features to be used for positioning. The keys $\bar{\mathbf{q}}^i$ of the relevant features per sub-region are the result of a randomized process and are thus random themselves. We have therefore carried out the feature selection and subsequent position estimation of the TPs three times independently. For each of these simulation runs the MSE of the estimated coordinates of the TPs is plotted in Fig. 4 as a function of the number h of selected features actually used for positioning. The difference of MSE of each run is caused by the randomization of LASSO-based feature selection. This figure only shows 3 out of the 200 randomized runs and the MSE path after majority voting from which, according to Table 1, the final feature selection is chosen. The figure shows that the accuracy of the estimated positions generally increases as the number of features used is increased from 1 to 40. However, the gain in accuracy is negligible if the number of features is increased above 6–10, in particular if the variability due to the randomized feature selection process is taken into account (different curves in Fig. 4).

Fig. 3 illustrates the positioning accuracy and processing time for different choices of parameters within the proposed approach. The accuracy is plotted in terms of cumulative positioning accuracy (CPA) i.e., cumulative density function of the positioning errors. When introducing the sub-region selection with 10 sub-regions into MAP-based positioning (but using all available features), the CPA is comparable to that of MAP-based positioning without sub-region or feature selection but the average processing time [4] for estimating the coordinates of one TP is 0.35 s (see Table 2),which is only about 1/4 of that of MAP-based positioning without sub-

---

[4]We used Python to implement the proposed method and evaluate the processing time using the *time* package (https://docs.python.org/3/library/time.html#module-time).

Table 2 The processing times

Methods	Time consumption (in s) for positioning one TP			
	Mean	Min	Max	Std
MAP (34 sub-regions, 399 features)	1.223	1.222	1.247	0.003
MAP (3 sub-regions, 399 features)	0.106	0.104	0.137	0.002
MAP (10 sub-regions, 399 features)	0.353	0.347	0.388	0.003
MAP (10 sub-regions, 6 features)	0.008	0.007	0.009	0.0002
MAP (10 sub-regions, 10 features)	0.012	0.011	0.013	0.0002

Fig. 4 Empirically determined accuracy (MSE) of TP coordinates estimated using the proposed approach with different number of selected relevant features

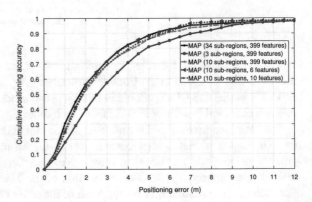

region or feature selection. Using only 3 sub-regions of course reduces the processing time further but leads to a considerable loss in accuracy.

By introducing both sub-region and feature selection the computational complexity and the data storage requirements can be reduced. Using 10 sub-regions and 10 features the average time of computing the coordinates of one TP is only 1% of that of MAP-based positioning with neither sub-region nor feature selection while the attained accuracy is virtually equal to the one obtained using all data. In agreement with the results depicted in Fig. 4 there is no significant loss in accuracy when using 6 features instead of 10, while the processing time decreases roughly by a factor of 2.

5   Conclusion

We proposed herein an approach to fingerprinting-based indoor positioning using the maximum a posteriori (MAP) principle for coordinate estimation. The main contributions are proposals to reduce data storage requirements and computational complexity in terms of processing times by segmentation of the entire RoI into sub-regions, identification of a few candidate sub-regions during the online positioningstage, and use of a selected subset of features instead of all available features for position estimation. Sub-region selection is based on a modified Jaccard index measuring the similarity between the features obtained by the user and those available within the reference fingerprint map (RFM). Feature selection is based on the randomized least absolute shrinkage and selection operator (LASSO) yielding a precomputed set of relevant features for each sub-region. The reduction of computational complexity is obtained both from the reduction of the number of candidate locations needed to analyze during online positioning and from the reduction of the number of features to be compared.

The experimental results corroborated the claim of reduced complexity while indicating that the positioning accuracy is hardly reduced by processing only 10 candidate subregions instead of the entire RoI and by selecting only 6–10 features instead of using all available ones. Given a fixed number of candidate sub-regions and a fixed, low number of features the computational burden of the entire algorithm is almost independent of the size of the entire RoI and of the number of available features across the RoI.

Further work will concentrate on increasing the stability of the feature selection via adaptive forward-backward greedy feature selection (Zhang 2011), on ranking features with respect to quality and impact, on taking into account user motion during sub-region selection and on handling temporal changes of the RFM. Furthermore, we are currently applying the proposed approach to larger and more complex datasets (Montoliu et al. 2017) and migrating to a mobile phone.

Acknowledgements The China Scholarship Council (CSC) financially supports the first author's doctoral research. Questions and proposals by three anonymous reviewers are acknowledged for contributing to improved quality of the paper.

References

Bekkali A, Sanson H, Matsumoto M (2007) Rfid indoor positioning based on probabilistic rfid map and kalman filtering. In: 2007 third ieee international conference on wireless and mobile computing, networking and communications, WiMOB 2007. IEEE, pp 21–21

Chen Y, Yang Q, Yin J, Chai X (2006) Power-efficient access-point selection for indoor location estimation. IEEE Trans Knowl Data Eng 18(7):877–888

Fastrich B, Paterlini S, Winker P (2015) Constructing optimal sparse portfolios using regularization methods. Comput Manag Sci 12(3):417–434

Feng C, Au WSA, Valaee S, Tan Z (2012) Received-signal-strength-based indoor positioning using compressive sensing. IEEE Trans Mobile Comput 11(12):1983–1993

Gu Y, Zhou C, Wieser A, Zhou Z (2017) Pedestrian positioning using wifi fingerprints and a foot-mounted inertial sensor, vol 1, pp 1–9. arXiv:1704.03346

Hazas M, Hopper A (2006) Broadband ultrasonic location systems for improved indoor positioning. IEEE Trans Mobile Comput 5(5):536–547

He S, Chan S-HG (2016) Wi-fi fingerprint-based indoor positioning: recent advances and comparisons. IEEE Commun Surv Tutor 18(1):466–490

Ingram S, Harmer D, Quinlan M (2004) Ultrawideband indoor positioning systems and their use in emergencies. In: 2004 Position location and navigation symposium, PLANS 2004. IEEE, pp 706–715

Jani SS, Lamb JM, White BM, Dahlbom M, Robinson CG, Low DA (2015) Assessing margin expansions of internal target volumes in 3d and 4d pet: a phantom study. Ann Nucl Med 29(1):100–109

Kasprzak S, Komninos A, Barrie P (2013) Feature-based indoor navigation using augmented reality. In: 2013 9th international conference on intelligent environments, pp 100–107

Kushki A, Plataniotis KN, Venetsanopoulos AN (2007) Kernel-based positioning in wireless local area networks. IEEE Trans Mobile Comput 6(6):689–705

Kushki A, Plataniotis KN, Venetsanopoulos AN (2010) Intelligent dynamic radio tracking in indoor wireless local area networks. IEEE Trans Mobile Comput 9(3):405–419

Lee C, Chang Y, Park G, Ryu J, Jeong S.-G, Park S, Park JW, Lee, HC, Shik Hong K, Lee, MH (2004). Indoor positioning system based on incident angles of infrared emitters. In: 2004 30th annual conference of IEEE industrial electronics society, IECON 2004, pp 2218–2222, vol 3

Madigan D, Einahrawy E, Martin, R. P., Ju, W. H., Krishnan, P., and Krishnakumar, A. S. (2005). Bayesian indoor positioning systems. In: Proceedings IEEE 24th annual joint conference of the ieee computer and communications societie, vol 2, pp 1217–1227

Meinshausen N, Bühlmann P (2010) Stability selection. J R Stat Soc Ser B (Stat Methodol) 72(4):417–473

Montoliu R, Sansano E, Torres-Sospedra J, Belmonte O (2017) Indoorloc platform: A public repository for comparing and evaluating indoor positioning systems. In: 2017 8th international conference on indoor positioning and indoor navigation, IPIN 2017. IEEE, pp 1–8

Niedermayr S, Wieser A, Neuner H (2014) Expressing location uncertainty in combined feature-based and geometric positioning. In: Proceedings European navigation conference 2014, EUGIN, pp 154–166

Padmanabhan VN, Bahl P (2000) RADAR: an in-building RF based user location and tracking system. In: Proceedings IEEE INFOCOM 2000. Conference on Computer Communications. Nineteenth Annual Joint Conference of the IEEE Computer and Communications Societies (Cat. No.00CH37064), vol 2(c), pp 775–784

Park JG, Charrow B, Curtis D, Battat J, Minkov E, Hicks J, Teller S, Ledlie J (2010) Growing an organic indoor location system. In: Proceedings of the 8th international conference on Mobile systems, applications, and services. ACM, pp 271–284

Pedregosa F, Varoquaux G, Gramfort A, Michel V, Thirion B, Grisel O, Blondel M, Prettenhofer P, Weiss R, Dubourg V, Vanderplas J, Passos A, Cournapeau D, Brucher M, Perrot M, Duchesnay E (2011) Scikit-learn: machine learning in python. J Mach Learn Res 12:2825–2830

Radu V, Marina MK (2013) Himloc: indoor smartphone localization via activity aware pedestrian dead reckoning with selective crowdsourced wifi fingerprinting. In: International conference on indoor positioning and indoor navigation, pp 1–10

Scott DW (2015) Multivariate density estimation: theory, practice, and visualization. Wiley

Tibshirani R (1996) Regression shrinkage and selection via the lasso. J R Stat Soc Series B (Methodol), 267–288

Wang S, Nan B, Rosset S, Zhu J (2011) Random lasso. Ann Appl Stat 5(1):468–485

Watson DF, Philip GM (1984) Triangle based interpolation. J Int Assoc Math Geol 16(8):779–795

Youssef M, Agrawala A (2008) The Horus location determination system. Wirel Netw 14(3):357–374

Youssef MA, Agrawala A, Shankar AU (2003) Wlan location determination via clustering and probability distributions. In: 2003 Proceedings of the First IEEE International Conference on Pervasive computing and communications, (PerCom 2003). IEEE, pp 143–150

Zhang T (2011) Adaptive forward-backward greedy algorithm for learning sparse representations. IEEE Trans Inf Theory 57(7):4689–4708

Part II
Mapping

Road Network Fusion for Incremental Map Updates

Rade Stanojevic, Sofiane Abbar, Saravanan Thirumuruganathan,
Gianmarco De Francisci Morales, Sanjay Chawla, Fethi Filali and
Ahid Aleimat

Abstract In the recent years a number of novel, automatic map-inference techniques have been proposed, which derive road-network from a cohort of GPS traces collected by a fleet of vehicles. In spite of considerable attention, these maps are imperfect in many ways: they create an abundance of spurious connections, have poor coverage, and are visually confusing. Hence, commercial and crowd-sourced mapping services heavily use human annotation to minimize the mapping errors. Consequently, their response to changes in the road network is inevitably slow. In this paper we describe `MapFuse`, a system which fuses a human-annotated map (e.g., OpenStreetMap) with any automatically inferred map, thus effectively enabling quick map updates. In addition to new road creation, we study in depth road closure, which have not been examined in the past. By leveraging solid, human-annotated maps with minor corrections, we derive maps which minimize the trajectory matching errors due to both road network change and imperfect map inference of fully-automatic approaches.

Keywords Map fusion · Map inference · Road closures

R. Stanojevic · S. Abbar · S. Thirumuruganathan (✉) · G. De Francisci Morales · S. Chawla
Qatar Computing Research Institute, HBKU, P.O. Box 5825, Doha, Qatar
e-mail: sthirumuruganathan@hbku.edu.qa

R. Stanojevic
e-mail: rstanojevic@hbku.edu.qa

S. Abbar
e-mail: sabbar@hbku.edu.qa

G. De Francisci Morales
e-mail: gmorales@hbku.edu.qa

S. Chawla
e-mail: schawla@hbku.edu.qa

F. Filali · A. Aleimat
Qatar Mobility Innovation Center, QSTP, P.O. Box 210531, Doha, Qatar
e-mail: filali@qmic.com

A. Aleimat
e-mail: ahide@qmic.com

© Springer International Publishing AG 2018
P. Kiefer et al. (eds.), *Progress in Location Based Services 2018*, Lecture Notes
in Geoinformation and Cartography, https://doi.org/10.1007/978-3-319-71470-7_5

1   Introduction

Map Fusion Problem: Generating accurate maps from geospatial data is an active area of research. A number of these works (Biagioni and Eriksson 2012; Cao and Krumm 2009; Chen et al. 2016; Edelkamp and Schrödl 2003) utilize crowd-sourced GPS data, e.g., from smartphones. An alternate strain of work tries to use other sources such as satellite images (Mnih and Hinton 2010). Despite considerable interest and effort by the research community, the existing automatic map inference solutions have a number of shortcomings, including: limited coverage, visually confusing layout, spurious roads, and imperfect turn restrictions. Hence, commercial maps such as Google Maps, Nokia HERE, and Apple Maps often use multiple sources of data information to generate initial maps, and then rely heavily on humans (both annotators and volunteers) to detect and correct the possible imperfections. However, the involvement of humans results in a very slow response in updating maps when a change in the road network occurs. In many cities in Asia and Africa, which are under heavy construction, this process results in substantial latency. One potential way to solve this issue is to automatically update the map using GPS traces given an existing map. However, most of those approaches are simple adaptations of classical map inference algorithms and suffer from the same disadvantages. In this work, we advocate for a new approach—Map Fusion—which automatically fuses two maps. One of the maps is a high-quality slowly updated map such as Open-StreetMap (OSM) (2017) or Google Maps (2017), while the other one is an automatically inferred map with incomplete coverage and imperfect topological structure. Our proposed system, `MapFuse`, synthesizes a new map that overcomes the deficiencies of the two maps discussed above. In the rest of the section, we enunciate this overall approach.

1.1   Challenges in Fully Automatic Map Inference

As mentioned above, there has been extensive work (see surveys (Biagioni and Eriksson 2012; Ahmed et al. 2015; Liu 2012)) on automatic map creation from GPS traces. However, these algorithms—both academic and commercial— face a number of important challenges. We now highlight three of the major ones.

- *Poor coverage*. The popularity of roads segments in the road network (measured, say, in number of trajectories which pass by the segment) is very skewed. While a few road segments (e.g., those lying on a highway) carry a massive number of trajectories, a large fraction of roads serves only a handful of cars. Hence, a vehicle fleet which opportunistically collects the GPS data needs to collect a massive amount of spatial samples in order to have a decent coverage of the road network. In the case of the fleet whose data we analyzed in this work, if we denote by L the total length of all the roads in Doha (L is in the order of 10 s of thousands of kilometers) our data, which corresponds to the trajectories with overall length

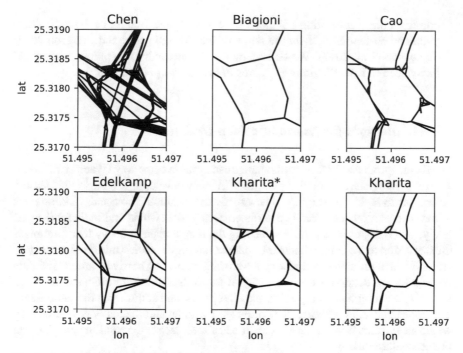

Fig. 1 Automatically inferred maps of 6 existing methods

of $175 \cdot L$, covers only about 48% of the road segments (see Fig. 3). In order to cover close to 100% of the road network with such opportunistic GPS probes, one would need to collect from one to two orders of magnitude more data, which in case of Doha would translate to 10 or 100 s of millions of kilometers of driving. Thus, independent of the map-inference method one utilizes, one needs to have an extremely high-volume of opportunistically collected GPS data in order to cover large portions of the road network.

– *Visually confusing outlook.* Most of the existing approaches do not control for the visual appearance of their maps, and hence the resulting maps have rather confusing look and are not visually appealing. In Fig. 1 we depict maps of a prominent "TV roundabout" in Doha derived by several well-known map-inference algorithms (Biagioni and Eriksson 2012; Cao and Krumm 2009; Chen et al. 2016; Edelkamp and Schrödl 2003; Stanojevic et al. 2017). Due to different nature of their inference process, they all have some unique features, yet they all have spurious or missing road segments, which can confuse the end-user and the navigation system which may utilize such maps.

– *Low topological accuracy.* Possibly the most serious concern regarding the existing map-inference methods is their low topological accuracy. Namely, due to the GPS noise as well as the inability to efficiently handle such noise, all existing methods often miss the connections between road segments or infer non-existing connections between road segments. Such topological inaccuracies are absolutely non-tolerable, yet existing solutions have topological Biagioni $F1$-score[1] (Biagioni and Eriksson 2012) in the range of 0.6–0.8 (Biagioni and Eriksson 2012; Stanojevic et al. 2017). We believe that a commercially acceptable map would likely need to have Biagioni $F1$-scores in the nearest proximity of 1.

1.2 Challenges for Automatic Map Updates

TomTom reports that 15% of roads change each year in some way (Wang et al. 2013). The road changes are particularly common in many developing countries in Asia and Africa due to rapid construction of new roads. For example, thousands of kilometers of new expressways have been constructed each year in China and India for the past few years (Wang et al. 2017). Automating the map update in a way that minimizes the disruption to the original map is of paramount importance. There has been extensive work on automatically updating an existing map using newly acquired GPS data (see Sect. 2 for details). However, many of these algorithms are often simple adaptations of existing batch map-inference algorithms, and suffer from the same issues mentioned above. In addition, they often start with an automatically generated map which also suffers from the issues mentioned above. Hence the resulting map is often of substandard quality.

1.3 Challenges for Hybrid Map Updates

According to the discussion so far, we believe that a hybrid method involving automatic algorithms along with humans is the way forward. The substandard quality of maps from purely automated means is often unacceptable for commercial map systems such as Google Maps, Apple Maps, Bing Maps, Nokia HERE, and Tom Tom. The creation of these maps is in many ways automated, however it requires human attention to examine possible places of interest. For example, Google Maps has a large team of so called operators who ensure the validity and consistency of the Google maps (Lookingbill and Weiss-Malik 2013) and hence any possible change in the road network needs to be approved by one of the operators. Similarly, the largest global crowd-sourced mapping effort OpenStreetMap (OSM) updates around 1M nodes per day. These maps have reasonably high accuracy in most cities with static road infrastructure.

However, even this approach has some fundamental limitations. Due to the human in the loop, they suffer from slow update response when changes happen (see Fig. 2). In many cities such as Doha, there are constant and large changes in road networks, that are not reflected in the maps in a timely manner. Conversely, automated

---

[1]Biagioni $F1$-score is a well known metric for measuring the topological accuracy of a map and lies in the range [0, 1] with 0 being absolutely wrong map, and 1 being a perfect map.

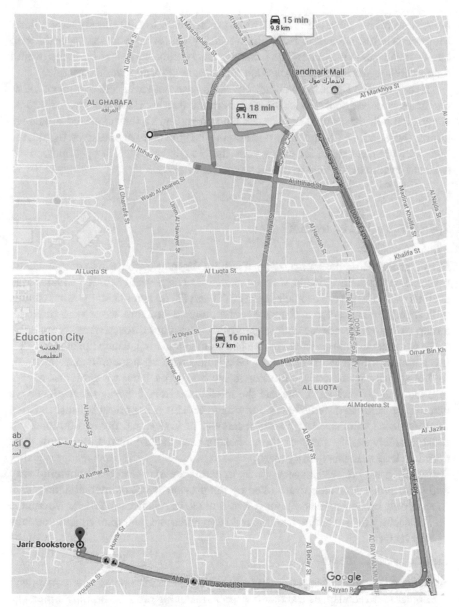

Fig. 2 Google maps route suggestion between two locations in Doha are almost twice longer (in length and duration) than the optimal route

algorithms often ignore the fact that most urban areas globally already have a fairly accurate map infrastructure. Not utilizing such great resource to construct the map (as most automatic map inference solutions do) is unfortunate and hurts the overall map inference process. Let us illustrate this effect with a real-world example.

In the city like Doha, with a very dynamic road network,[2] the quality of existing maps is rather poor. For example, when one queries Google Maps for a route suggestion between two points in west Doha (see Fig. 2), the suggested routes are almost twice as long (in both time and length) than the optimal one. Even though the optimal route has existed for over a year, the Google Maps has not yet updated the relevant portion of the map to reflect the current layout.

1.4   Proposed Approach

In this paper we propose `MapFuse`, a system for map fusion which automatically merges two maps. Specifically, we seek to fuse (1) a high-quality slowly-updated map such as OSM (2017) or Google Maps (2017) and (2) an automatically-inferred one, with incomplete coverage and imperfect topological structure. `MapFuse` produces a map which overcomes the deficiencies of the two maps discussed above.

In contrast with the existing approaches on map updating, which update the existing map (say OSM) by using a set of GPS trajectories via a specific map-inference tool, `MapFuse` is oblivious to the map inference approach one wishes to use to capture the road network segments and the interconnections between them. Hence we can fuse *any* map to the existing underlying map. This is important because existing map inference solutions suffer from a number of issues, and future solutions will most certainly rectify many of those. Fusing such better-inferred maps will most certainly lead to higher quality maps.

Finally, a very relevant aspect of map updating are road closures (both temporary and permanent) which are overlooked by the previous work on map updating, as it focuses only on new road additions (Shan et al. 2015; Wang et al. 2013). We use the GPS trajectory data to understand the road dynamics and infer road closures as soon as they happen.

Summary of Contributions:

- We introduce the problem of map fusion, which seeks to update a base map with another inferred map, as a geometric graph matching problem and show it can be treated as a minimal vertex cover problem on an appropriately-defined bipartite graph.
- Due to the size of the graphs representing the two maps (which can have hundreds of thousands of nodes) the polynomial solution to the bipartite vertex cover problem is not practical and we propose an efficient heuristic that fuses two maps.

---

[2]Influenced by a rapid construction of the city metro and a number of ongoing infrastructure projects.

- We suggest a new methodology for inferring closed road segments which utilizes dynamic statistics of the roads as well as a node centrality measure. As an unexpected advantage of our closure detection we identify the errors in the OSM maps (e.g., we can automatically pinpoint several roundabouts which are represented in the OSM as two-way roads, while they are obviously one-way only) which can be harmful to the navigation systems.
- Using a set of GPS trajectories from a fleet of vehicles in Doha we demonstrate that the fused map is more accurate than either of the two maps, and reduces the average/median/99th-percentile trajectory matching error by 30%.

2 Related Work

Map Inference: Constructing maps from crowdsourced GPS traces has been extensively studied (see surveys (Biagioni and Eriksson 2012; Ahmed et al. 2015; Liu 2012). K-Means based algorithms cluster the GPS points and link the resulting clusters into a routable map. Representative works include (Edelkamp and Schrödl 2003; Agamennoni et. 2011; Schroedl et al. 2004). Kernel density estimation (KDE) based algorithms such as (Chen and Cheng 2008; Davies et al. 2006; Shi et al. 2009) transform the GPS points into a density discretized image that are processed by image processing techniques to obtain maps. Trace merging based approaches start with an empty map and carefully add traces into it. Representative works include (Cao and Krumm 2009; Ahmed and Wenk 2012).

Maintaining Maps: Maintaining maps is closely related to map inference and often the algorithms for map maintenance are adaptations of those for map inference. Nevertheless, there are some subtle differences. While one can indeed obtain an updated map by re-running the entire inference pipeline, it is often efficient—in terms of both time and data—to treat it as a separate problem.

Recall that almost 15% of roads change every year in the US (Wang et al. 2013). This number is even higher in many developing countries in Asia and Africa due to rapid construction of new roads. For example, thousands of kilometers of new expressways are being constructed each year in China and India for the past few years (Wang et al. 2017). This necessitates research into work that maintain and update maps as and when new GPS data points arrive. Some representative work include (Ahmed and Wenk 2012; Schroedl et al. 2004; van den Berg 2015; Bruntrup et al. 2005; Wang et al. 2013; Zhang et al. 2010; Shan et al. 2015; Wu et al. 2015; Wang et al. 2017). However, most of these approaches do not have good practical performance and are very sensitive to differential sampling rates, disparity in data points, GPS errors etc. Often, these algorithms seek to directly extend one of the three approaches and suffer from bottlenecks arising from algorithmic step that is fundamental to it (such as clustering, density estimation, clarification, map matching) etc.

Additionally, while most of the prior work handle the simple case of new road additions, road closures are rarely addressed. CrowdAtlas (Wang et al. 2013) is exception that uses a simple heuristic in which each road segment is assigned an appropriate timeout proportional (3x) to the maximum time observed between the traversal of two successive vehicles in a training window. To cope with the cold start problem, no timeout is set for a segment until it has accumulated at least a week of data and at least five traces. Thus, most residential roads have no timeout established.

Graph Matching: Given two graphs, identifying if one graph is a subgraph of another is known to be NP-Complete (Garey and Johnson 2002). In fact, even identifying the minimal set of 'edits' to transform one graph to another is also NP-Complete (Zeng et al. 2009). However, it is possible to apply a number of heuristics for the case of road networks to solve this problem effectively. Matching of two road networks has been extensively studied due to its practical importance. The process of integrating different geospatial data to get new cartographic products is called map conflation. See (Ruiz et al. 2011) for a review of techniques used. Often, a wide variety of information including spatial features (such as distances, angles, shapes of the map) and topographical information (such as neighborhood) are used. For example, (Yang et al. 2013) proposed a heuristic probabilistic relaxation procedure to integrate multi-source geospatial data by using similarities between shapes. Recently, (Du et al. 2015) studied the problem of integrating authoritative geo-spatial data (such as OpenStreetMap) with crowdsourced GPS information. However, they use auxiliary information such as names and types of POIs that may not always be available.

3 Problem Formulation

A common representation of a map in the map-inference literature is a directed graph as following. A map is a geometric graph $G(V, E, L)$, where V is the set of vertices, $E \subseteq V \times V$ is the set of edges connecting pairs of vertices, and $L : V \rightarrow \mathbb{R}^2$ is a location function which assigns coordinates (latitude and longitude) to each vertex.

Given two instances of such graphs (maps), G_1 and G_2, our goal is to create a new fused graph $G_f = f(G_1, G_2)$ which preserves some properties of the source graphs. In particular, we wish for the connectivity of the fused graph to subsume the connectivity of the source graphs. However, we also wish to do so with the minimum number of edges, in order to avoid unnecessary and spurious ones.

In order to express the connectivity property, we consider the set of shortest paths π_i within each graph G_i. The fused graph G_f should be so that

$$\forall p \in \pi_i, \exists \hat{p} \in \pi_f \text{ s.t. } d(p, \hat{p}) \leq \theta, \ i \in \{1, 2\}, \tag{1}$$

where $d(\cdot, \cdot)$ is a suitable distance function between paths which takes into account their geometry, and θ is a user-specified tolerance parameter. In our paper we use the following distance function

$$d(p_0, p_1) = \min_{i=0,1} \max_{u \in p_i} v(u, p_{1-i})$$

where $v(u, p)$ is the minimum distance between a point u and path p measured in meters. Thus a small $d(p_0, p_1)$ indicates that one of the two paths can be matched onto the other.

In addition, we wish to find the "minimum" such graph, i.e., the one that minimizes the sum of the lengths of its shortest paths:

$$\arg \min_{G_f} \sum_{p \in \pi_f} \ell(p).$$

This problem formulation can be reconducted to a minimum vertex cover problem on a suitably-defined bipartite graph $H(\pi_1, \pi_2, F)$. The two sets of vertices in H are all the possible shortest paths in G_1 and in G_2 (π_1 and π_2, respectively). There is an edge (u, v) between two elements u and v if their distance is below the threshold, i.e.

$$(u, v) \in F \iff d(u, v) \leq \theta, \, u \in \pi_1, v \in \pi_2.$$

Finding a minimum vertex cover M on H is equivalent to finding a minimum set of shortest paths such that their union maintains the connectivity property of the two source graphs. Therefore, G_f can be build from the union of these paths $M \subseteq \pi_1 \cup \pi_2$.

Note that due to König's theorem, the minimum vertex cover problem on a bipartite graph is actually tractable in polynomial time (and not NP-hard as in the general case). However, the size of the problem is $\mathcal{O}(n^2)$, and that to materialize H naïvely we need to compute $\mathcal{O}(n^4)$ distances between pairs of shortest paths.

Graphs representing the OSM and inferred maps in a large city such as Doha have more than $n = \mathcal{O}(100\text{ K})$ nodes. Hence the polynomial solution we hinted above is impractical. Therefore in the following section we propose a simple and efficient heuristic for tackling map fusion problem.

4 New Roads Detection

A common approach used in the literature (Shan et al. 2015; Wang et al. 2013) to identify or detect new roads is the following. First, run a map matching algorithm between an existing map and a collection of GPS trajectories to identify the subset of trajectories that remain unmatched. Second, run some road creation algorithm on the collection of unmatched trajectories to identify the new roads. Finally, link the newly created road segments to the existing map. That is, at the heart of the process, an algorithm is required to create roads from GPS points, which is exactly what all map inference algorithms do. Thus, it is hard to understand the real added value of map updating algorithms compared to what map inference algorithms do. For instance, if we assume that the initial map is very sparse, then it becomes clear that map update algorithms will be creating most of the road network, just like map inference algorithms do. Another way to look at the issue is to consider an initial empty map: in this case the map update and map inference become equivalent problems.

In our work, we take a slightly different approach. We assume that two maps are given to us. One that represents the base map (e.g., OSM) and another one that is generated using GPS traces via one of the many map inference algorithms available. The problem is then redefined as merging these two maps.

The function *FindOutliers* takes as input two maps M_1 (original) and M_2 (inferred), and generates a set of outliers. Outliers are set of nodes in the map M_2 which are at distance at least θ (here we use $\theta = 20$ m). Mappings link nodes in M_2 to M_1, whereas outliers are those nodes in M_2 that have no correspondents in M_1. These nodes are considered as candidates to be part of new road segments not covered in M_1. Our road addition procedure (see Algorithm 1) works as follows.

Algorithm 1 `MapFuse`

1: **Input:** Base road map M_1, inferred map M_2
2: **Parameters:** collision radius (r, in meters)
3: $outliers = FindOutliers(M_1, M_2)$
4: $DRS = Subgraph(M_2, outliers)$
5: **for** each $o \in outliers$ **do**
6:     compute $distance(o, M_1)$
7: **end for**
8: $outliers = Sort(outliers)$ in decreasing order of distance to M_1
9: **for** each $o \in outliers$ **do**
10:     $sg = BFS(o, DRS)$
11:     **for** each node $n \in sg$ **do**
12:         **if** $distance(n, M_1) \leq r$ **then**
13:             $merge(n, argmin(n, M_1))$
14:         **end if**
15:         $outliers = outliers - \{n\}$
16:     **end for**
17: **end for**
18: **return** M_1

In line 4, the sub-graph of newly detected roads (DRS) in M_2 is generated from the outliers. In lines 5–8, the outliers are sorted in a decreasing order of their geometric distance to M_1. The intuition here is that the farther a node is from M_1, the more likely that node lays on a new road segment not covered by M_1. Outlier nodes are then processed in their order as follows. For each node o, we run a breadth first search (BFS) in M_2 starting from o until it reaches a leaf node or a node that is within a radius r (e.g., 2 m) from M_1. Leaf nodes are assumed to be dead ends of newly detected road segments whereas nodes within a radius distance r from M_1 are assumed to belong to M_1. Nodes in the latter case are then merged with their closest nodes in M_1 as per line 13.

It is not difficult to see that the output M_f of above algorithm satisfies the condition from the Eq. (1). All paths from M_1 are indeed in M_f and are obviously matched by paths of M_f, the nodes from M_2 which are more than θ away from M_1 eventually get merged into the M_f and clearly satisfy the matching requirement (1).

5   Closed Roads Detection

Recall that the input to our process is the original map M_1, GPS-level trajectory data and the automatically inferred map M_2. An important characteristic of the road network are road closures, which are sometimes permanent, but often temporary. Unfortunately, road closures have been overlooked by previous map-inference/map-update literature and in this section we propose two novel techniques for inferring road closures. The first one is 'static', in that it infers the road closures on a fixed input of trajectory data on the roads which have been closed prior to the start of the data collection. The second technique is more dynamic, as it observes the time series of the trajectories passing by a given road segment and by looking for anomalies is such time-series it effectively detects the road closures on the segments which have previously carried some trajectories in the data.

5.1   Cold-Start Road Closure Detection

As we hinted above, trajectory data collection inevitably has a starting point which is determined by either the functionality of the probe and the back-end system which stores the data, or by privacy regulations which may require sensitive trajectory data to be deleted after a period of time elapses.

What makes detection of closed road segments (from map M_1) difficult is the fact that there is a very high skew in the frequency of trajectories on different road segments: some segments (e.g., highways) carry a large number of trajectories while others in the capillary roads may not carry even a single trajectory. In Fig. 3 we show how many new segments are 'discovered' as more driving data is collected. If we denote by L the total length of the road network, after trajectories with total length of L, only about 10% of the unique road segments are touched by those trajectories. After the total trajectory length gets to $10 L$ they touch around 22% unique road segments. With all trajectories in our dataset with total length of $175 L$, we get to detect only about 48% of the road network.

Thus, therein lies a dilemma: is a segment from map M_1 which has not carried any trajectory a closed road segment or it simply did not see a trajectory due to its peripheral nature? To answer this dilemma we initially aimed to exploit the OSM meta-data of OSM road segments such as road type, speed limit, number of lanes or one-way tag. However, the OSM meta-data appears to be rather sparse and is unlikely

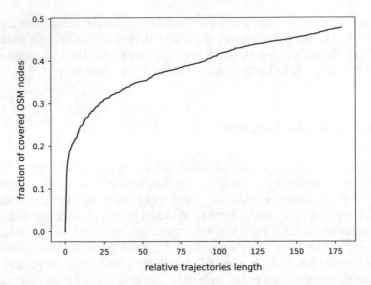

Fig. 3 Fraction of OSM nodes which are covered by at least one trajectory as a function of relative trajectory length defined as the ratio between the total length of all trajectories up to a point in time and total length of the road infrastructure. For close to 100% coverage one would need to have very, very, large trajectory dataset

to give us the relevant road importance score which would help answering the above dilemma.

We address the aforementioned question by evaluating the node betweenness centrality (BC)[3] in map M_1. The BC of a node acts as an indicator of the importance of the node in the graph M_1, and not-surprisingly we see a strong dependence between the centrality of a given road segment and the number of trajectories in our data that pass through it. As seen in Fig. 4, the trend is that the more trajectories a node has the higher BC and vice versa. In Fig. 5 we depict the empiric CDF of node BC for two classes of nodes: those who lie on at least one trajectory and those who do not. We observe that BC mean/median among the nodes which lie on at least one trajectory is an order of magnitude larger than among the nodes which are not carry any trajectory.

Based on these observations, we declare the road segment closed if it has no trajectories passing by it and its BC is greater than the threshold γ. We choose $\gamma = 0.01$ to shave off the tail of the BC distribution among the nodes with no trajectories. Such γ identifies a handful of roads which are closed which we confirm by inspecting each one of them. In addition to those closed roads which are a sequence of closed nodes (with $BC > \gamma$) there are several nodes which are candidates for closure but are isolated from the other candidates. In order to declare the road closed we require that

---

[3]We believe using another node-centrality measure would likely give similar results, though we do not evaluate the impact of the choice of centrality measure in this work. However, the use of betweenness is consistent with the problem definition in Sect. 3.

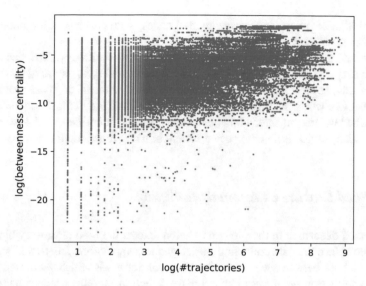

Fig. 4 Scatter plot OSM node betweenness centrality versus number of trajectories passing through each node (logarithmic scale)

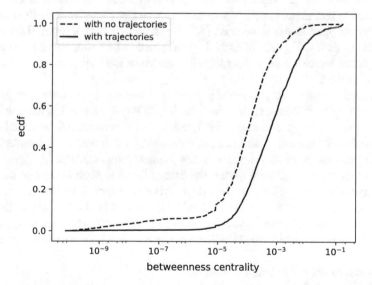

Fig. 5 Empiric CDF of node centrality for two classes of nodes: those who lie on at least one trajectory and those who do not. Node betweenness centrality is generally much smaller among nodes with no trajectories

at least 100 m segment (approximately 5–6 nodes) with corresponding nodes to be candidates for closure.

We would also like to point out that the proposed methodology allows us to infer inconsistencies between the OSM data and the traffic reality as captured by the GPS data. Namely, several roundabouts (formed by nodes with $BC > \gamma$) are represented in OSM as two way streets, however the clock-wise direction in those roundabouts is not matched by any trajectory and hence it is correctly identified as closed road (in that direction) which is an unexpected benefit of using the method described above.

5.2   Road Closure as Anomaly Detection

The method described in the previous section detects the road closures which have happened before the data collection started and it is applicable only to major roads - those with high betweenness centrality. However, for roads which get closed during the data collection we develop an anomaly detection module which monitors the traffic on each road segment and identifies "abnormal" gaps in the traffic stream.

For each node in the map M_1 we track the list of timestamps each time a trajectory is matched to that node. Note that sometimes a trajectory may have multiple records which are mapped to the same node (e.g., if the node is near a traffic light and the vehicle is static it will generate multiple data records which map to the same node in the map) and hence we only record the first match of the trajectory at the node and ignore the others.

As described previously, the road popularity (measured by number of trajectories which pass by it) distribution is rather skewed. In Fig. 6 we plot the number of trajectories that are mapped to every OSM node in our dataset and observe that a large fraction of nodes have only a handful of trajectories which pass by it. Consequently, detecting anomalies on such low-frequency roads is rather challenging.

To detect the road closure during the data collection, each node v in the OSM graph maintains $mean_v(t)$: the *average* inter-arrival time among all trajectories which have passed that node until time t. In addition to that it also maintains the time *elapsed* since the last trajectory: $e_v(t)$. Note that for optimization reasons, the time elapsed is also computed when needed such as a case where a route query is triggered.

We declare the node closed at time t if:

$$e_v(t) > \alpha \cdot mean_v(t)$$

where α is a parameter which determines how conservative we are when deciding to declare the road closed. Small values of α may declare roads closed prematurely, while with large α it may take a long time before a closed road is declared as such.

To understand what is the right choice of α in Fig. 7 we depict the histogram of the ratio between the maximum and the average trajectory inter-arrival time for all nodes which receive at least 2 trajectories per day, in average. We observe that the

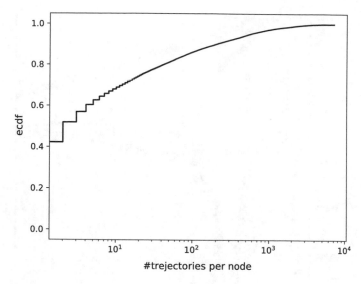

Fig. 6 Empiric CDF of the number of trajectories per node for all nodes in the OSM map. In our dataset only 18% of nodes have more than one trajectory per day in average

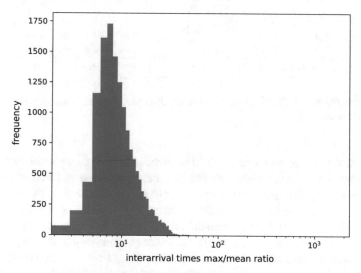

Fig. 7 The distribution of the ratio between maximum and average inter-arrival times for all nodes with at least 2 trajectory per day (in average). Most of the distribution falls in the range 1–40 with outliers corresponding to the nodes depicted in Fig. 8

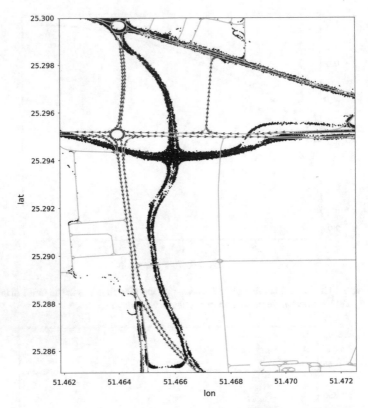

Fig. 8 Detected closed OSM road segments (red). OSM road network (yellow). GPS points after the road closure (black)

distribution of the max-to-average ratio is rather wide, and there is not clear cut-off point. However, most of the distribution is in the range between 1 and 40 with only a few nodes with the ratio greater than 40. Hence we choose $\alpha = 40$. Such choice results in only one closed road-section depicted in Fig. 8 during our 2-month long observation. It involves a closed roundabout and respective access roads.

Finally, note that choosing a smaller α is likely to identify temporary road closures. However, since we could not confirm whether or not such nodes correspond to actual road closure or they simply fall in the tail of the distribution we leave the detailed discussion of temporary closures to future work.

6   Evaluation

In this section we will exploit the GPS trajectory data to evaluate the quality of the fused map.

6.1 Data

As we discussed earlier, our map inference process uses data generated by a fleet of vehicles with GPS-enabled devices. In this paper we utilize the datasets from Doha (Qatar) with around 400 vehicles, 11 Million GPS points (sampled every 10 s). The dataset includes all GPS data points which fall into a rectangle (in lat, lon coordinates) of 6×8 km in an urban region in the city of Doha with a mixture of highways, high and medium volume roads, capillary streets, and roundabouts. Every data record contains: timestamp, latitude, longitude, speed, and heading of the moving direction of the vehicle. Heading is measured in angles against the North axis in degrees reporting values from 0 to 360°.

We preprocessed the data to eliminate those data points with speed ≤5 kmph which are known to have non-trivial noise when reporting location.

6.2 Using Trajectory Data to Evaluate Maps

In this section we analyze how well can we match trajectories to the maps. For a map $\mathcal{M}$ and a trajectory $\tau = (p_1, \ldots, p_k)$ we denote by $\delta(\tau, \mathcal{M})$ the maximum distance between the points on the trajectory τ and $\mathcal{M}$:

$$\delta(\tau, \mathcal{M}) = \max_{p_i \in \tau} \min_{(u,v) \in \mathcal{M}} v(p_i, (u, v))$$

where $v(p_i, (u, v))$ is simple distance to line segment in geo-distance, measured in meters.

In our data we split all the trajectories in two subsets: training and test. We use the training set for constructing map $\mathcal{M}_2$ and the test set of trajectories for evaluating the matching distance. Since many trajectories from the same driver coincide, we make sure that trajectories from the same driver do not fall into both training and testing data. To that end, we split the set of drivers into training/test drivers (75%/25% split) and assign all the trajectories from the training/test driver into training/test trajectory dataset, respectively.

For automatic map inference we use Kharita (2017), but note that using any other automatically inferred map (Biagioni and Eriksson 2012; Cao and Krumm 2009; Chen et al. 2016; Edelkamp and Schrödl 2003) could be used with relatively small (small, since only a handful of roads are being added to the map) impact on the final fused map.

For each trajectory in the test data we evaluate $\delta(\tau, \mathcal{M}_1)$, $\delta(\tau, \mathcal{M}_2)$, and $\delta(\tau, \mathcal{M}_1 \oplus \mathcal{M}_2)$, where $\mathcal{M}_1$ is the underlying (OSM) map, $\mathcal{M}_2$ is the automatically inferred map using the training trajectory data and $\mathcal{M}_1 \oplus \mathcal{M}_2$ is the merged map.

In Table 1 we report the mean, median and 99th-percentile trajectory matching distance for the three maps. All three metrics (mean, median and 99th-percentile) are minimized for the merged map and are around one third smaller than for automatic

Table 1 Trajectory matching distance

$\delta(\cdot, \cdot)$	mean (m)	median (m)	99th-% (mm)
OSM	40.3	9.3	333
automatic	12.3	9.1	70.4
merged	8.1	6.0	53.4

map. The improvements in trajectory matching come for two reasons. On one hand, trajectories which follow the new roads non-existing in the OSM map, but discovered by the automatic map, enjoy better matching in the merged map. On the other, the parts of the trajectories which correspond to the roads which are not covered in the training data, are likely to be covered in the OSM map and hence in the merged map.

7 Conclusion

In this paper, we proposed a new map update paradigm: map fusion. Instead using a customized map-inference algorithm when updating a map, we allow any map to be fused to the underlying (say OSM) map. Such fusion allows for quick map updates, with minimal changes to the high-quality underlying map. In addition to the map fusion, we also study in detail the road closure detection and propose two methods which efficiently detect road closure by comparing the statistical expectation of the traffic on a road segment against the actual traffic.

References

Agamennoni G, Nieto JI, Nebot EM (2011) Robust inference of principal road paths for intelligent transportation systems. IEEE Trans Intell Transp Syst 12(1):298–308

Ahmed M, Wenk C (2012) Constructing street networks from gps trajectories. In: European symposium on algorithms. Springer, pp 60–71

Ahmed M, Karagiorgou S, Pfoser D, Wenk C (2015) A comparison and evaluation of map construction algorithms using vehicle tracking data. GeoInformatica 19(3):601–632

van den Berg RP (2015) All roads lead to ROMA: design and evaluation of a robust online map-generation algorithm based on position traces. MS thesis, TU Delft

Biagioni J, Eriksson J (2012) Inferring road maps from global positioning system traces: survey and comparative evaluation. Transp Res Rec J Transp Res Board 2291(2291):61–71

Biagioni J, Eriksson J (2012) Map inference in the face of noise and disparity. In: ACM SIGSPATIAL

Bruntrup R, Edelkamp S, Jabbar S, Scholz B (2005) Incremental map generation with gps traces. In: Proceedings of intelligent transportation systems 2005. IEEE, pp 574–579

Cao L, Krumm J (2009) From gps traces to a routable road map. In: Proceedings of the 17th ACM SIGSPATIAL, pp 3–12

Chen C, Cheng Y (2008) Roads digital map generation with multi-track gps data. In: International workshop on geoscience and remote sensing. IEEE, vol 1, pp 508–511

Chen C, Lu C, Huang Q, Yang Q, Gunopulos D, Guibas L (2016) City-scale map creation and updating using GPS collections. In: Proceedings of the 22nd ACM SIGKDD international conference on knowledge discovery and data mining. ACM, pp 1465–1474

Davies JJ, Beresford AR, Hopper A (2006) Scalable, distributed, real-time map generation. IEEE Pervasive Comput 5(4)

Du H, Alechina N, Hart G, Jackson M (2015) A tool for matching crowd-sourced and authoritative geospatial data. In: International conference on military communications and information systems (ICMCIS). IEEE, pp 1–8

Edelkamp S, Schrödl S (2003) Route planning and map inference with global positioning traces. Computer science in perspective. Springer, pp 128–151

Garey MR, Johnson DS (2002) Computers and intractability, vol 29. WH Freeman, NY

Google (2017) Google maps. http://maps.google.com

Liu X, Biagioni J, Eriksson J, Wang Y, Forman G, Zhu Y (2012) Mining large-scale, sparse GPS traces for map inference: comparison of approaches. In: Proceedings of the 18th ACM SIGKDD international conference on Knowledge discovery and data mining. ACM, pp 669–677

Lookingbill A, Weiss-Malik M (2013) Project ground truth: accurate maps via algorithms and elbow grease, google i/o, 2013. https://www.youtube.com/watch?v=FsbLEtS0uls

Mnih V, Hinton GE (2010) Learning to detect roads in high-resolution aerial images. In: European conference on computer vision. Springer, pp 210–223

OpenStreetMap (2017) Openstreetmap. http://www.openstreetmap.org

Ruiz JJ, Ariza FJ, Ureña MA, Blázquez EB (2011) Digital map conflation: a review of the process and a proposal for classification. Int J Geog Inf Sci 25(9):1439–1466

Schroedl S, Wagstaff K, Rogers S, Langley P, Wilson C (2004) Mining GPS traces for map refinement. Data Min Knowl Discovery 9(1):59–87

Shan Z, Wu H, Sun W, Zheng B (2015) Cobweb: a robust map update system using gps trajectories. In: Proceedings of the 2015 ACM international joint conference on pervasive and ubiquitous computing. ACM, pp 927–937

Shi W, Shen S, Liu Y (2009) Automatic generation of road network map from massive gps, vehicle trajectories. In: IEEE ITSC 2009

Stanojevic R, Abbar S, Thirumuruganathan S, Chawla S, Filali F, Aleimat A (2017) Kharita: robust map inference using graph spanners. arXiv:170206025

Wang T, Mao J, Jin C (2017) Hymu: a hybrid map updating framework. In: International conference on database systems for advanced applications. Springer, pp 19–33

Wang Y, Liu X, Wei H, Forman G, Chen C, Zhu Y (2013) Crowdatlas: self-updating maps for cloud and personal use. In: Proceeding of the 11th annual international conference on mobile systems, applications, and services. ACM, pp 27–40

Wu H, Tu C, Sun W, Zheng B, Su H, Wang W (2015) Glue: a parameter-tuning-free map updating system. In: Proceedings of the 24th ACM international on conference on information and knowledge management. ACM, pp 683–692

Yang B, Zhang Y, Luan X (2013) A probabilistic relaxation approach for matching road networks. Int J Geog Inf Sci 27(2):319–338

Zeng Z, Tung AK, Wang J, Feng J, Zhou L (2009) Comparing stars: on approximating graph edit distance. Proc VLDB Endowment 2(1):25–36

Zhang L, Thiemann F, Sester M (2010) Integration of gps traces with road map. In: Proceedings of the second international workshop on computational transportation science. ACM, pp 17–22

Semantic Web Technologies Automate Geospatial Data Conflation: Conflating Points of Interest Data for Emergency Response Services

Feiyan Yu, David A. McMeekin, Lesley Arnold and Geoff West

Abstract Conflating multiple geospatial data sets into a single dataset is challenging. It requires resolving spatial and aspatial attribute conflicts between source data sets so the best value can be retained and duplicate features removed. Domain experts are able to conflate data using manual comparison techniques, but the task it is labour intensive when dealing with large data sets. This paper demonstrates how semantic technologies can be used to automate the geospatial data conflation process by showcasing how three Points of Interest (POI) data sets can be conflated into a single data set. First, an ontology is generated based on a multipurpose POI data model. Then the disparate source formats are transformed into the RDF format and linked to the designed POI Ontology during the conversion. When doing format transformations, SWRL rules take advantage of the relationships specified in the ontology to convert attribute data from different schemas to the same attribute granularity level. Finally, a chain of SWRL rules are used to replicate human logic and reasoning in the filtering process to find matched POIs and in the reasoning process to automatically make decisions where there is a conflict between attribute values. A conflated POI dataset reduces duplicates and improves the accuracy and confidence of POIs thus increasing the ability of emergency services agencies to respond quickly and correctly to emergency callouts where times are critical.

F. Yu (✉) · D. A. McMeekin · L. Arnold · G. West
Department of Spatial Sciences, Curtin University, Perth, WA 6845, Australia
e-mail: feiyan.yu@postgrad.curtin.edu.au

D. A. McMeekin
e-mail: D.Mcmeekin@curtin.edu.au

L. Arnold
e-mail: arnolds@iinet.net.au

G. West
e-mail: G.West@curtin.edu.au

D. A. McMeekin · L. Arnold
Cooperative Research Centre for Spatial Information, Carlton, Australia

© Springer International Publishing AG 2018   111
P. Kiefer et al. (eds.), *Progress in Location Based Services 2018*, Lecture Notes
in Geoinformation and Cartography, https://doi.org/10.1007/978-3-319-71470-7_6

Keywords Geospatial · Points of interest · POI · Data conflation
Semantic web · Ontology · RDF · SWRL · Rules · Reasoning ·

1 Introduction

Open Linked Data and Semantic Web technologies have been accepted widely by
the geospatial industry in the recent decade (Parekh et al. 2004; Patrick and Sven
2009; Janowicz et al. 2010; Zhang et al. 2013; Wiemann and Bernard 2016). The
Australian government has been working closely with W3C and OGC[1] to stan-
dardize information and technologies and promote best practice in the management
and use of spatial data on the web.[2] Australia has established its own government
linked data working group (AGLDWG)[3] to develop government standards and set
up Linked Data implementation techniques in response to its citizens and agencies'
needs. More recently, the Australian and New Zealand Cooperative Research
Centre for Spatial Information (CRCSI) published a white paper (Duckham et al.
2017) to propose moving traditional Spatial Data Infrastructures to a Next Generation Spatial Knowledge Infrastructure (SKI) which can automatically create, share,
curate, deliver and use data or information, as well as knowledge creation to support
decision making. Semantic Web technologies were identified as an essential element to support the SKI in connecting, integrating and analyzing data.

To be able to appreciate the benefit of data versatility as highlighted in the SKI
and embrace the advantages of Linked Data for knowledge acquisition, data conflation is an essential process for creating a single point of truth data set from
interrelated data sources, so that knowledge can be more easily derived.

Currently, duplicate geospatial data collection and maintenance exists across
Australian government agencies, leading to data management and processing
inefficiencies. Existing conflation processes are primarily manual and more automated conflation techniques are required (Yu et al. 2016).

The uniqueness of this research is the use of a SWRL Rule-based Data Conflation Framework to automatically match and link corresponding entities between
similar data sets and conflate these entities into a single dataset by selecting the
most accurate features while also removing duplicates without the need for human
intervention. The framework consists of four stages. Stage 1 is the creation of an
ontology based on a multipurpose data model. The multipurpose data model is one
that can be used by government agencies for various business purposes. Stage 2,
refers to the conversion of disparate source data sets into the RDF (Resource
Description Framework) format so they can link to the ontology during the conversion; and the development of SWRL rules to align attributes from the various

---

[1]http://www.opengeospatial.org/.

[2]https://www.w3.org/2015/spatial/wiki/Main_Page.

[3]http://linked.data.gov.au/index.html.

sources so they can be more readily compared and assessed in the latter stages of the conflation process. Stage 3 uses location proxy and other similarity measurements based on semantic descriptions to find matching candidates across data sets. Stage 4 uses a reasoning process to model how domain experts make decisions on which feature attribute values are the best or most accurate when they are considering various data sources.

In addition to the data sets to be conflated, SWRL rules reference other information and knowledge, such as building footprints data. The process is ordered sequentially according to the decision logic used by domain experts. This is an important step in the conflation methodology. Domain experts often refer to other data set(s) to compare attributes in candidate data sets, or look for information in the associated metadata to understand the level of accuracy of each source data set. In many cases, decisions are based on personal knowledge of an area and experience accumulated over time.

This paper explains the Data Conflation Framework and processes, and is organized as follows: Sect. 2 introduces the research background and related works. Section 3 presents the motivating example of conflating three government agencies' Points of Interest[4] (POI) data into a single authoritative for use in the emergency services response domain. Sections 4 and 5 demonstrate the implementation and evaluation of this research, respectively. The paper concludes with a summary of the research and describes a plan for future work.

2 Related Work and Background

It is well recognized in the spatial data domain that Lynch and Saalfeld (1985) were the first to make 'map conflation' a reality in 1985. Their approach to map conflation was to build a prototype using mathematical algorithms to perform geometric alignment between two vector datasets (e.g., census block boundary and road centerline map) (Saalfeld 1988; Kang 2001). This method is typically used to overlay and integrate map layers. The key is to correctly identify matched feature pairs from both base maps. They use the Delaunay triangulation algorithm to partition spaces based on data matches and a rubber-sheeting method to align datasets in each triangle. The process is repeated until all possible corresponding pairs are identified (Saalfeld 1988). Subsequent researchers have improved the efficiency of this method (Chen et al. 2004, 2006, 2008; Dongcai 2013).

However, as technology advances, ways to capture, store and present geospatial data have become more diverse. Geospatial data is recorded in more formats than traditional maps and the data required to support decision-making is often now distributed across the web. Over the past decades, researchers have made significant

---

[4]A wide-ranging definition of a Point of Interest (POI) is any feature or service that people wish to visit or know the location of, and is of value to the community (WALIS).

attempts to bring multiple interrelated geospatial data sets into the same data set to simplify analysis and create a unified view for better data visualization (Uitermark et al. 1999; Fonseca et al. 2002; Lutz et al. 2009; Zhang et al. 2013). The process is normally referred to as spatial data integration (Flowerdew 1991).

One barrier that has impeded spatial data integration is the heterogeneous nature of data. Data heterogeneity is classified into three categories: (1) syntactic heterogeneity, (2) schematic heterogeneity and (3) semantic heterogeneity (Bishr 1998). Syntactic heterogeneity is due to the use of different database systems (relational, object oriented etc.) and geometric representations (e.g., raster or vector representations). Schematic heterogeneity occurs when different data models are used to represent the same real world objects. Semantic heterogeneity arises when different disciplines or user groups have different interpretations for the same real world object. Naming heterogeneity is another form of semantic heterogeneity, such as the same real world object having multiple different names or the same name but referring to different real world objects. The heterogeneous nature of geospatial data makes it difficult to share and leads to data duplication problems.

A study by Lutz et al. (2009) shows that semantic heterogeneity can occur at the metadata level, schema level and data content level; each level blocks the discovery, retrieval, interpretation and integration of geographic information, respectively. They suggest ontologies as an appropriate mechanism to overcome these problems. Parekh et al. (2004) added semantics into metadata based on ontologies to improve geospatial interoperability efficiency and data discovery according to data content. Uitermark et al. (1999) developed a conceptual framework for ontology-based geographic data integration. Their work included generating domain ontology for certain disciplines, and application ontology for each geographic dataset. They also created abstraction rules to define the relationship between the concepts of domain ontology and application ontologies.

Based on the idea that concepts from different application ontologies are semantically similar if they refer to the same concepts or related concepts in the domain ontology, then corresponding object instances can be defined as semantically matched. Fonseca et al. (2002) proposed an ontology-driven geographic information system (ODGIS) in which ontologies are presented hierarchically with the Top-level Ontology at the highest level, Domain Ontology and Task Ontology at the middle level and Application Ontology at the bottom level. Their basic principle was to integrate what was possible and accept that some kinds of information will never be completely integrated due to their fundamentally different nature. They proposed that integration should always be done as the first point of intersection at the lowest level and then propagated upwards in the ontology tree.

As Semantic Web and Linked Data concepts become increasingly popular, more techniques have been studied in the geospatial integration process. There now exist ontologies designed to add semantics into the metadata through the Web Ontology Language (OWL) so computers can understand the meaning of the information and automatically operate actions on it (Parekh et al. 2004). Using the data integration system KARMA (Szekely et al. 2011; Zhang et al. 2013), geospatial data sets can be linked with design ontologies to transform various source formats into the RDF format so data being integrated can be published and reused with rich semantic descriptions on the Web. Zhang et al. (2013) also model integration steps using an ontology, so these processes can read RDF triples as input and also return results as RDF triples. As a result, the system is able to offer some meaningful match and link suggestions across data sets. A tool named FAGI-gis further explores semantic web technologies in the geospatial data domain (Giannopoulos et al. 2015). The input to the tool is two separate geospatial data sets converted to the RDF format and stored in PostGIS databases. SPARQL endpoints are used to pull linkages between entities from both data sets and their associated attributes. The tool uses Virtuoso as its RDF triple repository to store output and it supports GeoSPARQL[5] vocabularies so geospatial features are presented as GeoSPARQL WKT serialization and Basic Geo.

However, literature about spatial data integration has either focused on part of the integration processes, such as data discovery (Parekh et al. 2004), data retrieval (Walter and Fritsch 1999), data matching and linking separately (Sehgal et al. 2006; Wiegand and García 2007). Even when the processes have been studied as a whole, results only link the matched entities together and display all attribute values from each source (Zhang et al. 2013). The value conflicts between different sources for a same attribute haven't been resolved so the duplicate datasets still exist in silos.

There is more geospatial data conflation research required to combine overlapping geospatial data sources into a single source with richer attributes by reconciling conflicts and minimizing redundancy amongst source data sets while still retaining the best attributes from each source. Unlike traditional map conflation, once base maps for conflation are identified, much of the essential information required during the process is also known, such as, coordinate system, map scale, date created etc. So the conventional map conflation processes usually set the base map with higher geometry accuracy as the target map, then align each other map with the target map and transform attributes to the target map.

Contemporary spatial data conflation processes not only need to deal with all the difficulties associated with data integration, but furthermore to merge or fuse multiple data sets into a single data set. This involves decision making, such as "which data is most accurate?" and "which data is more up-to-date?" etc. However, the relevant information to support these kinds of decisions is usually vague.

Fusion can be further categorized. For example Szekely et al. (2011) merged point data with the latitude/longitude representing buildings or structures with address information from Yellow or White Pages. The connection between these datasets is the vector data attributed with street information. It uses latitude/longitude information for each vertex so it can calculate distance to point data. Having street names means it can compare with addresses extracted from Yellow or White Pages. Because each data set contains only one aspect of the real world object, the main challenge is finding matches. Once the nearest distance is identified and the name strings matched, the data sets can be fused. This method showcases

---

[5]The OGC GeoSPARQL standard supports representing and querying geospatial data on the Semantic Web. http://www.opengeospatial.org/standards/geosparql.

the 'attribute enrichment' aspect of data conflation, which involves combining the complementary properties.

The other part of the data conflation mission which is to resolve conflicts and reduce duplicates has not been well addressed. The work of Zhang et al. (2013) reduced data redundancy wherever attribute values from both data sets were exactly matched such as, exact name for a country/state or coordinates for a building. However, when the attribute value is different, the conflicts are not resolved. Instead they 'union' the attributes into a single list. Hence, there are multiple values for the same attribute in the resulting integrated list, such as two coordinate pairs representing the same building. The problem here is that two locations create confusion for a user when navigating to the building.

While matching and linking processes have been done semi-automatically or automatically using computer algorithms, the fusion process is difficult to automate with algorithms because it requires decision making not only to look at the data themselves but also requires reference to other information or knowledge. It is hard for the computer to do this because it needs domain expert's knowledge and intervention.

The fusion process requires holistic information, human logic and the sequencing of logic into a set of reasoning steps. Data sources that enable holistic reasoning include but not limited to, reference data, business rules, metadata, provenance, topological relationships or even domain expert's experience and knowledge stemming from years of work. The motivating example used in this research endeavors to replicate and sequence human logic through a series of automated reasoning steps and reference data sets to achieve a more holistic approach.

3 Motivating Example

The problem of duplication in the collection and management of spatial datasets is twofold. Firstly, duplication is costly for governments as it creates an unnecessary overhead in human and computing resources. Secondly, there is inconsistency between datasets meaning that the source of truth is not clearly understood and end-users may make decisions using incorrect or outdated information.

This is particularly a problem for emergency services. Incidents are often attended by more than one emergency service organization—ambulance, State and Federal police, fire and rescue, defense organisations and emergency volunteer associations. If each agency is using their own datasets there is a risk that information may be different leading to poor communication and coordination between first responders. For example, each organisation typically collects location data (points of interest), such as education institutions, pubs and clubs, pharmacies and civic places, to enable dispatch operations and incident management. However, these location features are often collected using different means, from distinct sources and at different times. The characteristics of these features are also recorded differently. Sometimes this is for unique and specific business purposes e.g., police record locations where licensed firearms are held, where restraining orders exist, and where violent behavior has occurred previously; whereas the fire department records the age and maintenance cycle of fire hydrants, location of arson and building floor plans. However, the more common reason why information is recorded differently is simply because there was no agreed standard for capturing and modeling information when these systems were first built.

Agencies are now coming to realise that collaborative data collection and shared resources is a more attractive alternative and one that makes incident management more effective. However, bringing multiple agency datasets together is problematic.

The data conflation case study used in this research is based on a project named LOC8WA, which was managed by Landgate (Western Australian Land Information Authority) in collaboration with WAPOL (Western Australian Police) and DFES (Department of Fire and Emergency Services). LOC8WA sought to conflate the POI data sets managed by each department into a single authoritative data set. The objective of LOC8WA was to improve the accuracy and confidence of emergency location information to increase the ability of emergency services to respond quickly and correctly to emergency callouts.

Identifying matched POIs across three datasets and conflating them into a single POI is a complex process. A scenario where all three POIs datasets related to a same region are combined is shown in Fig. 1. A point representing a shopping centre is highlighted inside a red circle. This point is from the Landgate data set and is represented by a small dot inside a building footprint. Whereas, the shopping centre is recorded in the DFES dataset as two red diamond shape points (within blue circles) located in a road intersection.

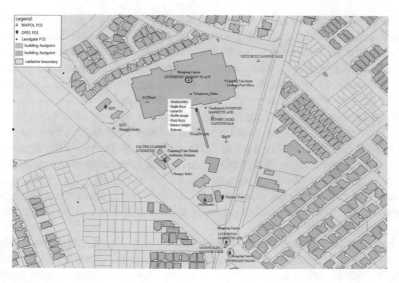

Fig. 1 POIs distributed around a shopping centre area

Noticeably, there are points inside the shopping centre with different categories such as supermarkets, bank branches and the post office. Around the shopping centre, there are other feature class points, bus stations, taxi ranks and fast-food outlets. The complexity or "confusion" in this situation is that some points are the same POI but their location is different. This is because they were sourced from different departments; or many POIs have the exact location but cannot be treated as the same POI as they have different names and attributes.

The LOC8WA project did not generate a conflated data set. Nonetheless, the importance of having an accurate POI data set for emergency services still remains and this has given rise to the importance of this research and the use of LOC8WA to case study automated conflation techniques using advanced semantic web technologies.

The amount of human effort required to complete the task was considered too great to correctly identify matches and make correct conflation decisions on a case-by-case basis. There are tens of thousands of POIs in total from these three agencies. Without the same ID to represent the same POI across agencies' data sets, the same POI's location varies from data set to data set, and there is no consistent naming convention. The research question is "How can it be known that the three points from the different data sets actually correspond to the same POI, which POI attributes (of each point) are the most correct and which points and attributes should be removed?".

4   Implementation

4.1   Stage 1: Ontology Development

Before ontology generation can be started, a fit for purpose output model should be defined which is able to satisfy multiple objectives and users. The data model represents the different models, each of which meets the business needs of each of the participating agencies. The choice of output model can affect the reasoning procedure design. For example, different models can define which data is ruled out and the final decision will consequently differ accordingly.

However, this research is not to define a completely new model from scratch; instead, the research will use existing models whenever possible (Yu et al. 2016). The LOC8WA project uses the Landgate's Points of Interest Data Model and participating agencies agreed that this model suited their business purposes. It was therefore adopted as the multipurpose mode for this study. The POI Ontology developed in this research is based on the Landgate data model and associated data dictionary. The POI ontology has potential to be adopted as a standard for all WA government agencies.

The essential knowledge in the data model was extracted and is shown in Fig. 2. It shows the classification system for the POIs which complies with a three-level hierarchy where red, blue and grey rectangles represent feature classes, feature

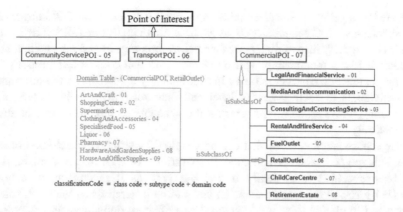

Fig. 2 A portion of Landgate POI data model

subtype and feature domains, respectively. A two-digit number following each hierarchy level value is the class code, subtype code and domain code, which together form a six digit classification code number for each POI.

The POI Ontology, designed according to the above structure, formally captures the scope of knowledge for Points of Interest using the Web Ontology Language (OWL), so it is machine-readable and reasoning can be done on the ontology. A part of the ontology corresponds to the same part of the data model demonstrated in Fig. 2 is shown in Fig. 3. There are three classes *POIClass*, *POISubtype* and *POIDomain* in the ontology and each represents a concept in the classification system, i.e., feature class, feature subtype and feature domain. On the right hand side of each class are their individuals or instances, an example is highlighted in red color at the bottom of the figure. The individuals showcased in *POIDomain* correspond to the "Domain Table" values in Fig. 2. They are all feature domains relating to *RetailOutlet* feature subtype; hence all *POIDomain* individuals are pointing to the *RetailOutlet* individual which is a subclass of *CommercialPOI* as

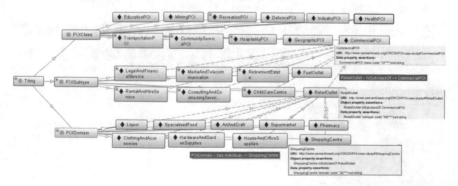

Fig. 3 OntoGraf (https://protegewiki.stanford.edu/wiki/OntoGraf) representation for classes and instances based on POI data model

indicated by a yellow pointer. All other individuals enumerated in *POISubtype* class are subclasses of *CommercialPOI* as well. Individual features also have a data property to specify its two digit code (see yellow box Fig. 3) and information about whether it has a relationship with another feature using an object property (see yellow pointer Fig. 3). The ontology in Fig. 3 clearly demonstrates the information for individuals in each hierarchy level and their relationship with others; more importantly, these relationships are machine-readable so inferences can be drawn automatically.

The classification code, which can be acquired by string concatenation of class code, subtype code and domain code, is an attribute of each feature domain. It has not been specified individually in the ontology as it is considered common knowledge for all the feature domains and can be inferred using a SWRL rule, as shown in Fig. 4. Consider the *ShoppingCentre* feature domain as an example. Its inferred classification code is inside the red rectangle. The rule together with all classes, instances for each class, object property and data properties presented are considered as the top-level ontology for Points of Interest (Fig. 4). The Top-level ontology includes the minimum information required to express the essential knowledge in this POI study area.

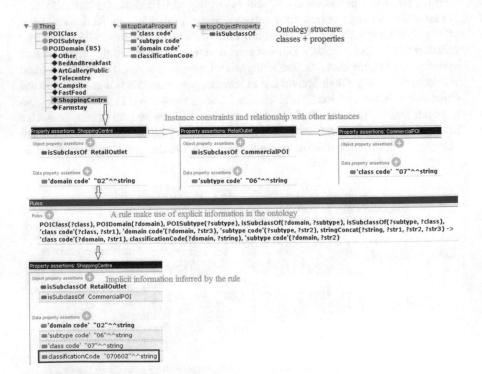

Fig. 4 POI top level ontology

4.2   Stage 2: Data Conversion and Alignment

When dealing with a specific project or application, the top-level ontology can be expanded to accommodate specific business needs. For example, the data property and object property lists are expanded so they can be used to transform the source data into RDF triples and used in reasoning processes (Fig. 5).

The three source datasets have quite different schemas including different levels of granularity. For example, even though the classification system for the POI was adopted by each source they represent it diversely. The WAPOL data set has three columns recording the POIs' feature class, feature subtype and feature domain values while DFES only contains the feature domain. The Landgate data set has six digital numbers to present the classification code. In order to automatically compare whether two POIs are in a same category, they need to all have a same attribute, either the feature domain value or classification code.

SWRL rules are used to read in the different kinds of classification attributes from each source and infer the missing information contained in the POI classification system so they can have the same attribute granularity. In the top-level ontology (Fig. 4), the 6-digit classification code has already been inferred for each feature domain. Hence, if a POI has a feature domain as "ShoppingCentre", its classification code can be retrieved from the ontology via a SWRL rule as well. This is because data is linked to the ontology during the RDF conversion process and therefore the data has the same semantic description as the ontology. Conversely, if a POI classification code is known, the relevant classification information can also be retrieved by a rule. The rules are shown in Fig. 6. Properties shown in yellow are inferred by the rules while the other data properties are drawn directly from RDF conversion. After alignment, the three example POIs shown below have the same attribute granularity.

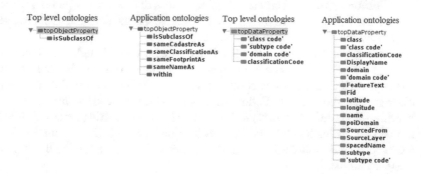

Fig. 5 Developed application ontology based on top level ontology

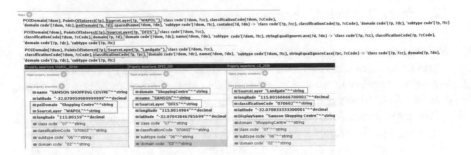

Fig. 6 Using SWRL rules to align disparate attributes

4.3 Stage 3 and Stage 4: Finding POI Matches and Attribute Conflation

The logic of finding matches and conflation is as follows:

1. Search points in buffer zone: The spatial (geographic location) characteristic is used as the first step in finding matches. For a selected POI, a buffer size is given by the user and used to calculate the distance between the POI and its surrounding POIs. Only points that fall inside the buffer zone of the selected point will be considered for conflation. This is because points that are close are more likely to be the same point than those further away. This is a mathematic calculation, so a rule is not used.

2. Compare classification code (Rule 1): the second step takes advantage of the POI classification system. As shown in Fig. 1, shopping centre, supermarket, fast food, bus station and taxi rank etc., they could all cluster within a buffer zone. However, each of them belongs to a different feature domain in the POI classification system so their classification code is different. Only points with the same classification code as the selected POI are considered as potential matches to be used in the next comparison step.

3. Compare by name string (Rule 2): For example, even though all POIs may belong to the *FastFood* feature domain, a POI named McDonalds[®] and another one named KFC[®] must not be conflated into a single POI because they represent different fast food stores. Following the classification code comparison, the matching list is further narrowed down by doing a name string measure. A POI named "KFC Cannington" and "Kentucky Fried Chicken Cannington" will be the matched points and a POI named "McDonald's Cannington" will not be in the matched list.

   Up to this point the matching and linking process is finished and a list of candidate POIs is ready to be conflated. The list normally contains two or three points, so the next step is to decide which point to keep.

4. Interrelated Relationships (Rule 3 and Rule 4): During the conflation stage, human intervention is normally required as human logic is currently more

efficient than comparison algorithm logic. Domain experts usually use contextual validation to decide which point to keep for each POI. For example, points representing a building are typically overlaid on top of aerial imagery to manually inspect which point is closest to the actual location of the building. In order for the system to perform this task automatically, this contextual validation process is replaced by intersecting POIs with two polygon data sets, i.e., cadastral boundary data and building footprints. The reason is because of the topological relationship they have with POI data. A building footprint must fall into a cadastral boundary, and if a point represents that building, theoretically it must fall into the footprint too. The point is less accurate if it is outside of the footprint but inside the cadastral boundary. It is even less accurate if it is outside the cadastral boundary. Using this logic, if only one point is within the building footprint, then it is considered the most accurate point. This is the point kept and the other physical points will be removed and their attributes conflated into this point. The next choice is the single point within the cadastral boundary.

5. User purposes (Rule 5): In the situation where there are still multiple points within the building footprint or none inside the footprint but more than one inside the cadastral boundary, experts usually decide which point to keep based on different purposes and these purposes can be formulated into rules. There are three rules generated in this study:

(1) *Provenance and Metadata Rule*: The order of reliability is determined by the combined information of metadata and interviews across agencies' experts. In the case study, the order is Landgate, WAPOL, and then DFES. The reason for selecting this option is the user wants to decide based on agencies authority.

(2) *Statistical Rule*: The centroid (mean location) of all the points in the candidate list determines the conflated point. The reason for selecting this option is when all data from the various sources is to be treated equally.

(3) *Random Rule*: Randomly select a point within the candidates list. The reason for selecting this option is when the location does not need a high level of accuracy, for example, for general navigation purposes.

According to the above logic, rules generated and are running in a sequential order, i.e., the result of previous rule will be used as a condition in the following rule, showcased in Fig. 7. It demonstrates a chain of rules to deal with the situation where multiple POIs are within a building footprint, the user makes a final decision based on *Provenance and Metadata rule* (Rule 5) and the result is output to a new class named *ConflatedPoint*.

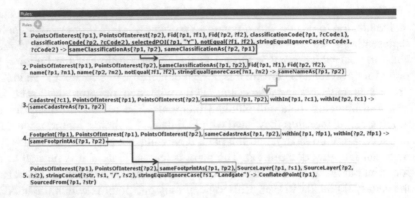

Fig. 7 Rule Chain for finding the best location based on provenance and metadata

5   Evaluation

5.1   Preliminary Testing

The methodology presented was tested with an example scenario shown in Fig. 8 and the process was run in Protégé.[6] A POI from the WAPOL dataset was selected (the blue point inside the basket icon) and a 250 m buffer around the point was calculated. Five points from Landgate, five points from WAPOL and one point from DFES (shown in yellow, blue and purple, respectively), all fall within the buffer zone.

The next stage compares the classification code of all points falling within the buffer zone. The selected WAPOL POI has the same code as one from DFES located in a roundabout and one from Landgate, which is located within the building footprint (represented by the green polygon). According to the conflation logic in Sect. 4.3, these three POIs will be conflated into a single point by taking the POI location from the Landgate dataset, shown using the star marker in Fig. 8.

All points in the example scenario and their relevant attributes were used in the reasoning processes listed in Fig. 9. These POIs were added to the same file as the designed POI ontology and SWRL rules so they could be run together with the Protégé reasoner. However, buffer distances are calculated using mathematical functions outside of Protégé. In addition, the comparison of POIs with the digital cadastre and building footprints is also pre-determined using methods, such as a layer intersection outside protégé. Here, the intersection results (listed in Fig. 9) show whether a POI is "within" a cadastral boundary or a building footprint (blue columns). The yellow columns represent data properties and the blue columns show the object properties.

---

[6]Protégé is a free, open-source platform that provides a suite of tools to construct domain models and knowledge-based applications with ontologies.

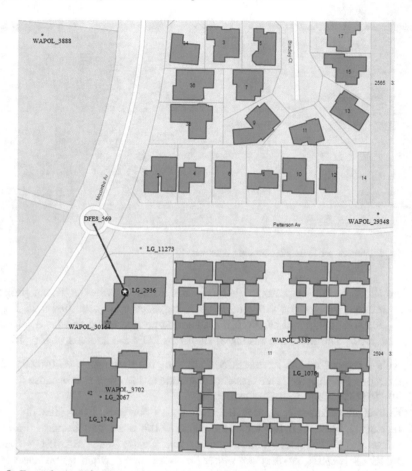

Fig. 8 Example scenario

		poiDomain	domain	classification code	name	DisplayName	FeatureText	Fid	logitude	latitude	SourceLayer	within	within
selected POI	WAPOL_30164	Shopping Centre			SAMSON SHOPPING CENTRE			191	115.801590	-32.070960	WAPOL	cad1	fp1
	WAPOL_29348	Park Reserve			OWEN FITZGERALD PARK SAMSON			23864	115.803330	-32.070370	WAPOL		
	WAPOL_3389	Park Reserve			SAMSON GARDEN PARKLANDS			24372	115.802760	-32.071010	WAPOL		
	WAPOL_3702	Community Centre			SAMSON RECREATION CENTRE			28809	115.801550	-32.071370	WAPOL	cad2	fp2
	WAPOL_3888	Toilet			SIR FREDERICK SAMSON RESERVE PUBLIC TOILET			28688	115.801160	-32.069420	WAPOL		
in Buffer Zone	LG_2936			070602		Samson Shopping Centre		34	115.801667	-32.070833	Landgate	cad1	fp1
	LG_2067			050201			SAMSON RECREATION CENTRE	1784	115.801549	-32.071372	Landgate	cad2	fp2
	LG_1742			070700			WANSLEA SAMSON OUTSIDE SCHOOL HOURS CARE	1578	115.801554	-32.071445	Landgate	cad2	fp2
	LG_11273			050102			SAMSON PUBLIC TELEPHONE	10120	115.801800	-32.070564	Landgate		
	LG_1076			070800			GARDEN PARKLANDS	1047	115.802842	-32.071265	Landgate	cad3	fp3
	DFES_569		ShoppingCentre		SAMSON			99	115.801498	-32.070438	DFES		

Fig. 9 Attribute list of example scenario POIs

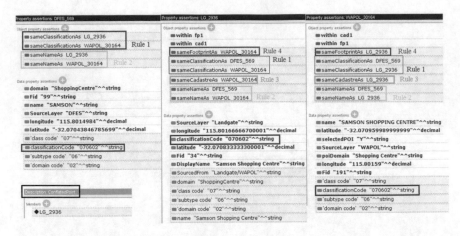

Fig. 10 Properties for POIs after running reasoner

The Protégé built-in reasoner Pellet[7] is run to check whether it can properly return inferred results for different POIs using each rule. As shown in Fig. 8, *DFES_569*, *LG_2936* and *WAPOL_30164* are supposed to be conflated into one, i.e., *LG_2936*. The inference results of the three POIs are showed in Fig. 10.

(1) Rule 1 returns results for the three POIs (see red rectangle). It correctly identifies one POI has the same classification code as the other two because they are all "070602" (see dark blue rectangle).

(2) Rule 2 also correctly returns inferred results for each POI. (See light blue rectangle). Each POI has the same name as the other two because the name values are "SAMSON", "Samson Shopping Centre" and "SAMSON SHOPPING CENTRE", so they are either an exact match when ignore case (e.g., "Samson Shopping Centre" and "SAMSON SHOPPING CENTRE") or one is contained within the other (e.g., "SAMSON" and "SAMSON SHOPPING CENTRE").

(3) Rule 3 and Rule 4 does not return any result for *DFES_569* because it is not within any cadastral boundary or building footprint. Both rules return a result for the other two POIs because they all within "cad1" and "fp1", so they have *sameCadastreAs* and *sameFootprintAs* with each other.

(4) Rule 5 returns the final result as *LG_2936*, which is an inferred member of *ConflatedPoint* class (see black rectangle in the lower left corner). This is the expected result for the test scenario based on the *Provenance and Metadata Rule*, i.e., Landgate data is more accurate than WAPOL data when two POIs from these two sources are both within a building footprint.

The inferred results for other points included in the test scenario are shown in Fig. 11. Because their classification codes are different than the selected POI, no

---

[7]Pellet is an open-source Java based OWL 2 reasoner https://www.w3.org/2001/sw/wiki/Pellet.

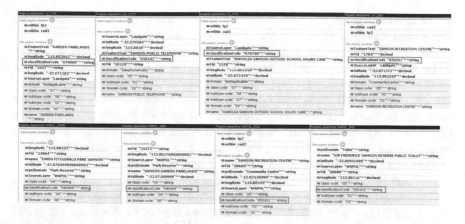

Fig. 11 Reasoning results for all other points

results are generated in Rule 1. Hence they are not carried any further in the reasoning process. This fulfills the expectation of the rules as only those candidates that meet the previous rules are carried into the next rule.

5.2 Proof of Concept Web Portal and Further Evaluation Data

The preliminary testing results demonstrate that the SWRL Rule-based Data Conflation methodology can model domain experts' decision making logic, thus enabling geospatial data to be conflated automatically. However, as Protégé is essentially an ontology and SWRL rule editor, there are many functions that cannot be performed, such as, calculate points within buffer zone, and intersect points with reference layers. Also, the example only demonstrates one scenario, which is two points within the same footprint and the final decision is based on *Provenance and Metadata Rule*. However, it is acknowledge that there could be other scenarios and different rules will come into play, such as a decision made by statistic rules or random rules, or if only one point is in a footprint, the point can be chosen automatically etc.

A Proof of Concept (PoC) web portal has been developed to integrate the aforementioned functions and automatically trigger different rules depending on the different situations.[8] The Data Conflation application server provides a visualisation layer so that the user can view the dataset points before and after conflation. The visualisation layer is developed using React JS. The user is able to access it through a common web browser such as Chrome and Firefox etc. The web application

---

[8]https://crcsi.amristar.com/automatedconflation; username: crcsi; password: l@ndg@te.

server also hosts the Apache Jena Semantic Web business rules engine that the web application interfaces to execute the conflation processes.

As the PoC web portal is capable of dealing with larger datasets and more complicated scenarios, a further evaluation was able to be performed. The evaluation is based on conflating *ShoppingCentre* feature domain points from the three sources including 351 POIs from Landgate, 255 POIs from WAPOL and 381 POIs from DFES. These POIs are well distributed across Perth metropolitan area. The reason for using this particular feature domain is that these points exist in all three datasets in the study area. The WA Police dataset and Landgate dataset cover most of the feature domains, whereas the DFES dataset only records *FastFood*, *Supermarket* and *ShoppingCentre* feature domains. However, the Landgate dataset does not contain enough samples in the *FastFood* and *Supermarket* feature domains with only 8 and 28 points in each feature domain, respectively. Furthermore, the points in these two Landgate feature domains occur outside the Perth Metro area where no building footprint data is available to compare. Therefore, the *ShoppingCentre* feature domain data in this case is the best test data to evaluate whether conflation decisions can be correctly made between the three sources.

The buffer size is set as 250 m is based on trial and error. A manual check on a few of the larger shopping centres in the metropolitan region showed that 250 m is sufficient to return relevant points and it is not too larger an area to decrease system performance. Nonetheless, in the PoC web portal, a user is able to select an area of interest rather than the whole dataset search area.

The building footprints and cadastral boundaries reference datasets are provided by Landgate, which is the recognised authoritative source.

5.3 Evaluation Criteria and Results

The evaluation focuses on two aspects; (a) whether the system can effectively reduce duplicate data; and (b) the accuracy of conflated results.

In terms of duplication, the number of conflated POIs is 493, whereas the number of POIs from the combined datasets is 987 (Fig. 12). This means that over half of the points are duplicated, and hence have been removed. At the same time, each source dataset has an increased number of POIs and thus coverage is improved. This is shown in Fig. 12 where Landgate has increased the number of valid POIs by 40%, WAPOL by 93% and DFES by 29%.

In order to examine how accurate the results are, manual validation was performed. Among the 493 conflated POIs, 283 points were generated from multiple points, i.e., either from more than one source or more than one point from the same source. Each of these 283 points were loaded into ArcMap and overlaid with the three source datasets and the two reference datasets to check whether or not the SWRL rule system effectively selected the best location for each scenario. The statistical results are displayed in Table 1.

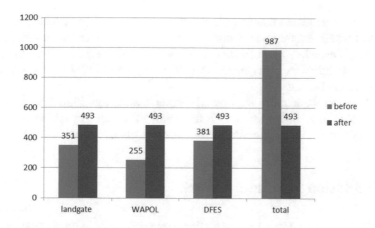

Fig. 12 Number of points before and after conflation for each source

Table 1 Evaluation result for conflate three datasets

Source	# Conflated POI				Total
	#Multi-sources			# Single source	
	Auto-select		Decided by rule		
	In footprint	In cadastre			
Landgate	58	2	156	60	276
WAPOL	15	4	24	63	106
DFES	15	0	9	87	111
Total	88	6	189	210	
					Total: 493

The test revealed that 88 points were conflated automatically because there was only one data source with the point inside the building footprint. There are 6 cases where no points were within a building footprint and only one point inside a cadastral boundary. The remaining 189 conflated points were decided by the *Provenance and Metadata Rule* as multiple source points existed in a same footprint or cadastral boundary. As the *Provenance and Metadata Rule* defines the Landgate dataset as the most accurate the result showing 156 points from Landgate source as the highest number of valid points was expected over the WAPOL (24 points) and DFES (9 points datasets. Changing the *Provenance and Metadata Rule* would achieve difference results.

Among the 283 conflated points, only 5 points were identified as incorrect and therefore, the conflation accuracy for *ShoppingCentre* POI is 98%.

There are 210 points in the conflated dataset, which were derived from a single source. However, 64 of these points should have match other points but were excluded due to the current name string method being too simple. The current string match method uses *SWRL Built-Ins for String*, which can only perform simple matches, such as match points with exactly the same name or where one name is contained within another *name* string. However, some name patterns such as full name (e.g., *Kentucky Fried Chicken*) and acronym name (e.g., *KFC*) will not return a result *as matched*. A better match method is required to deal with various name patterns across the datasets.

In future work, a more sophisticate string match algorithm will be used to generate custom Built-Ins for SWRL to improve the accuracy of the name string match in order to reduce the number of duplicate points further.

6 Conclusion and Future Work

Incidents are often attended by more than one emergency service organization. If each agency is using their own datasets there is a risk that information may be different leading to poor communication and coordination between first responders. A conflated single authoritative dataset is therefore desirable between agencies. This paper presents a new approach to data conflation where an ontology and RDF data conversion serve as the basis for the solution and SWRL rules are the core to automate the entire geospatial data conflation processes. By using a set of rules in a sequential order, human experts' logic can be used to find the most accurate or fit-for-purpose location and conflate the remaining attributes into the single location and removing duplicate features. In this way, the conflation processes can be run automatically without human intervention.

In the Proof of Concept web application, some other datasets are also used in the system, such as OpenStreetMap and BingImage. At this stage these are only used as based maps for visual reference and not included in the conflation process. Although the conflation with OpenStreetMap is not in the scope of this paper, including OpenStreetMap into the conflation reasoning process either as a reference dataset to facilitate decision making or used as a fourth source dataset to conflate into a single dataset is planned in the future work.

Acknowledgements The work has been supported by the Cooperative Research Centre for Spatial Information, whose activities are funded by the Australian Commonwealth's Cooperative Research Centre Program.

References

Bishr Y (1998) Overcoming the semantic and other barriers to GIS interoperability. Int J Geogr Inf Sci 12(4):299–314
Chen C-C et al (2004) Automatically and accurately conflating orthoimagery and street maps. In: Proceedings of the 12th annual ACM international workshop on Geographic information systems. ACM
Chen C-C et al (2006) Automatically conflating road vector data with orthoimagery. GeoInformatica 10(4):495–530

Chen C-C et al (2008) Automatically and accurately conflating raster maps with orthoimagery. GeoInformatica 12(3):377–410

Dongcai HE (2013) A study on theory and method of spatial vector data conflation. Res J Appl Sci Eng Technol 5(2):563–567

Duckham M et al (2017) Towards a spatial knowledge infrastructure Australia and New Zealand CRC for spatial information

Flowerdew R (1991) Spatial data integration

Fonseca F et al (2002) Using ontologies for integrated geographic information systems. Trans GIS 6(3):231–257

Giannopoulos G et al (2015) FAGI-gis: a tool for fusing geospatial RDF data. In: The semantic web: ESWC 2015 satellite events: ESWC 2015 satellite events, Portorož, Slovenia, May 31– June 4, 2015. Springer International Publishing, pp 51–57

Janowicz K et al (2010) Semantic enablement for spatial data infrastructures. Trans GIS 14 (2):111–129

Kang H (2001) Spatial data integration: a case study of map conflation with census bureau and local government data. http://www.cobblestoneconcepts.com/ucgis2summer/kang/kang_main. htm

Lutz M et al (2009) Overcoming semantic heterogeneity in spatial data infrastructures. Comput Geosci 35(4):739–752

Lynch MP, Saalfeld AJ (1985) Conflation: automated map compilation—a video game approach. Auto-Carto 7, Washington, DC, USA

Parekh V et al (2004) Ontology based semantic metadata for geoscience data. IKE

Patrick M, Sven S (2009) Data integration in the geospatial semantic web. J Cases Inf Technol (JCIT) 4(11):100–122

Saalfeld A (1988) Conflation automated map compilation. Int J Geogr Inf Syst 2(3):217–228

Sehgal V et al (2006) Entity resolution in geospatial data integration. In: Proceedings of the 14th annual ACM international symposium on advances in geographic information systems. Arlington, Virginia, USA, ACM, pp 83–90

Szekely P et al (2011) Exploiting semantics of web services for geospatial data fusion. In: Proceedings of the 1st ACM SIGSPATIAL international workshop on spatial semantics and ontologies. ACM, Chicago, Illinois, pp 32–39

Uitermark H et al (1999) Ontology-based geographic data set integration. In: Böhlen M, Jensen C, Scholl M (eds) Spatio-temporal database management, vol 1678. Springer, Berlin, Heidelberg, pp 60–78

Walter V, Fritsch D (1999) Matching spatial data sets: a statistical approach. Int J Geogr Inf Sci 13 (5):445–473

Wiegand N, García C (2007) A task-based ontology approach to automate geospatial data retrieval. Trans GIS 11(3):355–376

Wiemann S, Bernard L (2016) Spatial data fusion in spatial data infrastructures using linked data. Int J Geogr Inf Sci 30(4):613–636

Yu F et al (2016) Automatic geospatial data conflation using semantic web technologies. In: Proceedings of the Australasian computer science week multiconference, Canberra, Australia. ACM, pp 1–10

Zhang Y et al (2013) A semantic approach to retrieving, linking, and integrating heterogeneous geospatial data. In: Joint proceedings of the workshop on ai problems and approaches for intelligent environments and workshop on semantic cities, Beijing, China. ACM, pp 31–37

Topology Extraction from Occupancy Grids

Martin Werner

Abstract A fundamental problem in indoor location-based services is to compute the meaning of location with respect to an indoor location model. One specific challenge in this area is represented by the central tradeoff between two philosophies: a decent amount of the community tries to provide high-quality, high-fidelity models investing specialized knowledge and a lot of time in building such models for each building thereby increasing simplicity and quality of location-based services such as navigation or guidance. In contrast to that, other people argue that crowd sourcing and very simple representations of environmental information are the only way of generating indoor environmental information at scale. However, applications then have to tolerate errors and deal with oversimplified models. With this paper, we show for a specific widely accepted simple environmental model in which building floorplans are represented as black-and-white bitmaps, how we can provide algorithms for extracting higher order topological concepts from these trivial maps. We further illustrate how these can be applied to the hard problem of indoor shortest path calculation, indoor alternative path calculation, indoor spatial statistics, and path segmentation.

1 Introduction

Today, numerous location-based services enable new digital services and digital support for day-to-day life. Based on the wide availability of enabling technologies including GNSS positioning, satellite imagery, LiDAR scans, digital terrain models, and maps, many services have been developed especially for the outdoor space. The situation of Indoor Location-based services is, however, different (Werner 2014). First, it is not easy to derive a meaningful location from measurements of existing signals in the indoor area. Without investing much effort and money into a dedicated indoor location system, one is basically left with inertial sensory and signals that

M. Werner (✉)
German Aerospace Center (DLR), Remote Sensing Technology Institute (IMF),
Oberpfaffenhofen, 82234 Weling, Germany
e-mail: martin.werner@dlr.de

© Springer International Publishing AG 2018
P. Kiefer et al. (eds.), *Progress in Location Based Services 2018*, Lecture Notes
in Geoinformation and Cartography, https://doi.org/10.1007/978-3-319-71470-7_7

have not been deployed with positioning in mind. This renders the indoor positioning problem highly ambiguous and challenging. Second, the acquisition of indoor
location models is considerably harder as compared to the outdoor area. While for
outdoors, a simple graph model in which edges represent ways and vertices represent corners is sufficient for navigation, indoor environments are more complex and
cannot be represented by such a simple abstraction as a single graph. Instead, the
full freedom of movement of humans should be taken into account. Based on this, a
multitude of indoor location models has been defined ranging from a simple graph
modelling movement over set-based and hybrid models to complex models based
on occupancy grids or even 3D point clouds (Becker and Dürr 2005). The drawback
of this wide range of modeling methodologies is that there is no good compromise
to be reached: some of these models are extremely expressive and can be applied
in any environment (e.g., 3D point clouds), but are difficult to exploit computationally. Other models are extremely efficient for computations (e.g., set-based models,
graphs), however, are unable to express the full complexity of indoor navigation in
a natural way. Additionally, the modeling effort varies greatly. Drawing new map
information enhanced by a navigation graph consumes a lot of time. However, some
sort of map information is often available: floorplans and building blueprints.

In my opinion, one must rely on techniques that are simple enough such that a
majority of users understands these techniques and is able to model the environment themselves. One such model, though radical, is given by occupancy grids. An
occupancy grid is a map representation scheme in which indoor spaces are first split
into individual floors each of which is represented by a black-and-white floorplan
bitmap. In this bitmap, black pixels are partially or fully obstructed and white pixel
are walkable space. This model is simple enough such that even novice users are
able to create, understand and modify models based on such information as this only
needs basic image editing tools.

However, these representations have not been applied widely for complex queries
about the environment. Instead, they are mainly used to find connectivity by calculating shortest paths in free space ignoring that they will scrape along walls and to filter
erroneous sensor measurements in spatial particle filters. I believe that these simple
models are more powerful and show that it is possible to automatically extract a lot
of information from such occupancy grids, only.

The main contribution of this paper is the following: a method to extract higher
order topological information from these maps in which spaces such as rooms and
their interconnection (such as doors, hallways) are made explicitly available without
user intervention. Additionally, it is shown how to use reduced topological maps for
calculating shortest paths that do not scrape along geometry and, therefore, allow for
better postprocessing when, e.g., identifying visible landmarks for route description.
Third, the paper gives a scalable methods for extracting sets of sufficiently different
short routes between two points, which applications can use in order to optimize
complex additional criteria, e.g., find a sufficiently short path through the airport
such that I am able to change money along the way.

Note that algorithms for selecting alternative routes in street networks are not
applicable in this setting as there is a high number of similar paths sharing not

a single edge, which would be seen as "perfect" alternatives given the criteria used for selecting alternative routes in street networks.

The remainder of the paper is structured as follows: Sect. 2 shortly reviews related work with respect to building modelling and information extraction. Section 3 introduces the contraction pyramid and explains its relation to the Reeb graph of the contraction process. Section 4 introduces illustratory applications including shortest path, alternative routes, Wi-Fi positioning as a classification problem, spatial statistics, and turn-by-turn guidance in buildings. Finally, Sect. 5 concludes the paper.

2 Related Work

Extracting environmental information is a fundamental prerequisite to many location-based services. When providing services based on the location of an entity, the meaning of this location to the application needs to be understood. Therefore, most location-based services are designed by setting the location of a mobile device into the context of an environment and only very simple, just about trivial, services can be provided without environmental information.

A very early work towards the representation of indoor spaces for location-based services has been presented by Becker and Dürr (2005). They distinguish a set-based model in which the subset relation, for example, models rooms being on a specific floor, graph-based models, in which graph edges model neighborhood relations and graph connectivity models space connectivity in a navigational sense, and present combinations of both. Another approach to indoor modeling is explicitly based on the two concepts of locations (e.g., rooms) and exits (e.g., doors) (Hu and Lee 2004).

In the last decades, several new approaches for scalable extraction of environmental information have been proposed. One direction is the automatic extraction of higher order topological information like rooms or doors from building blueprints (Werner and Kessel 2012), another direction of research is towards building a community of people explicitly modelling topological information in open data platforms such as OpenStreetMap (Openstreetmap wiki—indoor mapping 2016).

However, both approaches have not seen wide adoption yet: the complexities of extracting meaningful information from drawings for humans is not sufficiently solved and the communities have not agreed on a single, sufficient way to represent indoor spaces.

Another track of research originates in the robotics domain. Here, the environmental models are limited not by the limitations of algorithms, but rather by the limited ability of mobile robots to acquire and analyze data. In this area, two-dimensional occupancy grids are often created in which white pixels model free space and black pixels model occupied space. This concept of modelling spaces by what devices actually measure, is quite successful. However, much information about indoor spaces is lost. Recent representatives of this approach are given by high-end SLAM systems such as NavVis (NavVis corporation 2016), which generate very large amounts of three-dimensional point clouds.

This paper proposes a method to extract meaningful topological information from occupancy grid maps which can be created by mobile robots or modern measurement devices. The most similar work in literature with respect to our work is given by Fabrizi et al., who extract information topological relevant objects such as rooms and corridors as a computer vision task on fuzzy occupancy grids (Fabrizi and Saffiotti 2000). Due to the nature of their approach, this does not easily generalize to complex environments as it is based on morphology operations using structuring elements, which have a chosen size and shape. This size and shape determines a lot of properties of the output.

The approach provides a set of features implicitly describing the building topology, which can be used to extract a topological map of rooms similar to the approach of Fabrizi. Most importantly, it is not needed to set parameters related to the expected size of topological features as doors or hallways. Additionally, it is shown that the given features can be used for other tasks with respect to the environment such as calculating realistically shaped short paths and reasonable families of alternative paths through space.

3   The Contraction Pyramid

Indoor maps usually come in two different flavors depending on how they have been constructed. Either they are vector drawings or they are bitmaps and occupancy grids (Werner 2014).

Geographic Information Systems (GIS) drawings are usually created by architects and technicians to build, enhance, understand, or improve the building while user-generated or machine-generated maps are often in the form of occupancy grids either by creating such representations using sensory or by using bitmap manipulation software in order to create an occupancy grid map of a building from floorplans.

When GIS drawings of the building are available and of usable quality or if maps for indoor navigation are created using GIS software, the indoor maps are most often collections of two-dimensional drawings of primitives including lines, circles, and arcs. A specific set of symbols is being used to draw special objects such as doors, escalators, elevators, windows, and other buildings objects. However, these drawings are often unclear about some details of these building objects which have to be modelled manually (Werner and Kessel 2010). For topology detection in the sense of this paper, such GIS drawings have to be preprocessed in order to model walkable space as empty space—that is, given a starting point inside the building, a recursive eight corner graph traversal can walk the complete walkable space. This can be done either manually by deleting organizational lines and doors. However, it can be automated to a high degree (Werner and Kessel 2012). In fact, this preprocessing results in a GIS drawing, essentially a set of lines, such that the walkable space can be extracted as an occupancy grid. These occupancy grids are transformed into an eight corner navigation graph by creating a vertex per walkable pixel and eight edges connecting neighboring white pixels. Note that this map could have more than one connected component, especially, when some rooms can only be accessed via building objects such as elevators or staircases. For floorplans given as a bitmap, it is usually quite easy to modify the image to contain white pixels for free walkable space and black pixel for the building and the surroundings. Note that the three-dimensional connections are not to be considered in the context of topology extraction, as they connect different topological parts of the building. Thereby, it suffices to perform topology extraction and detection in 2D and handle three-dimensional connections as connecting different parts from different topological objects.

The concept proposed with this paper is based on the idea of using the vertex degree as an indicator of the border of free space components. The degree of a vertex in an undirected graph is defined to be the number of adjacent edges. Given an eight-corner-system navigation graph spanning walkable space, two different types of vertices exist: full degree vertices, which are called inner vertices and vertices with lower degrees, which are called border vertices. Border vertices are generated near obstructive geometry, which hinders the generation of some edges. The degree of a vertex is used as a color in Fig. 1a.

You can clearly see inner vertices of full degree depicted in red, border vertices in green or yellow depending on their number of neighbors.

The proposed approach proceeds in iterations. In an initialization step, the connected components of the given navigation graph are calculated and all vertices are labelled with a number representing their connected component.

After initializing the data structures, the approach removes all border vertices from the graph and then again calculate connected components. This gives a set of connected components each of which is fully contained in a connected component of the previous iteration. However, a connected component from the previous layer can split into more than one component in the current iteration.

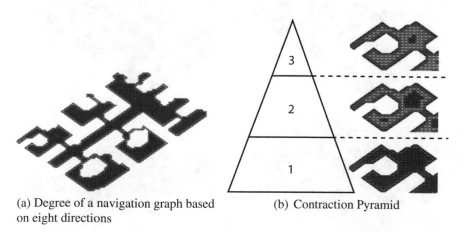

(a) Degree of a navigation graph based on eight directions

(b) Contraction Pyramid

Fig. 1 A degree map and the contraction pyramid of a simple example

Considering the iteration number as a vertical coordinate, one can create a stack of shrinking and eventually splitting connected components similar to what we know from Morse theory or Reeb graphs. This leads to a geometric objects as depicted in Fig. 1b in which the lowest layer 0 is the full graph, layer 1 is given by removing all border vertices of layer 0, layer 2 is given by removing all border vertices from the graph at layer 1, and so on.

Connected regions will soon break into several disconnected regions when enough border vertices are removed, e.g., the graph shrinks and connected components start splitting. Considering the step from layer 1 to layer 2 in Fig. 1b, you see the red marked room splits into a small kernel down and a slightly larger patch up.

When a given connected component splits in parts, all vertices that have been removed in this step and are adjacent to the created connected components are called topological borders. The process ends when connected components are only represented by a set of border vertices, which would disappear in the next iteration. These vertices are called kernel of a room or hallway or otherwise relevant topological object. Now, these kernels are expanded into all directions until they meet with topological border vertices. These connected regions are called a topological room.

Figure 2 depicts the process for a larger map as well as a three-dimensional visualization of the different steps stacked one upon each other.

Again, the three-dimensional representation of the topology is based on stacking the same graph iteratively reduced by removing border vertices vertically on top of the previous graph. This transforms the two-dimensional occupancy grid graph into a three-dimensional object of finite height.

Fig. 2 The process of iterative shrinking and the pyramid as a stack of layers

From this object, one can vertically connect all vertices which are above each other and thereby form a full three-dimensional object. This motivates the following formal definitions:

Definition 1 The *height* of a vertex v representing a white pixel in an occupancy grid map is defined to be the number of layers of the pyramid in which this vertex exists.

In math, Morse theory as well as Reeb graph theory are conceptually similar. They use a concept of sweeping along the height of objects realizing their topological structure. The Reeb graph is very similar to our construction, however, with one more level of abstraction. The Reeb graph of a manifold is constructed by sweeping along one axis (let it be the vertical one) and adding a vertex if and only if a connected component appears, disappears, or splits. These three operations are represented by vertices in the Reeb graph. An edge in the Reeb graph means that some object (e.g., a smaller connected component) appeared from another component during a split. This Reeb graph is combinatorially equivalent to the information in our pyramid, however, purely combinatorial. Hence, there is no access to the vertices and height after creating the Reeb graph. With our pyramid, we are using a slightly larger graph as compared to the Reeb graph. But in a loose sense, the Reeb graph of the contraction pyramid taken as a threedimensional surface is similar to the following definitions extracting topological objects from the pyramid.

From the pyramid, one can as well derive some other objects describing a less local topological feature around a vertex.

Definition 2 The *threshold-connected component* of an occupancy grid vertex v and an integer threshold τ is defined to be the connected component of the layer of the pyramid at height τ.

An example of such a threshold-connected component is given by Fig. 1b. The red vertices on the second layer build, for example, three connected components. The sets of vertices of these components are the threshold-connected components at height τ. Similarly, in Fig. 2, the black connected region in the top left is for height $\tau = 1$, as you can see from the non-black border vertices. To the right, one finds a different set of components by reducing all these components.

Unfortunately, for a fixed height, not all interesting components are realized. Small components die out earlier than large components. So in higher heights, the small components are not visible anymore.

This motivates the following construction of maximal or complete components, which first needs the term of a kernel to be defined more clearly:

Definition 3 The *kernel component* of a vertex is the highest connected component from the pyramid. That is the highest set of vertices such that all of these vertices (including the vertex itself) would be removed in the next iteration.

This leads us to the definition of maximal component:

Definition 4 The set of *complete* or *maximal* components is extracted from a kernel by taking the largest (e.g., lowest) component in the pyramid that fully contains the kernel and does not split in the process.

In other words, a maximal component is the largest connected region that will not split by iterating the described algorithm. In the following sections, the usefulness of the pyramid, threshold-connected components, and maximal components is highlighted in various application scenarios from the domain of indoor location-based services. Note that we do not claim that any of those applications taken for themselves are new or extraordinarily innovative. The aim of the section is to show that the innovation of extracting a topological environmental model without a single value ϵ related to the expected size of topological features is expressive enough for these problems. More clearly, that there is a *single and simple* data structure represented by the contraction pyramid in which these operations can be done.

4 Applications

Understanding the building topology is a quite general and important task for indoor positioning systems. Consequently, there are very many application areas for automatically extracted topological information. The following applications may serve as a examples for the vast applicability of topology as extracted using this framework.

4.1 Topology-Aware Shortest Paths

The extrusion of the map space into a pyramid as explained before is a powerful concept. As a first tool, the height of a vertex can be used to support better shortest paths for visualization and computation along paths without severe additional overhead. One can compute the heights h_i of each vertex and scale each edge weight $\omega_{i,j}$ in the navigation graph by a factor:

$$\tilde{\omega}_{i,j} = \frac{1}{h_i + h_j} \omega_{i,j}$$

This makes edges between higher heights shorter and therefore leads to paths favoring a location for visualization and calculation inside free space as opposed to along walls and corners. Figure 3 depicts an example of a shortest path using the modified weights $\tilde{\omega}_{i,j}$.

As you can clearly see, the shortest path avoids scraping along walls and is a reasonable tradeoff between preferring high vertex edges and short paths. Furthermore, the height map can be used to classify the topology along the path directly. Figure 4a

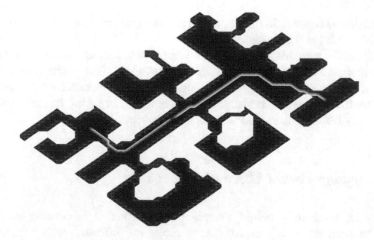

Fig. 3 A shortest path in the adapted weightmap keeping away from disturbing geometry

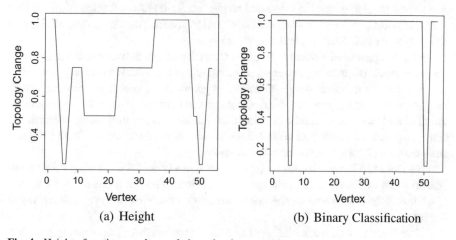

(a) Height   (b) Binary Classification

Fig. 4 Height of vertices as observed along the shortest path

depicts the height as observed when following the shortest path depicted in Fig. 3 and the binary classification of this height.

In the first figure, we follow the depicted path from the left to the right. You can see that we are somehow inside a room, then leave the room through a bottleneck at vertices around eight. Then follow a path of varying height (e.g., a hallway) before we enter a larger space from vertex 35–48 followed by a bottleneck and entering another room. The binary classification on the right shows only the bottlenecks. These two figures can be used to create hints of interest for route description engines creating sentences like: "Start at the given location, then leave the room, follow a medium-sized hallway. When the hallway opens up widely, turn right and go through the door." Essentially, each arbitrary threshold θ on the height returns in a binary classification as depicted in Fig. 4b, which marks significant spatial events along the path with respect to the surroundings.

Additionally, note that computations along the path are more sensible as the path is more similar to the path a human would actually follow. If one wants to describe the shortest path in a navigation application, one likely wants to identify landmarks that are visible from the path. The visible space is, however, larger for the path preferring to stay away from obstacles and walls as much as sensible.

4.2   Topology-Aware Alternative Routes

In the same situation as before, we might be interested in calculating alternative routes between two locations. Alternative routes are different routes between two locations that are reasonably short. It is quite complex to define and evaluate a good notion of different routes; the interested reader is referred to a definition of alternative routes in buildings (Werner and Feld 2014) and to general work regarding alternative routes (Dees et al. 2010; Bader et al. 2011).

Using the proposed pyramid, we can summarize routes between two locations by the set of labels of connected components the route visits. This is especially powerful, when a specific maximal height threshold is given. If we ignore all vertices whose height exceeds this given threshold, the connected components of this map can be labelled and two routes can be considered equivalent when they stride through the same sequence of connected components in this thresholded map. Figure 5 depicts such connected components for two example maps.

One can clearly see the rooms (and larger open spaces for the map taken from Starcraft) as red areas connected by blue corridors. The red areas are—for a constant threshold—the same as the connected components of the respective layer in the pyramid.

When it comes to the calculation of alternative routes, three important approaches can be identified. The most basic version of alternative route calculation is given by finding the top k shortest paths in a graph. This is relatively easy by first building a shortest path tree from the beginning and end of the intended route and then patching together a shortest path from the beginning, an edge (or a short sequence) that is not part of either shortest path tree, and the shortest path to the goal. Unfortunately, this approach requires calculation of two sufficiently complete shortest path trees which renders most optimizations of shortest path calculation useless and generates a huge amount of route candidates. Additionally, the patching nature of these routes makes many of those look quite unnatural.

There are two widely-used algorithms to reduce the amount of computation as well as the amount of candidates a bit. These two widely used algorithms can be enhanced by the extracted topology. One simple way to calculate alternative routes is given by the penalty method (Chen et al. 2007). This method is based on repeatedly calculating the shortest path between two points and increasing the weights of all edges on this shortest path by a certain amount. In this way, the shortest path gets

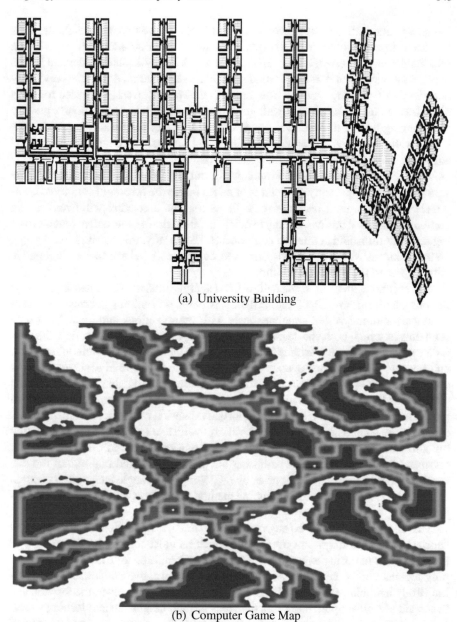

(a) University Building

(b) Computer Game Map

Fig. 5 Threshold connected components

longer and alternative routes become shortest routes in the updated weighting. This creates a large candidate set of alternative routes of increasing length from which a reasonable set of alternatives has to be selected. The sets of routes generated in this way are by an order of magnitude smaller than the sets generated with the previously described sidestepping method. The reason is that partial overlap between routes is disfavored as all edges along a path get a penalty. Thereby, the candidates quickly start ranging forth and back over space. One problem of the penalty method is the fact that the candidates keep quite similar for a number of iterations and "good" alternatives are first found after many iterations.

To this end, one can use the topology to not only increase the weights of the shortest paths, but possibly all weights of edges inside the threshold connected components. In this way, it becomes more likely that the algorithm will avoid such a component in few iterations leading to quick identification of alternative routes crossing different threshold connected components. Essentially, this behaves like a graph compression in which threshold-connected components behave like a single edge with respect to the penalty algorithm.

The second widely used algorithm for detecting alternative routes is given by the plateau algorithm (CVIT Ltd. 2016). It is based on building shortest path trees from start and end vertex simultaneously with optimizations and identifying path segments in which both shortest path trees overlap. For each such overlap, the candidate alternative route is extracted by routing from the start to the overlapping region and towards the end. Once overlap is detected, this is possible in constant time. As compared to the penalty algorithm, the plateau algorithm is able to quickly detect alternatives from a larger spatial area due to the uniform growing of shortest path trees. However, this is also the most important downside of this approach: the shortest path trees should be optimized to avoid explosion and quickly generate a solution but still have to cover a decent amount of space in order to find these overlapping regions. Usually, both shortest path trees are pruned at a fixed multiple of the distance between the start and end vertex by exploiting the triangle inequality. Note that this method is similar to the sidestepping method with a heuristic on which sidesteps to use first.

This algorithm can also be augmented by extracted topological information: the shortest path trees might not overlap too much as most search algorithms tend to keep left or right during expansion in free space. Therefore, quite unnatural overlapping regions will be generated and candidates for alternative paths will be counterintuitively irregular. By using a height-scaled weightmap, however, the forward and backward search are pulled towards the areas of high height vertices leading to collisions of both search trees and reasonable overlapping regions at high heights and high quality candidates. Again, this can be seen as a graph compression in which the connectivity of the graph is summarized by the high height vertices that are locally maximal. That is, as shortest paths in forward and backward search are pulled towards the same local maxima of height, they will collide there and each such local maximum (or plateau of such locally maximal vertices) represents the "kernel" of a threshold-connected component.

4.3 Selection of Alternative Routes

As already mentioned in the previous section, the most puzzling question for selecting alternative routes is a measure of alternativity of candidates.

It is quite easy to calculate very large sets of different routes between two points. As an easy approach, one calculates two shortest path trees, one from the start and one from the end vertex in the reverse graph. Then, each edge not in the shortest path trees creates an "alternative" by first going a shortset path from the start to this edge, then along this edge, and finally to the goal on a shortest path. However, these large sets of routes are not very useful for applications.

With the extracted topology, one can create a family of "alternative" routes, possibly even generated by mobile devices. Then, one can use the set of height components, these routes traverse, to select one and only one, e.g., the shortest, for each of these sets. Figure 6 depicts a result of selecting only one alternative route from a search based on the penalty algorithm in which two "alternatives" are considered equivalent, if they cross the same set of labels using a fixed height threshold. One clearly sees that only a limited set of routes is generated without much overlap. Hence, a "good" set of alternative routes: it is small subset of alternative route candidates, but covers many sensible examples. It is much more restrictive to use maximal connected components. Figure 7 depicts the filtering result in which maximal components have been used to define equivalence. This generates another, less complete, but smaller set of truly alternative routes between the two points.

It depends on the application, how much filtering of candidates is reasonable. Note that it is even possible to not only use set equality in filtering (reject candidates that are equal) but also to use set similarity provided for example by the Jaccard index for a fine-grained adjustment of the result set size and variety.

4.4 Indoor Spatial Statistics

When collecting large amounts of spatial data, several analytic approaches have been very successful. One elementary analysis for spatial and spatiotemporal datasets is given by the Getis-Ord G_i^* statistics. Basically, this statistic is based on comparing feature values of spatial cells (e.g., the number of events in a specific region) with those feature values of their neighbors identifying hotspots, i.e., locations, where this feature value is significantly larger than expected. Therefore, this statistics compares sums of features of regions and neighbors with the globally expected value for these sums.

The regions of Getis-Ord are often calculated by first aggregating feature values on a grid and then using the neighborhood relation on grid cells in order to find hotspots. However, when thinking about meaningful hotspots in buildings, this will be misleading.

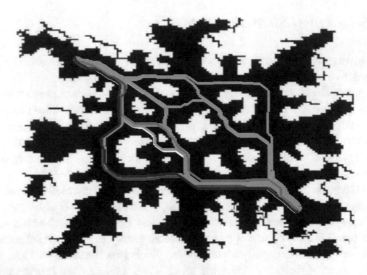

Fig. 6 Alternative routes selected using height components

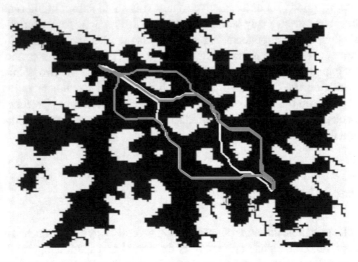

Fig. 7 Alternative routes selected using maximal height components

Assume, we are able to count the number of people in every room of the building with some sensors. A hotspot in this context should be a room in which many more people reside than one should expect from the average number of persons in a room. The grid-based approach, however, would detect people density as hotspots, that is rooms in which many people are near each other—a completely different question.

In this context, I propose to use the topological subdivision as provided by complete or threshold-connected components based on height in order to analyze hotspots in a topologically correct way. Consider Fig. 5a for an example of the spatial splitting using complete height components. Again, the most useful fact is here that the topology does not need to be generated by hand, but is extracted automatically from connectivity and does not depend on choosing a suitable threshold as in related work.

This approach can easily be extended to other spatial statistics methods in which spatial divisions have to be chosen including Moran's I and Geary's C for spatial autocorrelation.

4.5 Turn-by-Turn Guidance in Buildings

One of the hardest unsolved problems in indoor pervasive computing might be the computational generation of descriptions of movements and paths, that is a suitable analog of turn-by-turn guidance (Chewar and McCrickard 2002; Raubal and Winter 2002). In this area, two aspects interfere making a solution to the problem extremely hard:

The first aspect is about the availability of suitable map information needed to describe ways. This includes information about landmarks as well as topological information similar to what is extracted in this paper.

The second aspect is the unavailability of continuous user interfaces as well as the inability of humans to measure or estimate distances and to remember large sets of instructions. It has been discussed that the optimal number of instructions for describing a route through an airport would be roughly five (Ruppel et al. 2009). If users are given more than that, they tend to forget or—even worse—to confuse instructions.

In this context, the challenge is to generate descriptions with few instructions, the number of instructions independent from the length and complexity of the route, and still comprehensible for humans.

With respect to this problem, we envision a system that calculates and readily explains shortest paths in buildings in order to make an audio guide through buildings feasible and increase memorability of navigation instructions.

The hierarchical structure of the topology represented by the pyramid can be used to split any shortest path into flexible numbers of subpaths by increasing the height in the pyramid. While the path crosses one component on the base layer, this component splits several times and we expect these splittings to induce useful information for textual instruction generation. This can be seen as a Morse theory perspective on the navigation space: by increasing height, one can increase the number of components step by step and select a height in which the number of components is suitable for generating few instructions.

Additionally, as depicted in Fig. 4a, the height along the shortest path is full of information about the direct surroundings such as when the path is crossing small space such as doors or large spaces such as halls.

Furthermore, there is evidence that self-localization in large environments is better performed on schematic maps, while detailed information is better to be extracted from detailed floorplans (Meilinger et al. 2006). Such schematic maps can be constructed from our framework, for example, using the complete height components and highlighting all components crossed by the shortest path to be visually described allowing the mobile user to ignore irrelevant parts of the floorplan.

By fusing this information with information from other sources, I conclude that a lot of information for the optimization of textual instruction generation and shortest path visualization is made available. However, a deep investigation of this approach including a usability analysis is beyond the scope of this paper.

5 Conclusion

This paper has shown how to use an occupancy grid map in order to understand how the building topology is split into smaller pieces such as rooms and hallways. Additionally, it has shown the impact of observing the height in the contraction pyramid as a simple and powerful topological feature in order to annotate ways, calculate alternatives, and visualize shortest paths in buildings.

Additionally, a splitting of the topology into neighboring objects with sensible spatial extent allows for the application of spatial statistics such as the Getis-Ord hotspot statistics and similar spatio-temporal tools in indoor situations. This has not been the case without manually creating sensible neighborhoods which is a tedious, time-consuming and error-prone task.

Furthermore, we motivate another direction of applications in the area of compressed descriptions of complex paths through buildings. This is an area of ongoing research and we plan to explore this direction in future work.

References

Bader R, Dees J, Geisberger R, Sanders P (2011) Alternative route graphs in road networks. In: Theory and practice of algorithms in (computer) systems. Springer, pp 21–32

Becker C, Dürr F (2005) On location models for ubiquitous computing. Pers Ubiquitous Comput 9(1):20–31

Chen Y, Bell MG, Bogenberger K (2007) Reliable pretrip multipath planning and dynamic adaptation for a centralized road navigation system. IEEE Trans Intell Transp Syst 8(1):14–20

Chewar C, McCrickard DS (2002) Dynamic route descriptions: tradeoffs by usage goals and user characteristics. In: Proceedings of the 2nd international symposium on Smart graphics. ACM, pp 71–78

CVIT Ltd. (2016) Choice routing. http://www.camvit.com/camvit-technical-english/Camvit-Choice-Routing-Explanation-english.pdf

Dees J, Geisberger R, Sanders P, Bader R (2010) Defining and computing alternative routes in road networks. arXiv:10024330

Fabrizi E, Saffiotti A (2000) Extracting topology-based maps from gridmaps. In: IEEE international conference on robotics and automation, 2000. Proceedings. ICRA'00, vol 3. IEEE, pp 2972–2978

Hu H, Lee DL (2004) Semantic location modeling for location navigation in mobile environment. In: 2004 IEEE international conference on mobile data management, 2004. Proceedings. IEEE, pp 52–61

Meilinger T, Hölscher C, Büchner SJ, Brösamle M (2006) How much information do you need? schematic maps in wayfinding and self localisation. Spatial cognition V reasoning, action, interaction. Springer, pp 381–400

NavVis corporation (2016). http://www.navvis.com/

Openstreetmap wiki—indoor mapping (2016). http://wiki.openstreetmap.org/wiki/Indoor_Mapping#Tagging_proposals

Raubal M, Winter S (2002) Enriching wayfinding instructions with local landmarks. Springer

Ruppel P, Gschwandtner F, Schindhelm CK, Linnhoff-Popien C (2009) Indoor navigation on distributed stationary display systems. In: 33rd annual IEEE international computer software and applications conference, 2009. COMPSAC'09, vol 1. IEEE, pp 37–44

Werner M (2014) Indoor location-based services: prerequisites and foundations. Springer

Werner M, Feld S (2014) Homotopy and alternative routes in indoor navigation scenarios. In: Proceedings of the 5th international conference on indoor positioning and indoor navigation (IPIN 2014)

Werner M, Kessel M (2010) Organisation of Indoor Navigation data from a data query perspective. Ubiquitous positioning indoor navigation and location based service (UPINLBS), pp 1–6

Werner M, Kessel M (2012) A bitmap-centric environmental model for mobile navigation inside buildings. Int J Adv Netw Serv 5(1 and 2):91–101

Discovering and Learning Recurring Structures in Building Floor Plans

Andreas Sedlmeier and Sebastian Feld

Abstract Autonomous mobile robots show promising opportunities as concrete use cases of location-based services. Such robots are able to perform various tasks in buildings using a wide array of sensors to perceive their surroundings. A connected area of research which forms the basis for a deeper understanding of these perceptions is the numerical representation of visual perception of space. Different structures in buildings like rooms, hallways and doorways form different, corresponding patterns in these representations. Thanks to recent advances in the field of deep learning with neural networks, it now seems possible to explore the idea of automatically learning these recurring structures using machine learning techniques. Combining these topics will enable the creation of new and better location-based services which have a deep awareness of their surroundings. This paper presents a framework to create a data set containing 2D isovist measures calculated along geospatial trajectories that traverse a 3D simulation environment. Furthermore, we show that these isovist measures do reflect the recurring structures found in buildings and the recurring patterns are encoded in a way that unsupervised machine learning is able to identify meaningful structures like rooms, hallways and doorways. These labeled data sets can further be used for neural network based supervised learning. The models generated this way do generalize and are able to identify structures in different environments.

1 Introduction

Location-based services form a very interdisciplinary field of research ranging from Electrical Engineering over Computer Science to Social Science. Technological progress, especially the increase in computational power and the miniaturization of electronic devices and sensors, enabled the ideas of *Ubiquitous Computing*

A. Sedlmeier · S. Feld (✉)
Mobile and Distributed Systems Group, LMU Munich, Munich, Germany
e-mail: sebastian.feld@ifi.lmu.de

A. Sedlmeier
e-mail: andreas.sedlmeier@ifi.lmu.de

© Springer International Publishing AG 2018
P. Kiefer et al. (eds.), *Progress in Location Based Services 2018*, Lecture Notes in Geoinformation and Cartography, https://doi.org/10.1007/978-3-319-71470-7_8

(Weiser et al. 1991) and *Context Awareness* (Dey and Abowd 1999; Chen and Kotz 2000), both of which lead to the integration of location-based services into the daily life of many people. Built on this, mobile devices like smartphones or wearables contain several sensors for measuring movement (accelerometer), brightness (camera), volume (microphone), air pressure (barometer), position (GPS), and others. Thus, location-based services are basically context-aware services that incorporate spatial information (Küpper 2005).

Mobile robots can also be regarded to represent location-based services. Equipped with sensors like laser scanners, optical cameras, or tactile sensors they perceive and process their environment, resulting in the execution of simple tasks like the transportation of packets in storehouses. Further examples of research are mobile robots that lead tourist groups through an airport (Triebel et al. 2016) or serving as an assistance in housekeeping and everyday tasks (Rashidi and Mihailidis 2013). Due to recent advances in the fields of big data and machine learning, mobile robots get increasingly autonomous. Recent research focuses on cooperation, competition, and communication in order to solve more complicated tasks (Lowe et al. 2017; Mordatch and Abbeel 2017).

A related field of study deals with the visual perception of space. Since the end of the 1960s there are numerous empirical and experimental studies on the perception of architectural space. An early example is the work of Hayward and Franklin, who analyzed the influence of bordering elements like walls or trees on the perception of openness (Hayward and Franklin 1974). Today, there are different theories and tools to analyze spatial arrangements (Smith et al. 2007). The most basic term in this context is *Space Syntax*, summarizing mostly the acquisition of topological structures of an environment without geometric measurements (Hillier and Hanson 1984). A further concept in this area called *Isovist* has been introduced in (Tandy 1967) and describes the set of points in space that are visible from a specific vantage point. Based on this idea there have been presented a formal definition of isovists together with some analytical measures enabling a quantitative description of a spatial environment (Benedikt 1979).

Even if every building is different, one can still observe structures that recur constantly. Examples for such structural recurrences together with some semantics are rooms (small enclosed areas, often rectangular), corridors (long areas connecting rooms), or doorways (gaps in walls connecting rooms and corridors). Further exemplary structures are halls, staircases, or patios. The interesting part of such structures is that every room, corridor, and the like looks different, but they contain similarities that enable a (not necessarily distinct) recognition. Interestingly, this is a problem area in which huge progress was made in the last few years thanks to the advances in the field of deep learning with neural networks. Deep neural networks excel at the recognition of recurring structures in large data sets and the inference of underlying functions generating these structures.

The main idea of this paper is to investigate, whether the recurring structures inside buildings also have recurring isovist measures and whether such numerical features can be used to learn a model of such structures. Specifically this means that we incorporate unsupervised machine learning techniques of visual perception features to label a dataset consisting of geospatial trajectories through floor plans. This labeled dataset is utilized by supervised machine learning techniques to predict labels, thus structures, in unknown environments. The general use case of this idea is to create advanced spatial context for location-based services. A more concrete use case would be the problem of *Simultaneous Localization and Mapping* (SLAM), where a mobile robot has to build a map of its environment and estimate its pose simultaneously (Leonard and Durrant-Whyte 1991). Using the idea presented in this paper, a mobile robot would be able to independently learn a model of recurring structures inside buildings like, for example, rooms, hallways, and doorways. This model can then be reused in unknown buildings to recognize learned structures straight away. Alternative use cases are the off-line analysis or annotation of floor plans or the incorporation in computer games, such that non-player characters gain an additional understanding of altering surroundings.

The contributions of this paper are twofold. First, we present a framework that is able to automatically generate input data for learning a model of recurring structures inside buildings based on floor plans. The framework builds on the game engine *Unity* developed by Unity Technologies (Unity 2017) and uses the included navigation and route finding procedures to create a set of geospatial trajectories. Furthermore, our framework contains a custom isovist implementation in C# that is able to calculate isovist measurements for each time step of the trajectories of the data set. Second, we present a framework of machine learning techniques that can be used to train a model of recurring structures inside floor plans and to recognize such structures in unknown floor plans. We built upon existing scientific computing libraries written in the Python programming language (Pedregosa et al. 2011; Jones et al. 2001) as well as the open source neural network library Keras (Chollet 2015), which in turn uses TensorFlow (Abadi et al. 2015), a low level machine learning library developed by Google.

The remainder of this article is structured as follows: Sect. 2 describes the technical background for the further understanding of this paper together with related work. Section 3 incorporates the methodology for generating isovist measures along geospatial trajectories as well as the discovery, learning, and prediction of recurring structures in floor plans. In Sect. 4 we present our experimental results and present a detailed discussion. Section 5 concludes this paper.

2   Background and Related Work

This section contains the technical background for the further understanding of this paper together with related work. First, techniques for the analysis of visual perception as well as machine learning techniques are described. These are the main ingredients for the automatic generation of a model of recurring structures inside buildings. Furthermore, related work with respect to semantic annotation of floor plans are illustrated.

2.1 Analysis of Visual Perception

As already mentioned in Sect. 1, we utilize techniques that analyze the visual perception of space. There are numerous studies in the sector of cognitive psychology that address the behavior of people in typical buildings like hospitals (Haq and Zimring 2003), malls (Dogu and Erkip 2000), or airports (Raubal 2002).

Isovist Analysis is a concrete technique of *Space Syntax* that is used in many cases. As originally introduced in Tandy (1967) and more formally defined in Benedikt (1979), an isovist is the set of points in space that is visible from a specific vantage point.

The six isovist measures as defined in Benedikt (1979) are as follows:

1. A_x: the **area** describes the surface area of the isovist. The higher the value, the more space is visible from the vantage point. At the same time, this means that the vantage point can be observed from a large space.
2. P_x: the **real-surface perimeter** describes the length of the isovist's circumference that lies on visible obstacle surfaces, for example walls.
3. Q_x: the **occlusivity** describes the length of the isovist's circumference that lies in free space. With other words, these are the concealed radial borderlines that can be imagined as rays passing an obstacle and traversing through free space.
4. $M_{2,x}$: since the set of points in space that is visible from a specific vantage point can be calculated using rays sent out radially from the vantage point (Benedikt 1979), the **variance** is the second central moment of the rays' length.
5. $M_{3,x}$: the **skewness** is the third central moment of the rays' length.
6. N_x: the **circularity** is an isoperimetric quotient and evaluates the area against the perimeter. Basically, this is a numerical value that describes how similar a figure is in comparison to a circle. Circularity is calculated using $N_x = |\partial V_x|^2/4\pi A_x$, with $|\partial V_x|$ indicating the isovist's perimeter.

Isovist fields are likewise described in Benedikt (1979) as the set of isovists along a trajectory, or more complete, the set of isovists at all places of an investigated environment. Since a human is moving through an environment in a continuous manner, the isovist measures are also changing continuously. Thus, one can observe gradual changes in the isovist measures. This is the underlying idea of our approach: we calculate geospatial trajectories through floor plans, calculate isovist measures at every time step and analyze both, the absolute as well as the delta values to previous steps.

Although the framework proposed in this paper uses a 3D environment, the agent navigating through the building only walks on a 2D plane and thus creates 2D trajectories and corresponding 2D isovist measures. Nevertheless, there is literature that analyzes isovists in 3D space (Emo 2015).

The calculation of an isovist or rather the calculation of the isovist measures, as described above, is constrained by the environment's geometry and can potentially get complicated, since all corner points of visible walls and objects have to be determined and connected. Feld et al. (2016) showed that at least *area*, *variance*, *skewness*, and *circularity* can easily be approximated using a simple ray-scan algorithm.

The authors' motivation was to receive a preferably simple equivalent of isovist measures that can be applied on floor plans represented as occupancy grids via bitmaps. White pixels stood for walkable free space, black pixels represented obstacles like walls or other objects. Their experiments showed that there is a systematic error regarding the approximated and exact isovist measures, however, they show a strong correlation.

The ideas and solutions presented in this paper are using a similar ray-based approach.

2.2 Machine Learning

Machine learning can basically be regarded as a generic term for the generation of knowledge from experience. During a training phase a systems learns from examples and is able to make generalizations afterwards. Exactly this behavior will be utilized by the approach presented in this paper: we want to learn recurring structures inside floor plans of buildings that are as generic as possible in order to reuse the generated model on new and unknown floor plans. As our approach uses isovist measures for training, a necessary precondition is the assumption that recurring structures in buildings also have recurring structures in their isovist measures.

Generally, machine learning can be divided into several categories. **Unsupervised learning** methods use a set of unlabeled input data in order to infer a function that describes the data's inherent structure. In our case the input data consists of a large set of isovist measures forming time series that have been calculated along geospatial trajectories. As the input data is unlabeled, the algorithm has no explicit target values to learn and instead tries to determine a function that reflects patterns in the data.

A popular example of unsupervised learning is clustering, that is the automatic segmentation of data into groups of "similar" observations. Partitioning clustering techniques subdivide data into a predetermined number of k clusters. The assignment of observations to clusters will be modified until a certain error function is minimized. *k-means* is a widely used partitioning clustering technique (Lloyd 1982). Density-based clustering techniques arrange objects into groups which are close to each other, separated by areas with lower density. An example for such a technique is *DBSCAN* (Ester et al. 1996). The algorithm has got two parameters: ϵ representing the distance up to which two observations are reachable and *minPts* representing the minimal number of reachable observations that make an observation a cluster point.

A **supervised learning** algorithm, by contrast, tries to infer a function based on given pairs of input and corresponding known output labels. The idea is to train the system in order to create associations. In our case the input data again consists of a large set of time series of isovist measures calculated along geospatial trajectories. However, for each data point, there is a corresponding known ground truth, for example: "at this point the agent resides inside a room" or "at this point the agent traverses a doorway". A popular use case of supervised learning, where a lot of progress has been made in the last years, is the automatic classification of images using deep learning techniques with neural networks (Deng and yu et al. 2014). Given enough input data and the right structure, neural networks are able to learn arbitrary functions from labeled data sets. Using images of known classes, a model is trained that infers a function determining class boundaries. Afterwards, one can use this model to predict the classes of unknown images, or in other words: Observations that have not been used during the training phase. In our case a model is trained on floor plans where rooms, hallways, and doorways are known. Afterwards, this model can be used on unlabeled floor plans where no such information is available.

2.3 Semantic Annotation of Floor Plans

Map representations of spatial environments are an essential foundation for most location-based services. Even if the positioning of an object works without a map representation, further benefit can only be created using a map. Examples are road maps, touristic maps or floor plans of buildings.

Such map representations can include logical subdivisions. Road maps involve country roads, highways, crossroads, turns and more. Buildings, for example, can be subdivided into rooms, zones, units, and levels (Weber et al. 2010). Besides that, there are semantic subdivisions like rooms, hallways, and doors. This is the focus of the paper at hand.

There is extensive related work regarding semantic annotation of architectural floor plans. Samet and Soffer (1994) perform automatic interpretation of floor plans using statistical pattern recognition. Their work is distinct from ours as we do not detect concrete objects like tables or bathtubs explicitly marked in architectural plans. Ah-Soon and Tombre (1997) analyze architectural drawings using geometric analysis, symbol recognition, and spatial analysis. Again, our approach is not geometrical, but instead uses the numerical representation of visual perception. Dosch et al. (2000) aim to reconstruct the building in 3D based on architectural drawings. Using graphic recognition for image processing and feature extraction, the authors are able to recognize graphic layers, text layers, thick and thin lines as well as marked doorways, stair cases and more. Summarized, they try to identify marked semantics and transform this into 3D. In contrast, we try to identify semantics that are not explicitly marked. Lu et al. (2007) is a further work that tries to recognize typical structural objects and architectural symbols. Our approach works on floor plans that can be used by robots and not on architectural drawings. Weber et al. (2010) presents a system where a user can draw schematic abstractions of floor plans. Afterwards, the system searches for plans that are structurally similar. This is quite related to our approach, since they also seek for semantic relations. However, our focus is not on searching in databases, but on learning a model.

Further related work originates in the research field of mobile robots. What this work has in common, is that the ideas can be used for the problem of *Simultaneous Localization and Mapping* (SLAM) (Leonard and Durrant-Whyte 1991). This means, an autonomous robot has to examine an unknown area and tries to create a corresponding floor plan. Concurrently, the robot has to position itself. Thus, it makes sense to enrich the map just created with semantic information. The basic assumption is that the robot's perception, in most cases laser range scans, contains enough information about the environment. Basically, we indirectly follow this approach as well, since we use isovist measures based on rays. Buschka and Saffiotti (2002) describe a virtual sensor that can be used to detect rooms and to recognize already visited rooms in order to create a topological map of the environment. Our focus is wider than just detecting rooms, although we do not address topology. Anguelov et al. (2004) present a probabilistic framework for detecting and modeling doors. They use 2D laser range finders, but also panoramic cameras. Mozos and Burgard (2006) and Mozos (2010) extract the topology of buildings from geometric maps created by mobile robots using range data. The authors use supervised learning techniques in order to subdivide all points of the map into semantic classes. For this, they use the labels *room, corridor*, and *hallway* as the ground truth. This approach is very similar to the one presented in this paper, but the authors work only with supervised learning techniques and with different yet similar features. Goerke and Braun (2009) is also a similar related work that semantically annotates maps using laser range measurements of mobile robots. The authors follow two basic approaches. First, they use supervised learning techniques with the labels *doorway, corridor, freespace, room*, and *unknown*. Second, they use unsupervised learning techniques, but state that this approach did not produce satisfying results. Furthermore, the authors only work on a single floor plan, while our paper in particular addresses the aspect of generalization, which is why multiple maps are used. Chen et al. (2014) use deep learning techniques to identify doors, so that autonomous mobile robots are able to approach targets more accurately. Their focus in only on detecting doors visually, using cameras.

There is further related work on analyzing architectural space using isovist analysis. Bhatia et al. (2012) use 3D isovists in order to estimate salient regions in architectural and urban environments. Thus, the authors are able to detect regions that posses strong visual characteristics. Our approach focuses on recurring and not on salient structures. Feld et al. (2016) approximate four out of six isovist measures using a simple ray-casting approach while showing that the resulting error is systematically yet small, and the exact and approximated values show a strong correlation. Furthermore, they show with a few examples on a single map that trajectories of isovist measures potentially provide clues to identifying doors. The paper at hand goes much further and creates a model to recognize such structures. Feld et al. (2017) calculate isovist measures on 2D floor plans, cluster the values using archetypal analysis and interpret the results afterwards. They show that the identified clusters correspond to regions like streets, rooms, hallways, and the like. However, their approach is unsupervised learning with interpretation of relations, thus, they do not learn a specific model using which predictions can be made.

3   Methodology

This section is split in two parts: (1) It describes our framework for generating isovist measures along geospatial trajectories in a map-based simulation environment.
These measures provide the input for the following step, (2) the discovery, learning,
and prediction of recurring structures in floor plans. Unsupervised learning techniques are employed in the discovery phase, while supervised machine learning is
performed for the modeling and prediction tasks. Details regarding the exact implementation of these aspects can be found in Sect. 4.

3.1   Input Generation

Basic input for the framework is supplied as bitmap files representing building floor
plans. Walkable space is represented as white pixels, while black pixels depict obstacles like walls or furniture. Note that doors are excluded. In a first step, these bitmaps
are vectorized using a common vector graphics editor. The vector files are then
imported into Blender (blender.org 2017), an open-source 3D computer graphics
software, where a 3D-Extrusion is performed in order to generate a 3D map of the
building. These 3D maps serve as the basic asset for Unity (2017), a 3D game engine
and development environment. For each map, a navigation mesh (Snook 2000) is
generated in Unity to enable automatic navigation and pathfinding. Custom built
C# scripts then enable a player object (non-player character, NPC) to automatically
select a random point on the navigable area inside the map and move towards it
using Unity's built-in navigation algorithm. For each step of the NPC, another custom C# script was developed, which performs isovist measure calculations and logs
the results to disk. In order to generate the isovist, a configurable amount of rays
are cast, originating from the current position of the NPC, as can be seen in Fig. 1.

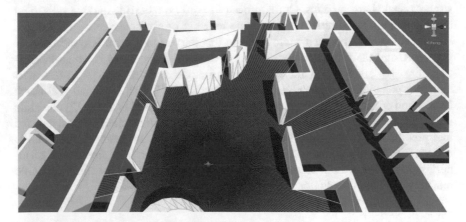

Fig. 1   3D view of a utilized floor plan, showing the non-player character casting 360 rays (red
lines) from it's current vantage point

Points in space, where the rays intersect with the map's mesh colliders (hitpoints) are detected and used to calculate the different isovist measures.

The isovist measures calculated are based on Benedikt (1979), as previously described in Sect. 2.1. As a discrete, ray-based isovist calculation is used, the calculated measures are only an approximation of the true isovist measures. The accuracy of the calculation can be adjusted, as the amount of rays cast is configurable.

One of the more challenging aspects to calculate is the differentiation between *real-surface* and *occlusivity* of the isovist. Benedikt states in Benedikt (1979) that the occlusivity of an isovist "measures the length of the occluding radial boundary R_x of the isovist V_x and indicates [...] the depth to which environmental surfaces are partially covering each other as seen from the vantage point".

In order to be able to differentiate occlusion from real-surface in our simulation's engine, we developed an algorithm which performs calculations based on the triangles that form the mesh of the environment. For every ray cast in a clockwise manner, a comparison with the previous ray's hitpoint on the environment's surface is performed. If the previous ray hit a triangle which shares none of it's edge coordinates with the currently hit triangle, we define the current ray to have hit an *un-connected triangle* (in respect to the previous triangle). The length of the line connecting the previous and current ray hitpoint in space is then counted towards the occlusion value of the isovist. If a *connected triangle* was hit, the length of the connecting line is counted towards the real-surface perimeter of the isovist. Figure 2 shows the

Fig. 2 In-engine view of the custom built algorithm's results for real-surface and occlusion isovist measure calculation. Red lines are the rays cast from the current vantage point, green lines visualize the meshes' triangles hit by the rays, blue lines denote real-surface while yellow lines denote occlusion

resulting lines calculated by our algorithm inside the Unity engine. Red lines are the rays cast from the current vantage point, green lines visualize the hit triangles of the meshes, blue lines denote real-surface, while yellow lines denote occlusion.

3.2 Unsupervised Learning of Unknown Floor Plan Structures

The first part of the learning framework is responsible for discovering hidden structures contained in the isovist measures. Goal of this step is to group input data into meaningful clusters, each representing a human-relatable concept, for example "isovists recorded in rooms" versus "isovists recorded in corridors". This can be achieved using unsupervised machine learning techniques.

In a first step of preprocessing, the logged isovist measures calculated during simulation time are vectorized in order to retrieve a suitable data set for unsupervised learning. This means that X and Y coordinates are removed so that a 6-dimensional vector remains, whereas each dimension represents one of the measured isovist features. We employ k-means (Lloyd 1982), a centroid-based clustering algorithm, as well as DBSCAN (Ester et al. 1996), a density-based clustering algorithm, both of which are implemented in the scikit learn python library (Pedregosa et al. 2011). The algorithm determines a configurable amount of cluster centers and assigns the data points to the nearest cluster center, by minimizing the squared distances from the clusters.

It is important to keep in mind that the input data given to the clustering algorithm as described above is static in nature. That is to say, each data point contains only the isovist measures of a single position along the trajectory trough the map. As there is no temporal component involved, the concept of movement and the dynamic change of isovists while moving along a path was not reflected in the analysis. The overarching idea of the next step, the inclusion of time, is to not only reflect the perception of "space" but the "changing of space perception" as caused by movement.

In order to tackle this idea, an additional data processing step was developed which reflects the temporal dimension of the data. For every feature of each data point, the delta of the current data point's feature x_c and the simple moving average (SMA) (Balsamo et al. 2013)—a method commonly employed in the statistical analysis of time series—of n previous data points' features, is calculated:

$$x_c - \frac{1}{n} \sum_{i=1}^{n} x_{c-i}$$

This way, the amount of features available to the machine learning algorithms is doubled from 6 to 12.

3.3 Supervised Learning of Known Floor Plan Structures

Using the method described in the previous section, labeled input data can be generated, which forms the basis for a following supervised machine learning step. The goal of this step is to learn a model representing the structures discovered in the data, by inferring a function which maps new unlabeled input data points to the respective cluster categories. This model can then be used, for example, in robots as a lightweight component enabling the robot to deduce the type of room it currently resides in, or whether it has just passed a doorway, by feeding it's current and previous isovist measures into the model. For our framework, we chose to implement a multi-layer feedforward neural network using the open source neural network library Keras (Chollet 2015), which in turn uses Tensorflow (Abadi et al. 2015), a low level machine learning library developed by Google, to execute it's calculations. Training data is provided by the labeled input data, as output from the unsupervised learning step. In order to verify the validity of the model generated using supervised learning methods, it is important to separate validation from training data. For this, we split the data into a left half and a right half, based on the data points' coordinates. Training was performed on the right half of the data, while validation was performed on the left half. As part of our evaluation of different neural network architectures, we found a rather small network of 5 fully-connected layers to be sufficient for our purposes. The input layer contains 12 neurons (one for each feature), connected to 3 hidden layers, each containing 64 neurons, followed by an output layer containing 4 neurons. Softmax activation is used on the 4-neuron output layer in order to build a classifier representing the 4 cluster labels, while rectified linear unit (ReLU) activation (LeCun et al. 2015) is used on all other layers. Categorical cross entropy is employed as the loss function while Adam (Kingma and Ba 2014) is used as the stochastic gradient descent algorithm. All in all, the network contains 9,412 trainable parameters. After the training step, the best model is selected based on the model accuracy score. In order to test the generalization capacity of the trained model even further, the model is then used to predict values from data captured on a different floor plan. The question to be answered by this is whether the model learned general abstractions (e.g. a concept of "doors") that capture underlying basic principles of the data which are independent of the specific floor plan layout.

4   Results and Discussion

For the evaluation, two distinct floor plans were chosen. The first one is a section of a university building of the Ludwig-Maximilians-Universität München (LMU). It features repeating structures of corridors and similar rooms. The second floor plan features the main hall and connected rooms of the Technische Universität München (TUM). By comparison, it contains a more irregular structure formed by large lecture halls and connecting hallways. Because of it's distinct and repeating structures, the

Fig. 3 Results of a k-means based clustering with $k = 3$ of static isovist measures on the LMU floor plan

LMU floor plan was chosen for training. We recorded more than 370,000 isovists along random trajectories on the LMU map and more than 220,000 isovists on the TUM map.

We compared two different unsupervised learning methods: k-means, a centroid-based clustering algorithm, and DBSCAN, a density-based clustering algorithm. As it is possible to define the amount of clusters to be found when using k-means, the results of using different values were compared. From all the values tested, we found a value of $k = 3$ to produce the most meaningful results on the LMU floor plan. As can be seen in Fig. 3, three different structures of the floor plan are separated into different clusters. A clear separation between the large horizontal corridor, the smaller vertical corridors, and rooms became apparent. It is important to keep in mind, that human concepts are not necessarily reflected by the clusters, which is why the meaning of a cluster is always subject to interpretation.

Besides the centroid-based clustering algorithm k-means, we also evaluated the density-based clustering algorithm DBSCAN. As the amount of clusters to be found is not to be specified in DBSCAN and can only be indirectly influenced by configuring two density parameters ϵ and *minPts*, it is a lot harder to produce a sensible amount of human interpretable clusters. For our data set, an ϵ value of 3 and *minPts* values between 1000 and 2500 produced meaningful results.

Compared to the clusters produced by the k-means algorithm, the resulting structures were less interpretable. This is why we decided to continue our analysis using the k-means based clustering.

After clustering the static data features and finding clusters that could be interpreted as rooms and floors, the delta of the current data point and the SMA of the isovist measures was calculated, in order to capture the temporal dimension of the data.

Fig. 4   Results of k-means based clustering of dynamic, SMA based isovist measures ($k = 3, c = 5$)

Using these "delta-features" as input to the k-means clustering, a completely different picture became visible: As can be seen in Fig. 4, a cluster now formed around passage ways, especially doorways.

This intuitively makes sense, as doorways are components in a building, often connecting structures of different shape, which is why movement through them leads to changes in the perception of space, in turn reflected in high changes of isovist measures.

By combining these static-data and dynamic-data clusters, we generated a merged set of data-labels containing four different clusters as shown in Fig. 5.

We interpret the clusters as follows:

Cluster-0   (blue): rooms
Cluster-1   (green): horizontal corridors
Cluster-2   (red): small vertical corridors and large rooms
Cluster-3   (purple): passage ways (e.g. doors)

Figure 5 also shows the training/validation split that was performed on the data. Training was performed on the right half of the data, while validation was performed on the left half. Good results could be achieved when training a 5-layer fully connected feedforward neural network using the 12-dimensional feature vector (static and SMA deltas) as input and the 4 cluster labels described above as targets. The best model showed an accuracy value of 0.9856 and validation accuracy value of 0.9188.

As can be seen in Fig. 6, the predictions produced by this model on the validation data match our interpretations of the clusters in the training data. This means, that the model was able to learn a function representing the structures bundled in the

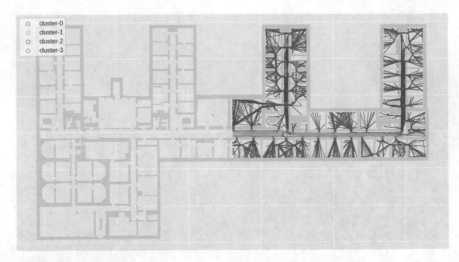

Fig. 5 Results of training/validation data split after merging static and dynamic cluster features. Only training data is visualized in this figure

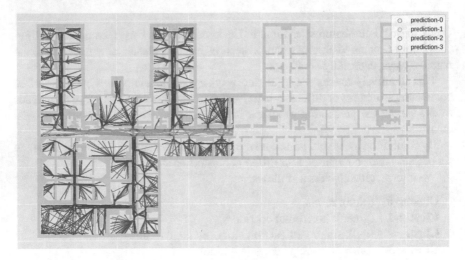

Fig. 6 Predictions of the 5-layer feedforward neural network on the validation data set

respective clusters. As it was able to predict meaningful results on previously unseen parts of the floor plan, it became obvious that the model generalizes to new data.

Figure 6 also shows that our previous interpretation of cluster-2 to be mainly comprised of small vertical corridors no longer holds. The small red colored corridor formed by the doors of interconnected rooms in the lower right part of the floor plan is clearly horizontal in nature. Apart from that, our previous interpretations of the clusters still hold.

Fig. 7 Predicted cluster memberships of all data points along random trajectories trough the TUM floor plan using the model trained on the LMU floor plan cluster membership dataset with 12 features (static and dynamic)

In order to test the generalization performance of our model even further, we used it to predict the cluster memberships on a completely different floor plan. As the room and corridor structures in the TUM plan are completely different from the LMU plan, on which the model was trained, it is a much more difficult task for the model to perform. Figure 7 shows the predicted cluster-memberships of all recorded points along the random trajectories.

Even though the maps have completely different layouts, a visual comparison with our previous cluster definitions provided a good match. Points on the trajectories in smaller rooms are almost completely predicted to belong to cluster-0 (blue), as they did on the LMU floor plan. Large rooms and corridors predicted to be cluster-2 (red) also match our expectations. As there is no directly corresponding structure to the

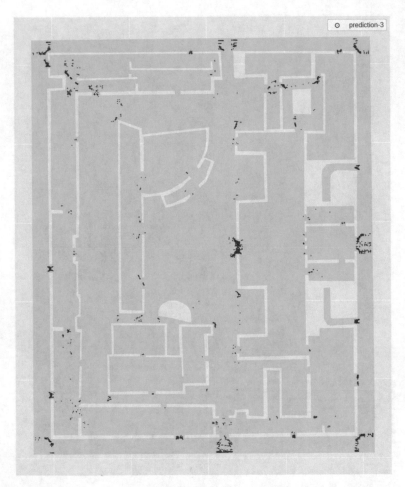

Fig. 8 Predicted membership of cluster-3 (passage ways and doors) using a confidence threshold of 99.99% on the TUM floor plan using the model trained on the LMU floor plan cluster membership dataset with 12 features (static and dynamic)

single large corridor (cluster-1: LMU) on the TUM floor plan, it is no surprise that labelings of cluster-1 (green) do not lend themselves to intuitive interpretations.

Most interestingly, structures labeled as cluster-3 (purple) match our definition of passage ways and doors almost perfectly. This becomes even more apparent when combined with a prediction confidence threshold. This is possible, because the output layer of the neural network does not produce binary label decisions but instead numeric values denoting the confidence that the current data point belongs to the respective label.

After increasing the threshold to 99.99%, doors and passage ways are marked largely correct, as can be seen in Fig. 8. What becomes apparent is that our definition of cluster-3 to contain only passage ways and doors might need to be expanded when

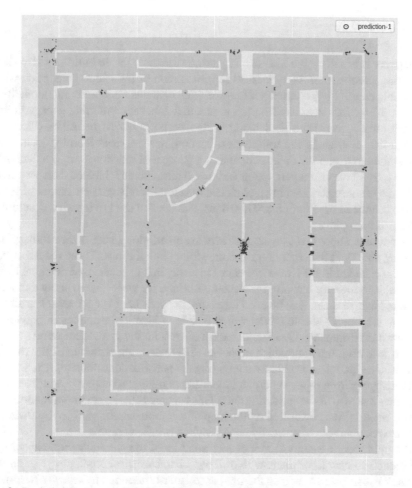

Fig. 9 Predicted cluster membership with confidence threshold of 99.95% on the TUM floor plan by a model trained using only the 6 dynamic SMA features on the LMU floor plan

applied to this floor plan. A more accurate description could be: Points where a different structural part of the building is entered. This structural part can, but need not, be explicitly separated by a door.

As a last step, we evaluated whether the model performance for this specific prediction of passage ways and doors could be further improved by excluding possibly irrelevant features for this task from the training data. A separate model having a smaller, 6-neuron input layer to take in only the 6 dynamic SMA features was trained on the LMU floor plan. The resulting predictions of now only two clusters are shown in Fig. 9. We think that these predictions fit our expectations even better, as the placement of labels is now more accurately inside the doors.

5 Conclusion

This paper is based on the idea that recurring structures inside buildings also show recurring structures in the numerical representation of the visual perception when traversing them. We presented a framework that contains three main functionalities. First, a 3D environment can be used to create a data set containing geospatial trajectories that traverse the floor plan together with 2D isovist measures calculated at each time step along the trajectories. Second, unsupervised learning techniques can be used to group the data set containing geospatial trajectories into meaningful clusters, based on visual perception features. Third, the now labeled data set can be utilized by supervised learning techniques to automatically create a model of recurring structures in the floor plan. This model can then be used to identify structure in unlabeled floor plans.

Our results show that isovist measures recorded along trajectories through the building do reflect the recurring structures found in buildings. These recurring patterns are encoded in the isovist measures in a way that unsupervised machine learning is able to identify meaningful clusters. Further, we were able to show that these clustered data sets can also be used for neural network based supervised learning in order to create a re-usable model which is able to identify structures in previously unknown environments. Good model accuracy results show, that the neural network is able to learn a function which represents the underlying structure of the training data. The validation score in turn shows that the network does not simply remember a 1:1 mapping from input to output, but abstracts general structures from the isovist measures that also fit the validation data. This becomes obvious in the validation step, where labeling was performed on the map of a completed different environment, as the network was able to correctly label previously unseen inputs.

As future work we envision a deeper analysis of the generalization capacity of the models to new floor plans with different characteristics, containing e.g. curved walls. We also plan to replace the ray-based isovist measures by exact isovist measures following the definition given in Benedikt (1979) in order to increase our data accuracy. Furthermore, we would like to analyze 3D floor plans of buildings using 3D isovist measures. Finally, extensive feature engineering can be conducted and different neural network architectures explored in order to improve model accuracy and generalization performance.

References

blender.org (2017) Home of the blender project—free and open 3d creation software. https://www.blender.org/. Accessed 23 July 2017

Unity (2017) Game engine. http://unity3d.com. Accessed 22 July 2017

Abadi M, Agarwal A, Barham P, Brevdo E, Chen Z, Citro C, Corrado GS, Davis A, Dean J, Devin M, Ghemawat S, Goodfellow I, Harp A, Irving G, Isard M, Jia Y, Jozefowicz R, Kaiser L, Kudlur M, Levenberg J, Mané D, Monga R, Moore S, MurrayD, Olah C, Schuster M, Shlens J, Steiner

B, Sutskever I, Talwar K, Tucker P, Vanhoucke V, Vasudevan V, Viégas F, Vinyals O, Warden P, Wattenberg M, Wicke M, Yu Y, Zheng X (2015) TensorFlow: large-scale machine learning on heterogeneous systems. http://tensorflow.org/

Ah-Soon C, Tombre K (1997) Variations on the analysis of architectural drawings. In: Proceedings of the fourth international conference on document analysis and recognition, 1997, vol 1. IEEE, pp 347–351

Anguelov D, Koller D, Parker E, Thrun S (2004) Detecting and modeling doors with mobile robots. In: ICRA'04. 2004 IEEE international conference on robotics and automation, 2004. Proceedings, vol 4. IEEE, pp 3777–3784

Balsamo M, Knottenbelt W, Marin A (2013) Computer performance engineering: 10th European workshop, EPEW 2013, Venice, Italy, September 16–17, 2013, Proceedings. Lecture notes in computer science. Springer, Berlin

Benedikt ML (1979) To take hold of space: isovists and isovist fields. Environ Plan B: Plan Des 6(1):47–65

Bhatia S, Chalup SK, Ostwald MJ et al (2012) Analyzing architectural space: identifying salient regions by computing 3d isovists. In: Conference proceedings. 46th annual conference of the architectural science association (AN-ZAScA), Gold Coast, QLD

Buschka P, Saffiotti A (2002) A virtual sensor for room detection. In: IEEE/RSJ international conference on intelligent robots and systems, 2002, vol 1. IEEE, pp 637–642

Chen G, Kotz D (2000) A survey of context-aware mobile computing research. Technical report TR2000-381, Dept of Computer Science, Dartmouth College

Chen W, Qu T, Zhou Y, Weng K, Wang G, Fu G (2014) Door recognition and deep learning algorithm for visual based robot navigation. In: 2014 IEEE international conference on robotics and biomimetics (ROBIO). IEEE, pp 1793–1798

Chollet F et al (2015) Keras. https://github.com/fchollet/keras

De Smith MJ, Goodchild MF, Longley P (2007) Geospatial analysis: a comprehensive guide to principles, techniques and software tools. Troubador Publishing Ltd

Deng L, Yu D et al (2014) Deep learning: methods and applications. Foundations and trends®. Signal Proces 7(3–4):197–387

Dey AK, Abowd GD (1999) Towards a better understanding of context and context-awareness. In: International symposium on handheld and ubiquitous computing. Springer, pp 304–307

Dogu U, Erkip F (2000) Spatial factors affecting wayfinding and orientation: a case study in a shopping mall. Environ Behav 32(6):731–755

Dosch P, Tombre K, Ah-Soon C, Masini G (2000) A complete system for the analysis of architectural drawings. Int J Doc Anal Recogn 3(2):102–116

Emo B (2015) Exploring isovists: the egocentric perspective. In: International space syntax symposium, pp 1–8

Ester M, Kriegel HP, Sander J, Xu X et al (1996) A density-based algorithm for discovering clusters in large spatial databases with noise. Kdd 96:226–231

Feld S, Lyu H, Keler A (2017) Identifying divergent building structures using fuzzy clustering of isovist features. In: Progress in location-based services. Springer, pp 151–172

Feld S, Werner M, Linnhoff-Popien C (2016) Approximated environment features with application to trajectory annotation. In: 6th IEEE symposium series on computational intelligence (IEEE SSCI 2016)

Goerke N, Braun S (2009) Building semantic annotated maps by mobile robots. In: Proceedings of the conference towards autonomous robotic systems, pp 149–156

Haq S, Zimring C (2003) Just down the road a piece: the development of topological knowledge of building layouts. Environ Behav 35(1):132–160

Hayward SC, Franklin SS (1974) Perceived openness-enclosure of architectural space. Environ Behav 6(1):37–52

Hillier B, Hanson J (1984) The social logic of space. Cambridge University Press

Jones E, Oliphant T, Peterson P et al (2001) SciPy: open source scientific tools for python. http://www.scipy.org/. Accessed 23 July 2017

Kingma D, Ba J (2014) Adam: a method for stochastic optimization. arXiv:1412.6980

Küpper A (2005) Location-based services. Fundamental and operation. Willey,

LeCun Y, Bengio Y, Hinton G (2015) Deep learning. Nature 521(7553):436–444

Leonard JJ, Durrant-Whyte HF (1991) Simultaneous map building and localization for an autonomous mobile robot. In: IEEE/RSJ international workshop on intelligent robots and systems' 91. Intelligence for mechanical systems, Proceedings IROS'91. IEEE, pp 1442–1447

Lloyd S (1982) Least squares quantization in pcm. IEEE Trans Inf Theory 28(2):129–137

Lowe R, Wu Y, Tamar A, Harb J, Abbeel P, Mordatch I (2017) Multi-agent actor-critic for mixed cooperative-competitive environments. arXiv:1706.02275

Lu T, Yang H, Yang R, Cai S (2007) Automatic analysis and integration of architectural drawings. Int J Doc Anal Recogn 9(1):31–47

Mordatch I, Abbeel P (2017) Emergence of grounded compositional language in multi-agent populations. arXiv:1703.04908

Mozos ÓM (2010) Semantic labeling of places with mobile robots, vol 61. Springer

Mozos OM, Burgard W (2006) Supervised learning of topological maps using semantic information extracted from range data. In: 2006 IEEE/RSJ international conference on intelligent robots and systems. IEEE, pp 2772–2777

Pedregosa F, Varoquaux G, Gramfort A, Michel V, Thirion B, Grisel O, Blondel M, Prettenhofer P, Weiss R, Dubourg V, Vanderplas J, Passos A, Cournapeau D, Brucher M, Perrot M, Duchesnay E (2011) Scikit-learn: machine learning in python. J Mach Learn Res 12:2825–2830

Rashidi P, Mihailidis A (2013) A survey on ambient-assisted living tools for older adults. IEEE J Biomed Health Inform 17(3):579–590

Raubal M (2002) Wayfinding in built environments: the case of airports. IfGIprints 14

Samet H, Soffer A (1994) Automatic interpretation of floor plans using spatial indexing. Prog Image Anal Process 3:233

Snook G (2000) Simplified 3d movement and pathfinding using navigation meshes. In: DeLoura M (ed) Game programming gems. Charles River Media, pp 288–304

Tandy C (1967) The isovist method of landscape survey. Methods of landscape analysis, pp 9–10

Triebel R, Arras K, Alami R, Beyer L, Breuers S, Chatila R, Chetouani M, Cremers D, Evers V, Fiore M et al (2016) Spencer: a socially aware service robot for passenger guidance and help in busy airports. In: Field and service robotics. Springer, pp 607–622

Weber M, Langenhan C, Roth-Berghofer T, Liwicki M, Dengel A, Petzold F (2010) a. SCatch: semantic structure for architectural floor plan retrieval. In: International conference on case-based reasoning. Springer, pp 510–524

Weiser M (1991) The computer for the 21st century. Sci Am 265(3):94–104

Part III
Landmarks and Mobility

Extracting Rankings for Spatial Keyword Queries from GPS Data

Ilkcan Keles, Christian S. Jensen and Simonas Saltenis

Abstract Studies suggest that many search engine queries have local intent. We consider the evaluation of ranking functions important for such queries. The key challenge is to be able to determine the "best" ranking for a query, as this enables evaluation of the results of ranking functions. We propose a model that synthesizes a ranking of points of interest (PoI) for a given query using historical trips extracted from GPS data. To extract trips, we propose a novel PoI assignment method that makes use of distances and temporal information. We also propose a PageRank-based smoothing method to be able to answer queries for regions that are not covered well by trips. We report experimental results on a large GPS dataset that show that the proposed model is capable of capturing the visits of users to PoIs and of synthesizing rankings.

1 Introduction

A very large number of searches are performed by search engines like Google or Bing each day. One source (Google 2016) reports that Google processes more than 7 billion queries per day. A recent study (Google 2014) of users' local search behavior indicates that 4 in 5 users aim to find geographically related information. It also shows that 50% of the users who conducted mobile search and 34% of the users who used a computer or tablet visit a point of interest (PoI) on the same day. These statistics indicate the importance of location-based web querying.

To support queries with local intent, the research community has proposed many different spatial keyword functionalities to find relevant nearby PoIs (Cao et al. 2012). A prototypical spatial keyword query takes a set of keywords and a location

I. Keles (✉) · C. S. Jensen · S. Saltenis
Department of Computer Science, Aalborg University, Aalborg, Denmark
e-mail: ilkcan@cs.aau.dk

C. S. Jensen
e-mail: csj@cs.aau.dk

S. Saltenis
e-mail: simas@cs.aau.dk

© Springer International Publishing AG 2018
P. Kiefer et al. (eds.), *Progress in Location Based Services 2018*, Lecture Notes
in Geoinformation and Cartography, https://doi.org/10.1007/978-3-319-71470-7_9

as arguments and returns a list of PoIs ranked with respect to a range of signals. Example signals include PoI ratings, properties of the neighborhoods of the PoIs, the distances of the PoIs to the query location, the textual relevances of the PoIs to the query keywords, and the relative expensiveness of the PoIs. These signals can be combined in multiple ways to obtain a ranking function. Most studies focus on indexing and efficient retrieval and thus evaluate the computational efficiency of proposed techniques. In contrast, the evaluation of the quality of the ranking functions is not covered well. We think that evaluation of the ranking functions is crucial since it is an important step towards increasing user satisfaction with location-based services; however, it is difficult to assess the quality of a ranking function when there is no yardstick ranking to compare against. The goal of this study is to propose a framework for constructing such baseline rankings that reflect the preferences of the users. Future studies will then be able to use the constructed rankings to evaluate the quality of different ranking functions.

A few studies (Yi et al. 2013; Chen et al. 2013; Stoyanovich et al. 2015; Keles et al. 2015) consider the use of crowdsourcing to synthesize rankings for objects and they can be used for spatial keyword queries. However, crowdsourcing-based approaches are expensive since workers need to be paid for each crowdsourcing task. They are also time consuming since there is a need to wait for the workers to complete the tasks. Further, it may be difficult to recruit workers who know about the spatial region of the query. Therefore, as a supplement to crowdsourcing, we focus on the use of GPS data to synthesize rankings.

We propose a method to build rankings for spatial keyword queries based on historical trips extracted from GPS data. We define a trip as a pair of consecutive stops extracted from a GPS trajectory. The stops represent the source and the destination of a trip, and we are interested in trips where the destination is a PoI. While the GPS data does not include spatial keyword queries, we can reasonably assume that a recorded trip to a PoI corresponds to issuing a spatial keyword query at the starting location of the trip with a keyword that is part of the textual description associated with the PoI. For instance, if a user visited a restaurant r starting from a location l, we assume that the user issued a spatial keyword query at l with the keyword "restaurant" and that r is the preferred restaurant. Further, a PoI is considered to be relevant to the users in a region if many trips starting in that region visit the PoI. To the best of our knowledge, this is the first study of using GPS data to synthesize rankings for spatial keyword queries.

To synthesize rankings, we first extract stops of users from available GPS data. Then, we assign the stops to the PoIs that were visited. Furletti et al. (2013) propose a PoI assignment method based on the distance between a stop and a PoI. We extend their method by taking into account temporal patterns of the users' visits to PoIs. Next, we extract all trips to PoIs.

Using the trips, we build a grid structure, where each cell records two values for each PoI, namely the number of trips from the cell to the PoI and the number of distinct users involved. To address the issue that some cells may have few or no trips, we adopt a personalized PageRank (Page et al. 1999) based algorithm to smooth the values. The intuition behind using PageRank is that nearby grid cells should have similar values just like web pages linking to each other should have similar values. The resulting grid structure is used to form a ranking for a given spatial keyword query. First, the grid cell that contains the query location is found. Then the PoIs are filtered with respect to the query keywords. Finally, the PoIs are ranked according to the number of trips and the number of distinct users. The resulting ranking reflects the preferences of the users for PoIs, and a ranking function that produces a ranking more similar to the synthesized ranking is more preferable. Although a given collection of GPS data is limited in its geographical coverage and its coverage of users, we are still able to produce rankings in the particular settings where the GPS data offers good coverage.

To summarize, the main contributions are: (i) A method for synthesizing rankings of PoIs from GPS data that is able to produce results for regions without GPS data and that employs the number of trips and distinct users to rank PoIs, (ii) A stop assignment algorithm that employs users' temporal patterns when assigning stops to PoIs, (iii) PageRank-based algorithm to smooth the values for grid cells, (iv) An evaluation using a dataset of some 0.4 billion GPS records obtained from 354 users over a period of 9 months.

The remainder of the paper is organized as follows. Section 2 covers preliminaries, related work, and the problem definition. The proposed model is covered in Sect. 3. Section 4 covers the evaluation, and Sect. 5 concludes and offers research directions.

2 Preliminaries

2.1 Data Model

The proposed method uses GPS records collected at one hertz from GPS devices installed in vehicles.

A GPS record G is a four-tuple $\langle u, t, loc, im \rangle$, where u is the ID of a user, t is a timestamp, loc is a pair of Euclidean coordinates representing the location, and im is the vehicle ignition mode. Even though im is not part of a GPS measurement, it is included in our dataset as a useful automotive censor measurement. An example GPS record is $\langle 5, 2014\text{-}03\text{-}01\ 13\text{:}44\text{:}54, (554025, 6324317), \text{OFF} \rangle$, where the coordinates of the location are given in the UTM coordinate system. Next, a trajectory TR of a user is the sequence of GPS records from this user ordered by timestamp t, $TR = G_1 \rightarrow \cdots \rightarrow G_i \rightarrow \cdots \rightarrow G_n$. We denote the set of all trajectories by S_{TR}.

We are interested in the locations where a user stopped for a longer time than a predefined threshold. We extract all such stops from S_{TR}. Specifically, a stop S is a three-tuple $\langle G, a_t, d_t \rangle$, where G is a GPS record, a_t is the arrival time at $G.loc$, and d_t is the departure time from $G.loc$. When we say the location of a stop, we refer to the location of G. Next, a point of interest (PoI) P is a three-tuple $\langle id, loc, d \rangle$, where *id* is an identifier, *loc* is a location, and *d* is a document that contains the textual description of the PoI.

We assume that a significant portion of users' stops are visits to PoIs, so when a user makes a stop, it is probable that the user did so to visit a PoI. We define an assignment A as a pair $\langle S, P \rangle$ of a stop S and a PoI P, indicating that a user stopped at the location of S to visit P. We are unable to assign all stops to PoIs, so only some stops have a corresponding PoI.

Having extracted all the stops of a user, we obtain the user's location history. In particular, the location history H of a user is defined as the sequence $H = S_1 \rightarrow \cdots \rightarrow S_i \rightarrow \cdots \rightarrow S_m$ of the user's stops ordered by a_t. A user's location history captures the user's trips. Specifically, a trip T is a pair $\langle S_i, S_j \rangle$ of a source and a destination stop. Given a trip $T = \langle S_i, S_j \rangle$ and an assignment $A = \langle S_j, P \rangle$, we say that T is a trip to PoI P.

Our goal is to use the trips extracted from GPS records to synthesize ranking of PoIs for spatial keyword queries.

Definition 1 (*Top-k Spatial Keyword Query*) Let S_P be a set of PoIs. A top-k spatial keyword query $q = \langle l, \phi, k \rangle$ on S_P is a three-tuple, where l is a query location, ϕ is a set of query keywords, and k indicates the number of results. The query q returns k PoIs from S_P that rank the highest according to a given ranking function. A frequently used ranking function is a weighted combination of the proximity of the PoI location to $q.l$ and the textual relevance of the PoI to $q.\phi$ (Cao et al. 2012).

Problem Statement. We assume a set S_G of GPS records obtained from vehicles and a set S_P of PoIs. Given a top-k spatial keyword query, we solve the problem of constructing a top-k ranking of PoIs included in S_P using S_G.

2.2 Related Work

Some studies propose crowdsourcing to obtain rankings of items (Yi et al. 2013; Chen et al. 2013; Stoyanovich et al. 2015; Keles et al. 2015). Yi et al. (2013) propose a method based on pairwise comparisons and matrix completion. Chen et al. (2013) also use pairwise comparisons and propose an active learning method that takes worker reliability into account to synthesize rankings. Stoyanovich et al. (2015) use list-wise comparisons and build preference graphs for workers and combine these to obtain a global ranking. Keles et al. (2015) propose a method based on pairwise comparisons in order to rank PoIs for a given query without assuming a total ranking on the PoIs. Crowdsourcing-based approaches are hard to apply in large-scale evaluations since they are expensive and time-consuming. In the context of spatial keyword queries, it is a challenge to recruit workers familiar with the relevant region and PoIs.

Some studies use GPS data to identify stops, visited PoIs, and interesting places (Alvares et al. 2007; Palma et al. 2008; Ashbrook and Starner 2003; Kang et al. 2004;

Zhou et al. 2004; Zheng et al. 2009; Cao et al. 2010; Montoliu et al. 2013; Furletti et al. 2013; Spinsanti et al. 2010; Bhattacharya et al. 2012, 2015). An important place is one where users stop for a while. In these studies, a stop is generally defined either as a single GPS record corresponding to the loss of satellite signal when a user enters a building or a set of GPS records where a user remains in a small geographical region for a time period.

Alvares et al. (2007) enrich GPS trajectories with moves and stops. They require a predefined set of possible stop places that is then used for annotating trajectories. Palma et al. (2008) enable the detection of stops when no candidate stops are available. They use a variation of DBSCAN (Ester et al. 1996) that considers trajectories and speed information. The main idea is that if the speed at a place is lower than the usual speed, the place is important. We use GPS data collected from vehicles, and we have a specific signal telling whether the engine is on or off. This simplifies the detection of stops.

Many clustering-based methods have been proposed to identify significant locations from GPS data. Ashbrook and Starner (2003) use a variation of k-means clustering to identify locations. Kang et al. (2004) propose a time-clustering method. Zhou et al. (2004) propose a density based clustering algorithm to discover personally meaningful locations. Zheng et al. (2009) propose a hierarchical clustering method to mine interesting locations. They employ a HITS (Kleinberg 1999) based inference model on top of the location histories of the users to define the interestingness of the locations by considering users as hubs and locations as authorities. Cao et al. (2010) employ a clustering method to identify semantic locations. They enhance the clustering using semantic information provided by yellow pages. They propose a ranking model that utilizes both location-location relations and user-location relations as found in trajectories. They also consider the stay durations and the distances traveled. Montoliu et al. (2013) propose time-based and grid-based clustering to obtain places of interest. We are not using clustering because we are not interested in regions; instead, we want to identify the specific PoIs that are visited by users.

Some studies use different strategies to extract significant places from GPS data. Bhattacharya et al. (2012) propose a method based on bearing change, speed, and acceleration for walking GPS data. In a recent study (Bhattacharya et al. 2015), they make use of density estimation and line intersection methods to extract places. Their work requires walking GPS data and polygon information for each PoI. Their method is not applicable in our setting.

Finally, methods have been proposed that identify visits to PoIs from GPS data. Given a stop, the goal is to identify a PoI. Spinsanti et al. (2010) annotate stops with a list of PoIs based on the distance between the stop and the PoIs and the average durations people spend at the PoIs. Their method requires average stay durations for each PoI, which are provided by experts. This information is not available in our setting. Furletti et al. (2013) propose a method that also forms a set of possibly visited PoIs by taking walking distance and opening hours into account. We extend their stop assignment strategy. Shaw et al. (2013) consider the use of learning-to-rank methods to provide a list of possible PoIs when a user checks in. They use historical check-ins to form a spatial model of the PoIs. They also make use of PoI popularity information and user-specific information like a user's check-in history and information about a user's friends that have already checked in at the PoI. Kumar et al. (2015) and Gu et al. (2017) model the geographic choice of a user taking into account the distance between the stop and a PoI as well as the number of possible PoIs and their popularity when multiple PoIs are possible. They use labeled data (direction queries and check-ins) to train their model. Since we have no personal information or check-in data, we are unable to use their method when assigning stops to PoIs.

3 Proposed Method

3.1 Overview

The method consists of two phases: model-building and ranking-building.

The model-building phase takes a set of GPS records and a set of PoIs as the input and outputs a regular grid that partitions the underlying geographical space. Each grid cell records two values for each PoI: the number of trips from the cell to the PoI and the number of distinct users making trips.

Using the GPS records, we first extract stops. Then we determine the home and work locations of the users and assign non-home/work stops to PoIs. In the next step, we extract the set of all trips to the PoIs and we compute the number of trips and distinct users for each cell and PoI. Finally, we smooth the values of the grid cells using an algorithm based on personalized PageRank (Page et al. 1999).

The ranking-building phase uses the grid structure constructed to synthesize rankings for top-k spatial keyword queries. Given a query, we first locate the cell that contains the query location. Then the PoIs that have values in this cell are filtered according to the query keywords. The remaining PoIs are sorted according to the scores produced by a scoring function that is a weighted combination of the number of trips and the number of distinct users of a PoI in the cell of a query. The first k PoIs constitute the output.

3.2 Stop Extraction

To extract stops, we use the ignition mode attribute. Similar to Cao et al. (2010), we employ a duration threshold parameter Δ_{th} to check whether an OFF record represents a stop. If the duration between consecutive OFF and ON records exceeds Δ_{th}, a stop is formed from the first GPS record. Arrival-time attribute a_t and departure-time attribute d_t of the stop correspond to the timestamp attributes of the OFF and ON records, respectively.

Since GPS readings might be inaccurate or missing, we augment the procedure with a distance threshold d_{th}. Only if the distance between the location of the ON record and the location of the OFF record is below d_{th} and the time difference exceeds Δ_{th}, the arrival record is classified as a stop.

To exclude stops at traffic lights but include short stops, e.g., to pick up kids at a kindergarten, Δ_{th} should be set to a value between 5 and 30 min. Parameter d_{th} can be set to a value in the range 100–500 m.

If the GPS dataset does not contain an ignition mode attribute, the stops can be extracted by the methods mentioned in Sect. 2.2. In other words, all the subsequent steps of the proposed method are applicable to GPS trajectories in general.

3.3 Determining Home/Work Locations

Home and work locations are not of interest to our study, so a first step is to eliminate stops that relate to such locations.

To determine home/work locations, we employ an algorithm based on DBSCAN (Ester et al. 1996), which is a density-based clustering algorithm with two parameters, *eps* and *minPts*. If a point p has more than *minPts* points in its *eps*-neighborhood, that is in the circular region centered at p with a radius of *eps*, it is a core point. The points in the *eps*-neighborhood of a core point p are said to be directly reachable from p. A point q is reachable from p if there is a sequence of points $\langle p_1, \ldots, p_n \rangle$ with $p_1 = p$ and $p_n = q$, where each p_{i+1} is directly reachable from p_i. The objects reachable from a core point forms a cluster.

Figure 1 shows an example cluster with *minPts* set to 4. Here, A is a core point since there are 5 points within its *eps*-neighborhood. Points B and C are directly reachable from A, and E is reachable from A since there is a sequence $\langle A, C, D, E \rangle$, where all of the preceding points of E are core points and each point in the sequence is directly reachable from the preceding point. All points reachable from A form the cluster.

The parameters are set to different values for each user since the total number of stops differs among users. For a given user, we set *minPts* to a value that is propor-

Fig. 1 An example
DBSCAN cluster

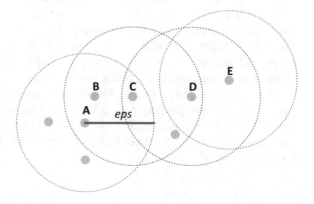

tional to the number of distinct days this user has stops, and we introduce a parameter p_{hw} as the constant of proportionality. For instance, a p_{hw} value of 4/7 means that the user should have at least four stops a week to consider clustering them into a home/work cluster.

To determine the *eps* parameter, we first compute the distances between the locations of each stop belonging to the user and its nth nearest neighbor stop with $n = minPts$. Then we sort these distance values and eliminate those that exceed the globally defined distance threshold parameter (dn_{th}). Finally, for each distance value v_i, we compute the percentage of increase of the next distance value v_{i+1}: $(v_{i+1} - v_i)/v_i$. The distance value with the maximum percentage of increase becomes the *eps* parameter.

Having found the DBSCAN parameters of a user, we cluster his/her stops with respect to the location and compute the average stay duration for each cluster. If the duration exceeds a threshold Δ_{hw}, we conclude that the cluster represents a home/work location, and we mark the stops in the cluster as home/work stops. The intuition behind using a duration threshold to determine home/work locations is that people typically spend a long duration at home and work.

3.4 Stop Assignment to PoIs

The next step is to assign the remaining stops to PoIs. The goal is to assign as many stops to PoIs as possible while being conservative, thus getting assignments that are true with high certainty. To achieve this, we propose two methods: *distance based assignment* and *temporal pattern enhanced assignment*.

Distance Based Assignment (DBA). Furletti et al. propose a stop annotation method (Furletti et al. 2013) that uses a maximum walking distance parameter and creates a list of PoIs that are within the maximum walking distance from the location of the stop and that have opening hours matching the time of the stop. Similarly, our DBA method searches for candidate PoIs in a circular region centered at the location of a stop with radius ad_{th}, a distance threshold that captures the maximum walking distance from the location of a stop to a PoI. In addition, the DBA method employs a parameter *lim* that sets an upper limit on the number of PoIs in the considered region. This avoids assigning a stop to a PoI when there are too many nearby PoIs. In such situations, it is not clear which nearby PoI was visited. Our goal is to make those assignments that we can make with relatively high certainty, so that the preferences of the users are captured while trying to avoid errors.

To assign a stop S to a PoI, we find the set of PoIs within the region defined by the location of S and parameter ad_{th}. Then we check whether the opening hours of the PoIs, if available, match with the arrival and departure time attributes of S. If the cardinality of the result set is below *lim*, we assign S to the closest PoI. Otherwise, we do not assign S to any PoI.

Temporal Pattern Enhanced Assignment (TPEA). The output of DBA might contain unassigned stops. These occur when there are either too many or no PoIs around the stops. We utilize temporal visit patterns to assign the unassigned stops.

For each user, we cluster non-home/work stops with respect to their locations using DBSCAN with $minPts$ equal to the lim and eps equal to ad_{th} from DBA. If a cluster contains stops that are assigned to PoIs, we construct a so-called *visit-pattern matrix* for the cluster.

In this 2D matrix, the first dimension represents different days, and the second represents different times of a day. The value in a cell is the number of PoIs that the user visited during the corresponding time period. We use three levels of groupings of weekdays: top, weekdays/weekends, day. At the top level, we do not use the day information and the matrix has only one row and groups PoI visits by periods of a day. At the weekdays/weekends level, the matrix contains one row for weekdays and one for weekends. At the day level, we build seven rows, one for each day of the week. An example matrix for the weekdays/weekends level is shown in Table 1.

Next, for each unassigned stop in the cluster, we check the number of PoIs for the corresponding cell starting from the top level of day grouping. If there is only one PoI, we assign the stop to this PoI. If there is more than one PoI, we proceed to the weekdays/weekends level and, finally, the day level. Otherwise, we conclude that we cannot assign the stop.

Example. Fig. 2a represents one of the clusters of a single user after the assignment with DBA. Yellow rectangles and red circles represent the assigned and unassigned stops, respectively. The corresponding PoIs are denoted by p_1, p_2, and p_3. We want to assign the stop in the ellipse, and we break a day into four 6-h periods.

We first check the matrix at the top level. We only have visits to PoIs in the time period between 12:00 and 18:00, and 3 distinct PoIs are visited. The top level matrix is shown in Fig. 2b. Since the relevant cell value (3) exceeds 1, we consider the week-

Table 1 Example visit-pattern matrix

	00:00 06:00	06:00 12:00	12:00 18:00	18:00 00:00
Weekdays	0	0	2	0
Weekends	0	0	1	0

(a)

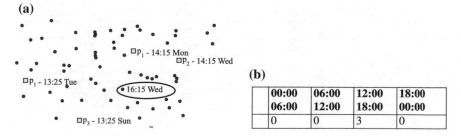

(b)

00:00 06:00	06:00 12:00	12:00 18:00	18:00 00:00
0	0	3	0

Fig. 2 Cluster of stops and visit-pattern matrix-top level

Table 2 Visit-pattern matrix-day level

	00:00 06:00	06:00 12:00	12:00 18:00	18:00 00:00
...				
Wednesday	0	0	1	0
...				

days/weekends level. We have 3 stops on weekdays with 2 different corresponding PoIs, and we have 1 stop on weekends. The matrix is shown in Table 1. The value of the relevant cell is 2, so we move to the day level, which is shown in Table 2.

Now the value of the relevant cell (Wednesday, 12:00–18:00) is 1, so the user visited only p_2 on a Wednesday during the time period containing 16:15. Therefore, we assign the stop to p_2.

3.5   Computing Values of Grid Cells

We use Danske Kvadratnet,[1] which is the official geographical grid of Denmark, as the underlying grid structure. The grid consists of equal-sized square cells of size 1 km^2 and it contains 111, 000 cells.

Initializing Values of Grid Cells. Next, for each PoI p_i, we form the set T_{p_i} of trips to p_i. Using these sets, we initialize a grid structure for each PoI. For each cell, the number of trips from the cell to the PoI and the number of distinct users making these trips are computed and recorded.

Smoothing the Values. After the initialization, many PoIs have sparse grids, where many cells have no trips to the PoIs. Table 3 shows the number of cells with non-zero values for top-5 PoIs. Only 3 PoIs have more than 100 cells with non-zero values after initialization. Sparse grids are a problem since this reduces the number of spatial keyword queries that we can construct rankings for. If neighboring cells of an empty cell have non-zero values, it is reasonable to assume that trips starting in the these cells are also relevant for the empty cell. So, the neighboring cell values

Table 3 Number of cells out of 111,000 cells with non-zero values for top-5 PoIs

PoI ID	Before smoothing	After smoothing
1	148	2,021
2	142	1,521
3	115	2,123
4	98	2,148
5	98	1,652

---

[1] http://www.dst.dk/da/TilSalg/produkter/geodata/kvadratnet.

can be used for smoothing to address the sparsity problem. The smoothing also helps reduce noise in cell values.

As the smoothed grids of multiple PoIs will be used in the ranking building phase, a smoothing method should have two properties. First, for a PoI, a smoothing algorithm should not change the sum of all the values in the grid for that PoI. Inflating or deflating the sum of values would unfairly promote or demote the PoI in relation to other PoIs in a constructed ranking. Second, the ordering of the values for all PoIs in a specific cell before and after smoothing should be similar in order to reduce distortion of the original spatial popularity data for the PoIs.

The literature contains some smoothing and interpolation methods for spatial grid based data. The inverse distance weighting (IDW) method (Shepard 1968) is proposed to interpolate missing values using distance-based weighting. This method builds on the intuition that the effect of a cell's value on the value of an originally empty cell should depend on how close the cell is to the empty cell. However, IDW does not contain a smoothing method, and it changes the sum of values in the grid. Therefore, we do not utilize IDW in our work. Kernel-based methods (Hastie et al. 2009) have also been proposed for smoothing. For these, it is possible to preserve the sum of the values since they produce a probability distribution as output. The sum of the values can then be distributed according to this distribution. However, they might introduce changes to the ordering of grid cells since kernel-based methods yield continuous functions that might not reflect the original properties of the data.

We use a smoothing algorithm based on personalized Pagerank (Brin and Page 1998) to interpolate values for cells with no trips. The PageRank algorithm was proposed for web graphs, where web pages are the vertices and hyperlinks are the edges. The algorithm assigns a page rank value to each website to indicate the relative importance of it within the set. A web page is considered important if other important web pages link to it. The algorithm can be described as a random walk over a directed graph $G = \langle V, E \rangle$. A random walker starts from a randomly chosen vertex. Then, with probability $1 - \alpha$, it follows an outgoing edge, and with probability α, it teleports to another randomly chosen vertex $y \in V$, where α has the same value for each web page and $0 < \alpha < 1$. The PageRank of a vertex is the probability that a random walker will end up at the vertex.

Personalized PageRank (Page et al. 1999) was proposed in order to incorporate personalized preferences. This is achieved by changing the uniform probability distribution of teleportation to a random web page to a personalization parameter that is basically a distribution based on user preferences. We use this parameter to utilize the initial values of the grid cells while smoothing the values.

The PageRank algorithm is a good candidate for smoothing, since, if a cell is close to another cell, they should have similar values just like the page rank values for the web pages linking to each other. The main idea is that if a PoI is of interest to drivers leaving from a cell, it might also be of interest to drivers leaving from nearby cells.

We first convert the underlying grid into a directed graph. For each cell, we introduce a vertex. Then, we add edges from a "cell" to the neighboring "cells" with weight $w = 1/d^2$, where d denotes the distance between the centers of the cells.

The edge weights define how the page rank value of a vertex should be distributed to the adjacent vertices. In the initial version of PageRank, it is equally distributed. In our case, we use weights based on distance to make sure that the page rank value is distributed inversely proportional with the distance between the grid cells corresponding to the vertices. Then we apply PageRank to this graph for both number of trips and number of users values. We use the initial cell values obtained after the initialization to determine the personalization parameters. The probability of teleportation to a vertex is set proportional to the actual value of the corresponding grid cell.

The procedure yields a probability distribution that indicates the likelihood that a random walker will end up at a particular vertex. We distribute the total number of trips and the total number of distinct users to the cells proportional to the output probability distribution. For instance, assume that we are smoothing the numbers of trips and that the total number of trips to the PoI is 100. A cell with probability 0.23 then gets the value 23. Note that this smoothing procedure is done for each PoI. Table 3 shows that smoothing provides a significant increase in the number of cells with non-zero values.

Example. Let G be a grid with cells $c_1, c_2, c_3, c_4, c_5, c_6$. The grid structure is shown in Fig. 3a. The first value represents the number of trips from each cell to a PoI before smoothing.

The graph representing the grid is given in Fig. 3b. Each vertex has an edge to each vertex that represents a neighboring cell. Each edge is assigned a weight as explained above. For instance, the distance between c_1 and c_2 is 1 unit, and the distance between c_1 and c_5 is $\sqrt{2}$ units, so the weight of edge (c_1, c_5) is $1/d^2 = 0.5$.

Then, we apply personalized PageRank using the initial values as the personalization input. The second value of each cell in Fig. 3a represents the resulting probability of the cell.

Finally, we distribute the sum of the values according to the pagerank values. The third value of each cell in Fig. 3a represents the smoothed value. Here, c_5 has the second largest value because it is closer to the cell with the largest value (c_2) than c_4 and c_6, and it has more edges than c_3 and c_1 since it is in the middle column. The

(a)

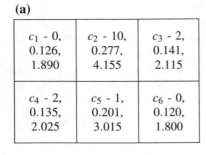

c_1 - 0, 0.126, 1.890	c_2 - 10, 0.277, 4.155	c_3 - 2, 0.141, 2.115
c_4 - 2, 0.135, 2.025	c_5 - 1, 0.201, 3.015	c_6 - 0, 0.120, 1.800

(b)

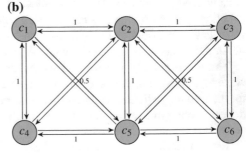

Fig. 3 Grid structure and corresponding graph

effect of the number of edges is not an issue when smoothing is applied on a large grid since the grid cells, except the cells on the boundary of the grid structure, have the same number of edges.

3.6 Extracting Rankings for Queries Using the Model

To form a ranking for a given top-k spatial keyword query, we use the grid model.

Algorithm 1 The Algorithm for Ranking-building Phase

Input: q—top-k spatial keyword query, *model*—the grid model
Output: r_k—a ranked list of PoIs
1: $c \leftarrow$ the corresponding grid cell for $q.l$ in *model*
2: $p \leftarrow$ the set of PoIs that have values in c
3: $p_f \leftarrow$ the PoIs in p which are filtered using the the query keywords $q.\phi$
4: Rank the PoIs in p_f with respect to their values in c and assign it to r_{all}
5: **if** r_{all} has more than k elements **then**
6:    $r_k \leftarrow$ the first $q.k$ elements of r_{all}
7:    **return** r_k
8: **else**
9:    **return** r_{all}
10: **end if**

The algorithm, given in Algorithm 1, first finds the cell that contains the query location l. Then it filters the PoIs with values in the cell with respect to the query keywords to exclude the PoIs that do not contain query keywords. Here, we assume that all PoIs with descriptions that do not contain any of the query keywords will not be among the popular ones for this query. The ranking is computed using the ranking function given in Eq. 1, where n denotes the number of trips, d denotes the number of distinct users, and β is the weighting parameter.

$$s = \beta \times n + (1 - \beta) \times d \text{ where } 0 \le \beta \le 1 \tag{1}$$

If there are more than k relevant elements in the ranking then the top k elements form the output of the algorithm, as shown in Lines 5–7. Otherwise, the ranking is the output, as shown in Line 9. It is important to note that the output ranking might contain fewer than k elements. Although the algorithm does not produce a complete ground-truth ranking, the partial ranking is still useful for evaluation purposes as it provides valuable information about the expected result.

4 Experimental Evaluation

In Sect. 4.1, we report on studies aimed at understanding effects of parameter settings. In Sect. 4.2, we compare the stop assignment methods with a baseline method. Finally, we study the effect of smoothing and weighting parameters on the output rankings in Sect. 4.3. We do not present a complexity analysis since we think that it is not a real concern due to the fact that model-building is performed only once.

In the experiments, we use 0.4 billion GPS records collected from 354 cars traveling in Nordjylland, Denmark during March–December 2014. The PoI dataset used in the experiments contains around 10,000 PoIs of 88 categories. All of the PoIs are located in or around Aalborg.

4.1 Exploring the Parameters

To explore the effects of changing the parameters on the outputs of the different steps, we vary one parameter at a time while fixing other parameters to their default values. The parameters are described in Table 4.

4.1.1 Stop Extraction

Here, we study the effect on stop extraction of varying Δ_{th} and d_{th}.

Table 4 Parameters

Notation	Step	Explanation	Default value
Δ_{th}	Stop extraction	Minimum duration between two GPS records to consider the first one as a stop	10 min
d_{th}		Maximum distance between two GPS records to consider that their locations match	250 m
Δ_{hw}	Determining home and work locations	Minimum average stay duration in home/work locations	240 min
p_{hw}		Minimum fraction of days in a week a person is expected to visit home/work	Three days a week (3/7)
ad_{th}	Stop assignment to PoIs	Maximum distance between a PoI and the location of a stop	100 m
lim		Maximum number of PoIs within the region bounded by the location of a stop and the ad_{th} parameter	5 PoIs

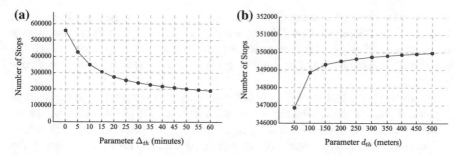

Fig. 4 Effects of parameters Δ_{th} and d_{th}

As shown in Fig. 4a, the number of stops decreases as Δ_{th} increases, as expected. The decrease is smooth. In order to capture meaningful stops from GPS data, parameter Δ_{th} should be set to a value in the range of 5–30 min since this setting would exclude quite short stops which might not be a visit to a PoI and as we see from the figure, there are a lot of quite short stops.

As can be seen in Fig. 4b, the number of stops increase when d_{th} increases, which is as expected. Although this parameter has some effect on the number of stops extracted, the increase in the number of stops is negligible. This parameter is introduced to eliminate inaccurate GPS readings, and the results suggest that there are only few inaccurate readings in our dataset.

4.1.2 Determining Home/Work Locations

Here, we analyze the effect of p_{hw} and Δ_{hw} on determining home/work locations.

As shown in Fig. 5a, both the number of home/work locations and the number of stops assigned to home/work locations decrease when p_{hw} increases, because of the fact that more weekly stops are required to form a cluster. The decrease in the former is sharper than the decrease in the latter which is a consequence of that parameter

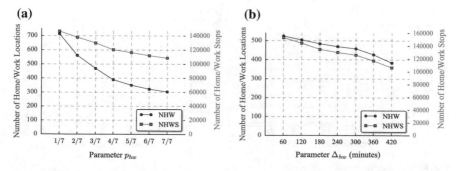

Fig. 5 Effect of parameters p_{hw} and Δ_{hw}

p_{hw} is used to limit the number of stops in a cluster to be considered as a home/work location. The clusters that are left out when p_{hw} increases are the clusters with a small number of stops. This is why the decrease in the number of stops assigned to home/work locations are not as sharp as the decrease in the number of home/work locations.

Figure 5b shows that the numbers of home/work locations and stops decrease as Δ_{hw} increases, as expected. Unlike when increasing p_{hw}, the patterns of decrease are quite similar for both when increasing Δ_{hw}. This suggests that the average durations that users spend inside clusters are not correlated with the number of stops in the clusters.

4.1.3 Stop Assignment to PoIs

Here, we analyze the effect of parameters ad_{th} and lim on the assignment of stops to PoIs. In addition, we assess the effect of ad_{th} on the distance between stop location and the assigned PoI for TPEA. Figure 6 shows the effect of varying ad_{th} on the number of stops that can be assigned to PoIs and the number of PoIs that receive assignments of stops when lim is set to 5 or 10.

Figure 6a shows that for both lim values, the number of assigned stops increases up to a point and then decreases when ad_{th} increases. When ad_{th} is small, it is impossible to assign some stops since there are no PoIs in the region defined by the location of the stop and the parameter. However, when ad_{th} increases, the number of PoIs within the bounded region increases as well. At some point, the number of PoIs starts to exceed the value of lim, and we are unable to assign the stop to a PoI. When this occurs, the number of stops starts to decrease.

Figure 6b shows that the number of PoIs with stops assigned to them follows a very similar pattern. The decrease after an increase is also the result of having too many PoIs ($>lim$) inside the region bounded by parameter ad_{th}.

Figure 7 shows the effect of varying ad_{th} on the distribution of the distance between the stop location and the assigned PoI for TPEA when lim is set to 5. The

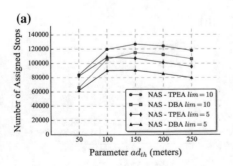
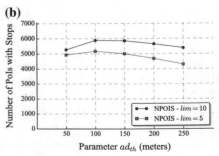

Fig. 6 Effect of parameter ad_{th}

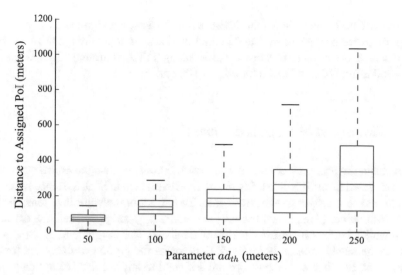

Fig. 7 Effect of parameter ad_{th} on distance distribution for TPEA

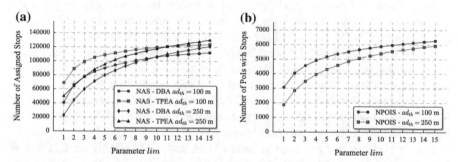

Fig. 8 Effect of parameter lim

red lines show the medians, and the green lines show the means of the distance values. For this experiment, we only consider the stop locations assigned by TPEA. The figure shows that the distance between the stop and the assigned PoI increases when ad_{th} increases, as expected since parameter eps is set to ad_{th} parameter in the density based clustering. This figure also shows that although no specific distance threshold parameter is employed by TPEA, the distance between the stop and the assigned PoI does not exceed a few multiples of ad_{th} since it employs density-based clustering to form the visit pattern matrices. In other words, the assigned PoIs when TPEA is employed are still within reasonable distance from the stops.

Figure 8 shows the effect of varying parameter lim on the number of assigned stops and the number of PoIs in assignments when ad_{th} is set to 100 or 250 m. As expected, both the number of stops and the number of PoIs increase when lim increases because the algorithm is able to assign stops that it could not assign for smaller values of lim.

Figures 6 and 8 also show that TPEA is able to assign additional stops using the temporal patterns of the users found in the initial assignment with the DBA method. For instance, the number of stops assigned using TPEA is around 110,000 while it is 90,000 using DBA in Fig. 6a for $ad_{th} = 100$ and $lim = 5$.

4.2 Evaluation of Stop Assignment

To evaluate the accuracy of stop assignment methods, we use the home/work locations extracted from GPS data using the method explained in Sect. 3.3 since we do not have access to a proper ground-truth data. To extract home/work locations, we use the default values given in Table 4 for parameters Δ_{hw} and p_{hw}. Then, the extracted home/work locations are inserted to the PoI database. The assignments of home/work stops to the newly inserted home/work PoIs forms the ground-truth dataset for this experiment. In other words, no stops are assigned to any regular PoI in this ground truth. However, we use the set of all PoIs (regular PoIs and home/work PoIs) in this experiment. We assign the complete set of home/work stops using the proposed methods and compare our methods with the closest assignment method (CA) that assigns the stop to the closest PoI regardless of the number of PoIs around it.

In particular, we report the precision and recall (Baeza-Yates and Ribeiro-Neto 1999) for the stop assignment methods. The true positives are the stops marked as home/work stops that are assigned to the correct home/work PoI, and the false positives are the non-home/work stops that are assigned to a home/work PoI. The false negatives are the home/work stops that are assigned to a PoI different from the ground-truth PoI or are not assigned at all.

Figure 9a shows that precision values of DBA and TPEA are higher than that of CA. The precision of DBA is slightly higher than that of TPEA since utilization of the temporal visit patterns of the user can introduce false positives. The precision, at 0.93, indicates that DBA and TPEA are able to assign home/work stops and the remaining stops almost correctly.

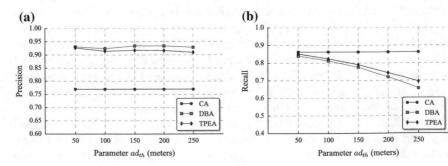

Fig. 9 Precision and recall

Figure 9b shows that CA has better recall than DBA and TPEA that cannot assign all the home/work stops due to the constraints set by parameters ad_{th} and lim. Since the unassigned home/work stops are false negatives, DBA and TPEA have lower recall than CA. For $ad_{th} = 50$ and $ad_{th} = 100$, DBA and TPEA achieve a recall above 0.8. TPEA achieves a slightly higher recall because of an increase in the true positives and a decrease in the false negatives compared to DBA.

4.3 Exploring the Effect on Output Rankings

We proceed to study the effect of the grid smoothing method and the weighting parameter of the ranking function (β) on the output top-k rankings.

Smoothing, described in Sect. 3.5, changes the original values of the grid as well as introduces non-zero values in cells that lack data. This, in effect, extrapolates the available data to wider geographical areas, but it may also distort the original data. To observe the effect of the smoothing, we compute the top-10 PoIs for the grid cells that have initial values for at least 10 different PoIs, before and after smoothing, and we report the distribution of the Kendall tau distance (Fagin et al. 2004) between them. The top-10 lists are formed according to the ranking function in Eq. 1.

To explore the effect of the weighting parameter (β) in Eq. 1, we present the average Kendall tau distance between the rankings constructed for top-k spatial keyword queries using different β values. The set of queries used in this experiment consists of top-k queries with $k = 10$ and $k = 15$. The set of keywords used in top-k queries is {"restaurant", "supermarket", "store"}, and the set of locations include the centers of grid cells that contain values for at least k PoIs.

Kendall Tau Distance. The distance is defined in Eq. 2 (Fagin et al. 2004), where R_1 and R_2 denote the rankings that are compared and P is the set of pairs of the PoIs.

$$K(R_1, R_2) = \frac{\sum\limits_{(p,q)\in P} \bar{K}_{p,q}(R_1, R_2)}{|P|} \tag{2}$$

Function $\bar{K}_{p,q}$ is given in Eq. 3. If R_1 and R_2 agree on the ranking of PoIs p and q, the function evaluates to 0; otherwise, it evaluates to 1.

$$\bar{K}_{p,q}(R_1, R_2) = \begin{cases} 0 \text{ if } R_1 \text{and } R_2 \text{agree on } p, q \\ 1 \text{ if } R_1 \text{and } R_2 \text{ do not agree on } p, q \end{cases} \tag{3}$$

4.3.1 Effect of Smoothing

We report the Kendall tau distance distributions between top-k rankings obtained before and after smoothing for different β values.

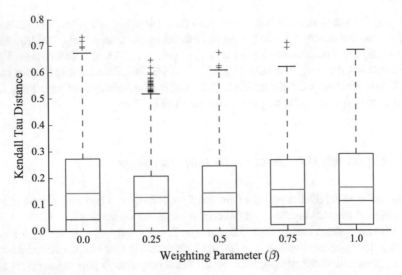

Fig. 10 Kendall Tau distance distribution

The results are shown in Fig. 10, where the red lines show the medians and the green lines show the means of the distance values. The points denoted by a plus sign shows outlier distance values. On average, we achieve a Kendall tau distance around 0.15, which means that we can capture 85% of the relations between pairs after smoothing. We can also see that for all β values, the resulting distribution is right-skewed. We achieve a Kendall tau distance less than 0.1 for half of the grid cells and a Kendall tau distance around 0.3 for 75% of the grid cells. Further, for β values 0, 0.25, and 0.5, the smoothing does not introduce any changes in top-k PoIs for at least 25% of the grid cells.

4.3.2 Effect of Weighting Parameter

Figure 11 reports the average Kendall tau distance between the top-k rankings produced using different β values for top-10 and top-15 queries. For instance, in Fig. 11a, the green bar on the group $\beta = 0$ indicates that the average Kendall tau distance between the rankings produced with $\beta = 0$ and $\beta = 0.25$ is around 0.12. The distances between rankings produced with different β values are less than 0.2. This suggests that the proposed model to extract output rankings is not overly sensitive to the weighting parameter.

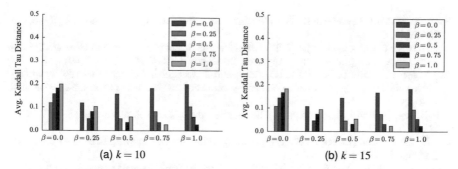

Fig. 11 Avg. Kendall Tau distance for top-k queries

5 Conclusion and Future Work

The paper proposes a model with two phases, model-building and rank-building, to synthesize rankings for top-k spatial keyword queries based on historical trips extracted from GPS data. We propose a novel stop assignment method that makes use of the distances between the locations of the stops and the PoIs as well as temporal information of the stops to obtain the trips. We also propose a Pagerank-based smoothing method in order to extend the geographical coverage of the model. Experiments show that the model is able to produce rankings with respect to the visits of the users, and that the output rankings produced by the model are relatively insensitive to variations in the parameters.

In future work, it is of interest to use the methods proposed here for evaluation of the ranking functions for spatial keyword queries, as this is the motivation behind this work. Another future direction is to explore probabilistic stop assignment in order to contend better with dense regions since the proposed methods use a conservative distance based approach when assigning stops to PoIs. In other words, we assign a stop to a PoI if it is highly probable that the visit occurred. As a result, it is not possible to assign stops in regions with many PoIs. It is also of interest to try to employ more advanced home/work identification methods to be able to determine home/work locations more accurately. It is also of interest to combine data sources like geo-coded tweets and check-ins with GPS data to form rankings for spatial keyword queries.

References

Alvares LO, Bogorny V, Kuijpers B, de Macedo JAF, Moelans B, Vaisman A (2007) A model for enriching trajectories with semantic geographical information. In: SIGSPATIAL GIS'07, pp 22:1–22:8

Ashbrook D, Starner T (2003) Using GPS to learn significant locations and predict movement across multiple users. Pers Ubiquitous Comput 7(5):275–286

Baeza-Yates RA, Ribeiro-Neto B (1999) Modern information retrieval. Addison-Wesley Longman Publishing Co, Inc

Bhattacharya T, Kulik L, Bailey J (2012) Extracting significant places from mobile user GPS trajectories: A bearing change based approach. In: SIGSPATIAL GIS'12, pp 398–401

Bhattacharya T, Kulik L, Bailey J (2015) Automatically recognizing places of interest from unreliable GPS data using spatio-temporal density estimation and line intersections. Pervasive Mob Comput 19:86–107

Brin S, Page L (1998) The anatomy of a large-scale hypertextual web search engine. In: WWW' 98, pp 107–117

Cao X, Chen L, Cong G, Jensen CS, Qu Q, Skovsgaard A, Wu D, Yiu ML (2012) Spatial keyword querying. In: ER' 12, pp 16–29

Cao X, Cong G, Jensen CS (2010) Mining significant semantic locations from GPS data. Proc VLDB Endow 3(1–2):1009–1020

Chen X, Bennett PN, Collins-Thompson K, Horvitz E (2013) Pairwise ranking aggregation in a crowdsourced setting. In: WSDM'13, pp 193–202

Ester M, Kriegel HP, Sander J, Xu X (1996) A density-based algorithm for discovering clusters in large spatial databases with noise. In: KDD'96, pp 226–231

Fagin R, Kumar R, Mahdian M, Sivakumar D, Vee E (2004) Comparing partial rankings. SIAM J Discrete Math 20:47–58

Furletti B, Cintia P, Renso C, Spinsanti L (2013) Inferring human activities from GPS tracks. In: UrbComp'13, pp 5:1–5:8

Google: Understanding consumers' local search behavior (2014). https://goo.gl/7TrZNl

Google: Annual search statistics (2016). https://goo.gl/fxTWZs

Gu Q, Sacharidis D, Mathioudakis M, Wang G (2017) Inferring venue visits from GPS trajectories. In: SIGSPATIAL GIS'17

Hastie T, Tibshirani R, Friedman J (2009) Kernel smoothing methods. Springer, New York

Kang JH, Welbourne W, Stewart B, Borriello G (2004) Extracting places from traces of locations. In: WMASH'04, pp 110–118

Keles I, Saltenis S, Jensen CS (2015) Synthesis of partial rankings of points of interest using crowdsourcing. In: GIR'15, pp 15:1–15:10

Kleinberg JM (1999) Authoritative sources in a hyperlinked environment. J ACM 46(5):604–632. https://doi.org/10.1145/324133.324140

Kumar R, Mahdian M, Pang B, Tomkins A, Vassilvitskii S (2015) Driven by food: modeling geographic choice. In: WSDM'15, pp 213–222

Montoliu R, Blom J, Gatica-Perez D (2013) Discovering places of interest in everyday life from smartphone data. Multimedia Tools Appl 62(1):179–207

Page L, Brin S, Motwani R, Winograd T (1999) The pagerank citation ranking: bringing order to the web. In: TR 1999-66, Stanford InfoLab

Palma AT, Bogorny V, Kuijpers B, Alvares LO (2008) A clustering-based approach for discovering interesting places in trajectories. In: SAC'08, pp 863–868

Shaw B, Shea J, Sinha S, Hogue A (2013) Learning to rank for spatiotemporal search. In: WSDM'13, pp 717–726

Shepard D (1968) A two-dimensional interpolation function for irregularly-spaced data. In: ACM'68, pp 517–524

Spinsanti L, Celli F, Renso C (2010) Where you stop is who you are: Understanding peoples' activities. In: BMI'10, pp 38–52

Stoyanovich J, Jacob M, Gong X (2015) Analyzing crowd rankings. In: WebDB'15, pp 41–47

Yi J, Jin R, Jain S, Jain A (2013) Inferring users' preferences from crowdsourced pairwise comparisons: A matrix completion approach. In: HCOMP'13, pp 207–215

Zheng Y, Zhang L, Xie X, Ma WY (2009) Mining interesting locations and travel sequences from GPS trajectories. In: WWW' 09, pp 791–800

Zhou C, Frankowski D, Ludford P, Shekhar S, Terveen L (2004) Discovering personal gazetteers: an interactive clustering approach. In: SIGSPATIAL GIS'04, pp 266–273

Towards a Dynamic Isochrone Map: Adding Spatiotemporal Traffic and Population Data

Joris van den Berg, Barend Köbben, Sander van der Drift
and Luc Wismans

Abstract This research combines spatiotemporal traffic and population distribution data in a dynamic isochrone map. To analyze the number of people who have access to a given area or location within a given time, two spatiotemporal variations should ideally be taken into account: (1) variation in travel time, which tend to differ throughout the day as a result of changing traffic conditions, and (2) variation in the location of people, as a result of travel. Typically, accessibility research includes neither one or only variation in travel time. Until recently, we lacked insight in where people were located throughout the day. However, as a result of new data sources like GSM data, the opportunity arises to investigate how variation in traffic conditions and variation in people's location influences accessibility through space and time. The novelty of this research lies in the combination of spatiotemporal traffic data and spatiotemporal population distribution data presented in a dynamic isochrone web map. A case study is used for the development of this isochrone map. Users can dynamically analyze the areas and people who can reach various home interior stores in the Netherlands within a given time, taking into account traffic conditions and the location of people throughout the day.

J. van den Berg (✉)
Faculty of Geosciences, UU—Utrecht University, Utrecht, The Netherlands
e-mail: jorisvdberg@live.nl

B. Köbben
Faculty of Geo-Information Science and Earth Observation, ITC—University of Twente,
Enschede, The Netherlands
e-mail: b.j.kobben@utwente.nl

S. van der Drift · L. Wismans
DAT.Mobility, Deventer, The Netherlands
e-mail: svddrift@dat.nl

L. Wismans
e-mail: lwismans@dat.nl

L. Wismans
Engineering Technology Faculty, CTS—University of Twente, Enschede, The Netherlands

© Springer International Publishing AG 2018   195
P. Kiefer et al. (eds.), *Progress in Location Based Services 2018*, Lecture Notes
in Geoinformation and Cartography, https://doi.org/10.1007/978-3-319-71470-7_10

Keywords Isochrone map · Dynamic · Spatiotemporal · Traffic
Population distribution · GSM · Accessibility

1 Introduction

Accessibility is an interdisciplinary concept that has been used increasingly in a variety of disciplines and studies (Li et al. 2011; Cascetta et al. 2016). In this research, accessibility refers to passive accessibility which is described as *'the ease with which an activity can be reached by potential users in the study area'* (Cascetta et al. 2016 p. 45). Accessibility is a spatiotemporal phenomenon, meaning that it changes through space over time (Andrienko et al. 2013). On the one hand, travel times increase and, as a result, the area that can effectively be reached within a given time changes throughout the day. And on the other hand, in order to determine the number of potential users in a study area, you need to know the location of these people through time.

One way of analyzing accessibility is through isochrone maps. Such a map displays isochrones which are the points, lines or areas that can be reached from a given location, within a given time (Bauer et al. 2008; Efentakis et al. 2013; Marciuska and Gamper 2010). Besides visualizing points, lines or areas that are within reach from a given location within a given time, the number of people within an area can be determined by combining the isochrone area with population distribution data. This determines the number of people that could theoretically reach or be reached from a given place within a given time.

A common problem with contemporary isochrone maps is that static travel speeds are assumed when calculating isochrones. As we argued before, traffic conditions, and therefore accessibility, changes significantly over space and time (Li et al. 2011). Using static travel times in accessibility studies and isochrone maps mean that significant variations in accessibility through time and space would be ignored. Moreover, earlier research determining the number of people in isochrone areas (Efentakis et al. 2013; Innerebner et al. 2013) fall short on one crucial point: Spatiotemporal variation in population distribution and movement of people is completely absent.

Currently, mobile data, being GPS tracking, mobile phones and locational media worldwide, provide new opportunities for research into the movement of individuals, the dynamics of traffic and population distribution (Zook et al. 2015). By combining isochrone maps with mobile data, a whole new range of interesting questions can be answered and a more accurate insight into accessibility can be achieved. These mobile data can be used to determine travel times and population distribution data for determining the number of people within areas at a given time. Also, combining spatiotemporal traffic and population distribution data allows knowing where and at what times many people are at the same location, which allows businesses to adjust their opening hours, schedule of events and optimal location (Steenbruggen et al. 2015). The rise of alternative ways to track the movement of people, and the spatiotemporal distribution of populations is particularly interesting in accessibility studies, and until recently has not been used (Järv et al. 2016). The relative newness of these data sources means that no best-practices have been developed yet (Zook et al. 2015).

To dynamically visualize isochrone maps incorporated with spatiotemporal data means a more efficient visualization of spatiotemporal change in accessibility (Innerebner et al. 2013). Besides, it provides a possibility to display interactive statistics allowing easier interpretation of the presented results. As Ullah and Kraak (2015) mention, there is a need for interactive geovisual analytical representation of the produced spatiotemporal data in order to produce useful insights and to make sense out of the data.

Despite these benefits spatiotemporal data might have when implemented in isochrone maps, this implementation can cause new problems. Problems both technically, how to calculate isochrones using vast amounts of spatiotemporal data, as well as how to visualize dynamics in isochrone maps. More data does not necessarily mean more accurate or better results. Ironically, more (spatiotemporal) data means more complications and more effort to conduct useful research (Zook et al. 2015). Although a lot of effort has been put in developing visualization methods that meet the needs to analyze and understand spatiotemporal data (Zeng et al. 2014), options that effectively deal with temporal data in cartography still have not been developed sufficiently (Andrienko et al. 2010; Li and Kraak 2008).

This research aims to tackle the problems mentioned, by combining spatiotemporal traffic and population distribution data in a dynamic isochrone map. The main question in this research is: How can spatiotemporal traffic and population distribution data be incorporated in a dynamic isochrone map? First, we will briefly discuss related work in Sect. 2 before continuing with the methodology in Sect. 3, where we discuss how we combined the data and visualized it in a dynamic isochrone map. In Sect. 4 we discuss the results and end with a discussion and conclusion in Sect. 5.

2 Related Work

The strength of isochrone maps is to visualize accessibility (Doling 1979; O'Sullivan et al. 2000). Still isochrone maps have been used infrequently in the literature and are often absent from well-known studies on accessibility (O'Sullivan et al. 2000). First applications use static (i.e. fixed) spatiotemporal information on travel times and location of people (mostly based on census data). Assuming static travel speeds and static population distribution has various consequences: users of isochrone maps can make (investment) decisions or interpretations based on overly simplified or erroneous images of the realities of accessibility (Tenkanen et al. 2016). Isochrone map users, like urban planners, would carry a risk of over—or underestimating accessibility or the number of people within reach in peak hours. Social equity is another related field where problems could occur when using static traffic data (Li et al. 2011; Shaw 2006). People who live relatively close to facilities but suffer from traffic congestion have more difficulties accessing certain facilities than others who are not experiencing traffic congestion. This is especially true in urban areas (Melhorado et al. 2016). Errors could also occur in non-residential areas which are crowded, like airports or business areas. Because officially no one is registered to live in these areas, using static population distribution data in accessibility studies would assume that no one is present in those areas, whereas in the real world significant numbers of people travel to these places. Using spatiotemporal traffic data already proved to be successful in several accessibility studies (Jariyasunant et al. 2010; Jihua et al. 2013; Innerebner et al. 2013; Li et al. 2011; Marciuska and Gamper 2010).

Problems associated with using static travel speeds when calculating isochrones are identified in different studies (Miller et al. 2009; Shaw 2006). These studies conclude that using static travel time in accessibility studies and isochrone maps mean that significant variations in accessibility through time and space would be ignored. Traditional work focused particularly on space (locational) constraints whereas time constraints have been mostly disregarded (Li et al. 2011). A proposed solution is to use spatiotemporal traffic data to calculate isochrones as done by Lee et al. (2009). While research on spatiotemporal traffic data has gained attention in route computation research, spatiotemporal traffic data for calculating isochrones have not received the same consideration (Baum et al. 2015). Efentakis et al. (2013) presented one of the few studies that used static and dynamic traffic data to research differences between the two. They concluded that spatiotemporal traffic data have a '*huge*' impact on informed business intelligence decisions and showed that the number of potential customers varied between the twenty and forty percent depending on traffic, these variations are quite significant and could be even more significant when taking into account spatiotemporal population distribution. Jihua et al. (2013) created accessibility profiles to display variations in accessibility throughout the day. By plotting the isochrone area in square kilometers versus different hours a day, a better insight in the accessibility of a location was realized. Although the dynamics in travel time also results in variation of people who can reach a location within a certain travel time threshold, there is little to no research in which the dynamics in people's location is taken into consideration as well.

3 Methodology

Figure 1 shows the methodology developed in this research to construct a dynamic isochrone map. First, a network dataset needs to be prepared (1). Secondly, we calculate the isochrone network using a database network extension (2). Using the

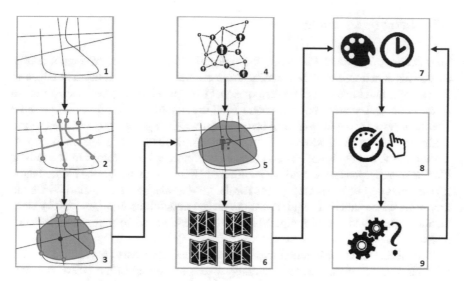

Fig. 1 Workflow construction isochrone map

isochrone network calculated in the previous step we can calculate isochrone areas using a specific buffer (3), this is discussed in Sect. 3.3. The next step is to prepare the population distribution dataset (4). We then combine the isochrone areas with the population distribution data to determine the number of people within the isochrone areas (5). This results in a series of images that display areas, and estimated numbers of people within these isochrone areas (6). The remaining steps consist of visualizing the data processed and calculated in previous steps (7). After adding interactive elements (8), the isochrone map is tested (9) and the visualization process is repeated if necessary.

3.1 HERE Network

This research uses HERE Traffic Patterns data (HERE 2017). HERE Traffic Patterns offers extensive average traffic speed data for 83 different countries, including the Netherlands. These patterns are constructed based on billions of Floating Car observations. The HERE traffic patterns data used in this research contains average driving speeds for every 15 min for every road in the Dutch road network. These speed patterns do not exceed the maximum allowed speed on that specific road. This is because HERE traffic patterns are mostly used for navigation systems and trip planners, and these should not encourage users to exceed the maximum allowed speed. These data are stored and processed in a PostgreSQL database.

3.2 Driving Distance

To calculate an isochrone network, the pgRouting function pgr_drivingdistance is used. This function calculates, starting from a given point, all nodes and edges in the network that have costs less than or equal to a given cost. In our case, costs are given in time. A starting point is entered and the function calculates in all possible directions how far the network can be traversed within a given cost. We used a case study to test our methodology. First, we determined the points from which to calculate isochrones, being the coordinates of different home furniture stores. For illustrational purposes, we used the location of IKEA stores within the Netherlands. These stores served as starting points for the driving distance calculation. Since the driving distance function requires a node on the network as input, the IKEA store locations (retrieved from Google Maps 2016) are snapped to the closest network node.

The driving distance function requires a field that represents the cost and reverse cost per road segment. Since the average speeds are stored in the HERE network data and the road length can be calculated, we created a function that calculates the time it takes to traverse a road segment. In our case, the costs required by the PGrouting function are given as time in minutes. The time it takes to traverse a road can be calculated by combining average driven speeds and the road length. Normally, the driving distance function calculates what parts of the network can be reached from a given point. However, in this research, we are interested in how many people can potentially drive to a store. So instead of stores being a start point, we rather want them to be an end point. This is achieved by switching the cost and reverse cost values. This ensures that the calculation uses the costs for roads towards the IKEA store only.

After the driving distance calculation is completed, a table is created. This table is joined with the output table of the original road network table. The result is a table of all roads on which one can reach an IKEA store, containing a geometry field which can be visualized accordingly.

Despite the fact that pgRouting is fit for handling complex routing computations on extensive network datasets, there are two limitations. When calculating driving distances, pgRouting uses data from one input column for costs. Since one column represents one time step, inaccuracies can occur. If for example average speeds from 9 o'clock with a maximum cost of one hour are used in a driving distance calculation, the 9 o'clock data is used during the entire calculation. Ideally, the calculation would have a Time Dependent Dynamic Shortest Path algorithm (TDDSP) meaning the cost field used in the calculation changes according to the time already driven.

Another limitation is that the pgr_drivingdistance function by default returns nodes and edges which do not exceed the maximum input cost. Some road segments are relatively long. If for example they have a node at each end of which one exceeds the maximum cost and one does not, only the latter is returned whereas in reality part of the road segment could still have been traveled before exceeding the maximum cost. Obe and Hsu (2017) have identified this problem as well and describe a possible solution called '*node injection*'. Both of these problems cannot be solved with the current default pgRouting functions.

3.3   Variable Distance Buffer

Since the isochrone area is used to calculate the number of people in reach, it is important to draw an isochrone area around the isochrone network which is as accurate as possible. Marciuska and Gamper (2010) discuss a variety of methods, each with different accuracies. We tested a variety of methods as well. The concave hull, convex hull, alpha shape, link—and surface based approach (Marciuska and Gamper 2010) and different buffers.

However, each of these methods in—or excluded areas that were actually in or out of reach. This means that when intersecting the calculated isochrone areas with the number of people, large under—or overestimations could occur. We developed a method which in our case is most accurate: the variable distance buffer. Here, a buffer is drawn according to the aggregated costs of the isochrone network. In other words: time already spent can be taken into account when drawing a buffer. The further driving time increases, the smaller the buffer size. Figure 2 illustrates this solution. The pgr_driving distance calculation explained earlier, stores the aggregated time for each road segment. The variable distance buffer is calculated using a simple IF-THEN statement. If the amount of driven time exceeds a certain value, the buffer size is adjusted.

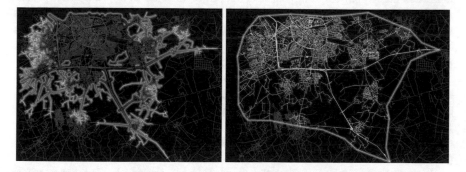

Fig. 2 Variable distance buffer versus concave hull approach

3.4 GSM Data

This research uses GSM data as a way to measure population distribution. This passive mobile positioning data is aggregated and provided by Mezuro. It originates from Call Detail Records (CDR) collected by a single network provider which facilitates between 30–40% of the Dutch mobile phone usage. This means that by accessing these data we can derive travel information of about one-third of the total Dutch population. No other data source is known that gives travel information on a national scale at a level this high. This data is preprocessed using a validated rule-based algorithm to approximate and classify the number of people within areas based on phone activity for every hour of the day. Because of the known issues regarding the spatial accuracy of determining the location using CDR data (e.g. described in Bonnel et al. 2015), the location data is aggregated at the level of villages. As a result, the Netherlands is split into 1.261 areas for which the number of people is made available, where each city or village is a separate zone. This area definition is the result of earlier analysis by Mezuro of the dataset (CDR and cell tower plan properties) provided. The largest cities in the Netherlands are split into city districts. Mezuro uses a complex algorithm to translate the sample into total estimated number of people, classified in different groups, within these areas. This algorithm takes into account different factors such as the number of people within a GSM area subscribed to the network provider, the number of active phones of subscribers per area and the number of inhabitants per area.

GSM data is privacy sensitive. In theory, it is possible to track someone's movements. To secure the privacy of mobile phone owners, the CDR's are anonymized. This means that data of individual phones remain with the network operator. Also, it is not possible to track or filter individual phones out of the provided data. At least 16 phones have to be in the same area before they are registered in the final dataset (Meppelink et al. 2015). As a result of the area definition and by means of aggregating data over multiple days (i.e. all days within a certain month), the impact of this "rule of 16" is minimized (i.e. analysis shows that the impact of this rule is less than 1%). Although the algorithms are tuned for the accuracy of location determination and bias of the available sample, there possibly still remains a bias in the determined number of people in areas.

The aggregated GSM data, from here on simply referred to as GSM data, consists of two tables. The first table contains the administrative areas used. This table does not contain any information on the number of people in that area yet. It merely serves as a spatial reference to the GSM areas. The second table holds the estimated numbers, including further classified population groups in areas for a given month. These two tables are linked through matching area ID's.

The population groups within the GSM data are classified using observations from the given month. Assumed inhabitants for example, are people that have been observed in the same area during most of the nights in the given month. Regular visitors are observed in a specific area at least 10 times a month. There is a possibility that these people visit these areas because of their job or school. Frequent visitors are people which are observed 3–9 times in a GSM area per month. It is hard to determine the goal with which frequent visitors travel to certain areas. It could, for example, be visiting friends or families once a week. The same goes for incidental visitors; People which are observed one or two times a month in a GSM area.

For this specific research, we are interested in the number of people which are located in an isochrone area at a specific time during the day. A differentiation is made only between inhabitants and visitors within the isochrone areas. To determine the number of inhabitants and visitors within isochrone areas at a specific time of day, different calculations were used for inhabitants and visitors. This is because we can further increase spatial accuracy for inhabitants using PC6 points. PC6 points are the centroids of all postal code areas in the Netherlands containing the (static) number of residents in that postal code zone. These points can be used to more accurately determine the distribution of inhabitants within GSM areas. Since PC6 points only contain static information on inhabitants, the method cannot be applied for visitors.

The method for inhabitants intersects the PC6 points with the GSM areas to determine the static total number of inhabitants within a GSM area (p_t). By dividing the number of static inhabitants for each individual PC6 point (p_p) by the total static number of inhabitants in the GSM area the fraction of the total number of inhabitants per PC6 point is determined. Using these fractions the estimated inhabitants located in the GSM areas are distributed more accurately. Assuming that this distribution remains the same through time, we can multiply the fraction of each PC6 point which is located in the isochrone area with the associated GSM area dynamic number of measure inhabitants (I_{gsm}) which results in the dynamic number of inhabitants present per PC6 point. The sum of the dynamic inhabitants for these PC6 points (i.e. located within the isochrones area) is the total dynamic number of inhabitants in the isochrone area (I_{DT}).

$$I_{DT} = \sum \left(\left(\frac{p_p}{p_t} \right) \times I_{gsm} \right)$$

I_{DT} Total number of dynamic inhabitants in isochrone area.
p_p Population PC6 Point.
p_t Total PC6 population GSM area.
I_{gsm} Dynamic number of inhabitants in isochrone per GSM area.

The method used for calculating inhabitants using PC6 points cannot be applied to visitors since their exact location within the GSM areas is unknown. An alternative approach is to use the isochrone and GSM areas' surface area. First, isochrone areas are intersected with the GSM areas. The share of the isochrone area (A_{isa}) within the GSM area (A_{gsm}) can be used to calculate the relative number of visitors in that particular area (V_{gsm}). The major assumption in this method is that the visitors are distributed evenly throughout the GSM area. For example, when

50% of the isochrone area intersects with the GSM area, we assume 50% of the total visitors in the GSM area are in the isochrone area. By summing up the visitors within each share of the isochrone area, the total number of visitors in the isochrone area is determined (V_T).

$$V_T = \sum \left(\left(\frac{A_{iso}}{A_{gsm}} \right) \times (V_{gsm}) \right)$$

V_T *Total number of dynamic Visitors.*
A_{iso} *Area isochrone within GSM area*
A_{gsm} *Total area GSM area*
V_{gsm} *Dynamic visitors GSM area*

3.5 Visualization

The two major components which are visualized in the web map are the isochrones and the population distribution data. Using the QGIS TimeManager (QGIS TimeManager 2017) isochrone areas were visualized based on the time attribute in our data. The TimeManager filters out, visualizes and exports specific times in the data as images. Using this function, 96 static maps for every 15 min of the selected day are exported. Using JavaScript, we created a web page with an interactive interface, to control the playback of the animation.

We have chosen to visualize the population distribution data as an animated line graph, to allow the user to see trends throughout the day. The number of inhabitants, visitors and a total number of people were visualized as a line graph using Microsoft Excel, animated using Microsoft PowerPoint and then added as a video to the web page, synchronized to the map animation using JavaScript code.

We deliberately chose to visualize the population distribution data and the isochrone areas separately to maintain the simplicity of an isochrone map. An alternative would, for example, be to visualize the number of people in a third dimension. We argue that this would overcomplicate the map thereby making it less useful.

4 Results

The resulting Ikea isochrone web map (Fig. 3) can be seen online at http://kartoweb.itc.nl/students/isochroneswebmap/nederland.html.[1] It displays an animation of accessibility throughout the day. Users can 'slide' through time using the

---

[1]If the website animation does not run smoothly, we advise using the step-by-step button.

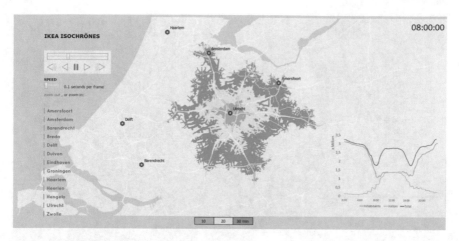

Fig. 3 Isochrones webmap IKEA Utrecht

provided time slider and can see the change in area size during different hours of the day. They can zoom into a specific IKEA store to also include the dynamic number of people within the area throughout the day.

As we can see in Fig. 3, the difference in the number of people that can reach IKEA Utrecht changes quite noticeably throughout the day. The number of inhabitants (orange line) decrease in the early morning while the number of visitors (green line) increase. This can potentially be explained by the fact that people start commuting. This pattern happens during the end of the afternoon in reverse. Around 16:00, commuters start heading back home. We observe significant drops in the total number of people (black line) around the typical rush hours (06:00–09:00 and 16:00–19:00). The variation of the number of people who can reach this location (black line) shows that differences are possible of over 1 million people, roughly resulting in a variation of 30%. Such large differences are certainly relevant for determination of service areas, e.g. choosing the best location for opening a new store.

Another interesting result is that the number of people that can reach an IKEA store located in regions not as much affected by congestion stays relatively stable. Although not visible in Fig. 4, but well visible once animated, you can see that the isochrone areas around IKEA Groningen change less in size compared to the isochrones areas around busier IKEA stores such as Utrecht and Amsterdam. For IKEA Groningen we see a relatively stable black line representing the total number of people (Fig. 4) which is mainly the results of number of inhabitants and visitors compensating each other resulting in a relatively stable number of people within the isochrone areas. This could mean that the influence of traffic (i.e. travel time variation) has a larger impact on the total number of people that can reach an IKEA store than the movement of people through time.

All in all, the differences in the area size and the number of people which can reach an IKEA store within given times justifies the use and need for spatiotemporal

Fig. 4 Isochrones webmap IKEA Groningen

traffic and population distribution data in accessibility studies. With differences of up to 1 million people, it would be a mistake to use static traffic and population distribution data.

5   Conclusion and Discussion

We have proven that spatiotemporal traffic and population distribution data can be combined in a dynamic isochrone map to research accessibility. The method used in this research can be used for similar cases without the need to redevelop the methodology. The combination of spatiotemporal traffic and population distribution data is particularly interesting for calculating dynamic service areas which can be used in different fields. A specifically interesting potential use would be the optimization of potential locations for new stores or facilities. One point of attention: some of the calculations were rather time-consuming. Calculating all isochrone areas during a day for the Netherlands lasted approximately 2 days. This should be taken into account when planning to calculate even bigger isochrone areas.

The results of this research are promising although some points can be improved in future research. First of all, pgRouting by default is more focused on the use of static input data for network calculations. Time Dependent Dynamic Shortest Path Algorithms (TDDSP) and a method to increase the accuracy of the nodes returned by pgRouting, for example, would increase the overall accuracy of the isochrone calculations and thus the calculated number of people within these areas. These functionalities can be added by editing the default pgRouting functions. Also, the interactivity currently offered in the isochrones web map could be extended. It would be nice if users could search for their own address and pan and/or zoom the map.

Besides these methodological points, new, more accurate spatiotemporal data sets might become available in the near future. It would be interesting to calculate isochrone areas using actual traffic speeds of one single day instead of the Traffic Patterns used in this research. Also, more research into GSM data should be conducted to draw better conclusions on the accuracy and usability of these data.

Moreover, the dynamic isochrone map should be tested with actual end-users in future research to evaluate the usability and other potential benefits or shortcomings compared to traditional static isochrone maps. In this research, we claim that adding spatiotemporal dynamics to isochrone maps lead to a better and more accurate insight in accessibility but the potential need and use for such an application are not researched. Another interesting application would be to analyze day-to-day dynamics. This research only focused on a single day but there are significant differences in accessibility between different days as well.

Although there is always room for improvement, especially regarding the visualization of the results, we hope our work encourages further research into dynamic isochrone maps using spatiotemporal traffic and population distribution data. Besides potentially improving the methodology presented in this research, we hope to see relevant new case-studies in which the benefits of a dynamic isochrone map, as presented in this research, are shown.

References

Andrienko G, Andrienko N, Bak P, Keim D, Wrobel S (2013) Visual analytics of movement. Springer, Berlin, Heidelberg. https://doi.org/10.1007/978-3-642-37583-5

Andrienko G, Andrienko N, Dykes J, Kraak M, Schumann H (2010) GeoVA (t)—geospatial visual analytics: focus on time. J Location Based Serv 4(3):141–146. https://doi.org/10.1080/17489725.2010.537283

Bauer V, Gamper J, Loperfido R, Profanter S, Putzer S, Timko I (2008) Computing isochrones in multi-modal, schedule-based transport networks. In: Proceedings of the 16th ACM SIGSPATIAL international conference on advances in geographic information systems—GIS'08, 2. https://doi.org/10.1145/1463434.1463524

Baum M, Buchhold V, Dibbelt J, Wagner D (2015) Fast computation of isochrones in road networks, 1–27. https://doi.org/10.1007/978-3-319-38851-9

Bonnel P, Hombourger E, Olteneanu-Raimond AM, Smoreda Z (2015) Passive mobile phone dataset to construct origin-destination matrix: potential and limitations. Transp Res Procedia 11:381–398

Cascetta E, Cartenì A, Montanino M (2016) A behavioral model of accessibility based on the number of available opportunities. J Transp Geogr 51:45–58. https://doi.org/10.1016/j.jtrangeo.2015.11.002

Doling J (1979) Accessibility and strategic planning. Centre for Urban and Regional Studies, University of Birmingham, Birmingham

Efentakis A, Grivas N, Lamprianidis G, Magenschab G, Pfoser D (2013) Isochrones, traffic and DEMOgraphics. In: GIS: Proceedings of the ACM international symposium on advances in geographic information systems, pp 538–541. https://doi.org/10.1145/2525314.2525325

Google maps (2016) Ikea. https://www.google.nl/maps/search/ikea/@52.246146,4.7822063,8.48z Accessed 16 Dec 2016

HERE (2017) Traffic analytics. https://here.com/en/products-services/products/here-traffic/here-traffic-analytics Accessed 26 Jan 2017

Innerebner M, Böhlen M, Gamper J (2013) ISOGA: a system for geographical reachability analysis, pp 180–189

Jariyasunant J, Mai E, Sengupta R (2010) Algorithm for finding optimal paths in a public transit network with real-time data. In: Transportation research board 90th annual meeting, pp 1–14. https://doi.org/10.3141/2256-05

Järv O, Tenkanen H, Salonen M, Toivonen T (2016) Dynamic spatial accessibility modelling: access as a function of time. AGILE 2016:14–17

Jihua H, Zhifeng C, Guangpeng Z, Ze H (2013) A calculation method and its application of bus isochrones. J Transp Syst Eng Inf Technol 13(3):99–104. https://doi.org/10.1016/S1570-6672(13)60111-7

Lee W-H, Tseng S-S, Tsai S-H (2009) A knowledge based real-time travel time prediction system for urban network. Expert Syst Appl 36(3):4239–4247. https://doi.org/10.1016/j.eswa.2008.03.018

Li Q, Zhang T, Wang H, Zeng Z (2011) Dynamic accessibility mapping using floating car data: a network-constrained density estimation approach. J Transp Geogr 19(3):379–393. https://doi.org/10.1016/j.jtrangeo.2010.07.003

Li X, Kraak M (2008) The time wave. a new method of visual exploration of geo-data in time. Cartographic J 45(3):193–200. https://doi.org/10.1179/000870408X311387

Marciuska S, Gamper J (2010) Determining objects within isochrones in spatial network databases, 392–405

Melhorado AMC, Demirel H, Kompil M, Navajas E, Christidis P (2016) The impact of measuring internal travel distances on self-potentials and accessibility. Eur J Transp Infrastruct Res 16(2):300–318

Meppelink J, Van Langen J, Siebes A, Spruit M (2015) Know your bias : scaling mobile phone data to measure traffic intensities

Miller HJ, Bridwell SA (2009) A field-based theory for time geography. Ann Assoc Am Geogr 99(1):49–75. https://doi.org/10.1080/00045600802471049

Obe RO, Hsu LS (2017) pgRouting: a practical guide. Locate Press, Chugiak

O'Sullivan D, Morrison A, Shearer J (2000) Using desktop GIS for the investigation of accessibility by public transport: an isochrone approach. Int J Geogr Inf Sci 14(1):85–104. https://doi.org/10.1080/136588100240976

QGIS (2017) QGIS. http://www.qgis.org/en/site/ Accessed 13 Jan 2017

QGIS TimeManager (2017) QGIS python plugins repository. https://plugins.qgis.org/plugins/timemanager/ Accessed 13 Jan 2017

Shaw S-L (2006) What about "time" in transportation geography? J Transp Geogr 14:237–240. https://doi.org/10.1016/j.jtrangeo.2006.02.009

Steenbruggen J, Tranos E, Nijkamp P (2015) Data from mobile phone operators: a tool for smarter cities? Telecommun Policy 39(3–4):335–346. https://doi.org/10.1016/j.telpol.2014.04.001

Tenkanen H, Saarsalmi P, Järv O, Salonen M, Toivonen T (2016) Health research needs more comprehensive accessibility measures: integrating time and transport modes from open data. Int J Health Geogr 15(1):23. https://doi.org/10.1186/s12942-016-0052-x

Ullah R, Kraak M (2015) An alternative method to constructing time cartograms for the visual representation of scheduled movement data. J Maps 11(4):674–687. https://doi.org/10.1080/17445647.2014.935502

Zeng W, Fu CW, Arisona SM, Erath A, Qu H (2014) Visualizing mobility of public transportation system. IEEE Trans Visual Comput Graph 20(12):1833–1842. https://doi.org/10.1109/TVCG. 2014.2346893
Zook M, Kraak M, Ahas R (2015) Geographies of mobility: applications of location-based data. Int J Geogr Inf Sci 29(11):1935–1940. https://doi.org/10.1080/13658816.2015.1061667

Continuous Trajectory Pattern Mining for Mobility Behaviour Change Detection

David Jonietz and Dominik Bucher

Abstract With the emergence of ubiquitous movement tracking technologies, developing systems which continuously monitor or even influence the mobility behaviour of individuals in order to increase its sustainability is now possible. Currently, however, most approaches do not move beyond merely describing the status quo of the observed mobility behaviour, and require an expert to assess possible behaviour changes of individual persons. Especially today, automated methods for this assessment are needed, which is why we propose a framework for detecting behavioural anomalies of individual users by continuously mining their movement trajectory data streams. For this, a workflow is presented which integrates data preprocessing, completeness assessment, feature extraction and pattern mining, and anomaly detection. In order to demonstrate its functionality and practical value, we apply our system to a real-world, large-scale trajectory dataset collected from 139 users over 3 months.

Keywords Mobility · Trajectory mining · Anomaly detection
Sustainability · Behavior change

1 Introduction

Human mobility is ubiquitous in modern societies and represents an integral part of our daily behavioural routines. At the same time, however, there are numerous undesirable effects, such as traffic jams or increased fossil fuel consumption (Taaffe 1996). With regards to Switzerland, for instance, roughly a half of the total CO_2 emissions are contributed by the transportation sector (including international aviation), with motorized individual mobility being responsible for around two thirds of these emissions (Bundesamt fuer Umwelt 2014). If no major changes occur in the

D. Jonietz (✉) · D. Bucher
Institute of Cartography and Geoinformation, ETH Zurich, Stefano-Franscini-Platz 5, 8093 Zurich, Switzerland
e-mail: jonietzd@ethz.ch

D. Bucher
e-mail: dobucher@ethz.ch

© Springer International Publishing AG 2018
P. Kiefer et al. (eds.), *Progress in Location Based Services 2018*, Lecture Notes in Geoinformation and Cartography, https://doi.org/10.1007/978-3-319-71470-7_11

transport system, these numbers are widely expected to rise in the coming decades (Boulouchos et al. 2017).

Recently, the significance of emerging technologies which enable ubiquitous monitoring as well as real-time regulation and management of human mobility has been emphasized as potential game changing aspect for increasing the sustainability of travel behaviour (Boulouchos et al. 2017). Indeed, current developments in the field of location-acquisition technologies such as Global Navigation Satellite Systems (GNSS), Wireless Local Area Networks (WLAN), or Global System for Mobile Communications (GSM) allow to monitor and record human movement at an exceptional level of detail and at relatively low cost and effort (Feng and Zhu 2016). Due to the widespread use of modern smart phones, as well as a general trend towards digitalization in the transportation and mobility sector, Big Mobility Data are now widely available and ready to be utilized for gaining unprecedented insights into the fundamental mechanisms that guide human mobility (Brunauer and Rehrl 2016).

In fact, since the late 1990s, human movement trajectories, i.e. series of chronologically ordered x, y-coordinate pairs with time stamps (Andrienko et al. 2016), have increasingly been used for travel surveys (Shen and Stopher 2017). Apart from notable exceptions (e.g. Schlich and Axhausen 2003; Stopher et al. 2013), however, these studies have mainly applied a snapshot approach (e.g. Schüssler 2008; Kohla and Meschik 2013), with the center of interest being put on inter-personal variability (differences in the behaviour of different persons) rather than intra-personal variability (different behaviour of one person from day to day) (Schlich and Axhausen 2003). What has often been neglected, therefore, is analysing the dynamic dimension of mobility behaviour, i.e. behaviour changes such as trying out new travel alternatives, or forming new mobility habits.

Especially today, however, it would be worthwhile to be able to automatically detect and analyse such changes in mobility behaviour. On the one hand, in contrast to merely surveying mobility behaviour, there are now systems which move further by aiming to directly influence people's mobility behaviour towards more sustainable transport alternatives (cf. Banister 2008), e.g. by using mobile applications which continuously record the movements of users, stream the data to a server, and utilize them to provide their users with feedback or even suggest more sustainable travel options (Froehlich et al. 2009; Montini et al. 2015). To the best of our knowledge, currently none of these systems apply strategies for automatically detecting behaviour change, but instead require manual checking of the data for evaluating the effectiveness of the conducted persuasive measures. A fully automated system which continuously monitors movement behaviour based on a stream of trajectory data, and detects behavioural changes, however, could take over this tedious task and even trigger dynamic reactions to users based on their behavioural changes, e.g. encourage sustainable mobility behaviour adaptations and discourage in the opposite case. On the other hand, apart from application scenarios where behaviour change is actively induced, the development of methods for detecting such variations in movement data would also be useful for general transportation research and planning purposes. Thus, for instance, insights are still needed in terms of evaluating and predicting peoples reactions to today's novel mobility options, such as shared mobility,

mobility as a service, electric mobility and autonomous vehicles. Being confronted with these, one can expect numerous people to adapt their mobility behaviour, e.g. by testing novel alternatives and even forming new travel habits (Boulouchos et al. 2017). In order to accurately understand these behavioural changes, travel surveys are needed which involve tracking numerous participants over a long period of time. In addition, a set of suitable methods are necessary to analyse the collected data and be able to accurately understand these behavioural changes.

For developing such methods, however, a practical problem is posed by insufficient data quality. It is especially data incompleteness which represents a critical challenge for GNSS-based travel surveys, since it comprises missing records for parts of trips, one or more full trips, or even one or more full days of the recording period (Hecker et al. 2010). These gaps can have various causes, e.g. the cold start problem at the start of movement, bad signal reception, participants leaving the device switched off, or other technological problems (Shen and Stopher 2017). While shorter gaps can often be handled by means of map matching techniques (see Sect. 2.1), longer ones can heavily distort or bias the results of the following analyses. In the context of automated behaviour change detection, for instance, the occurrence of missing movement data could lead to misleading calculations, e.g. drastically lower values for CO_2 emissions produced during the respective week of recording. In this case, a system might erroneously interpret this drop in numbers as a behaviour change, whereas it is in fact merely the result of missing data. To avoid such misdetection of behaviour changes, methods need to be sensitive to recording gaps, i.e. distinguish them from cases where observed changes are actually due to changed mobility patterns.

Before this background, this study proposes a method for identifying and evaluating changes in human mobility behaviour by first detecting and quantifying spatio-temporal recording gaps in a stream of movement trajectory data, and then continuously mining it for anomalies with regards to various mobility features, i.e. a subset of variables which can be extracted from movement data, and describe selected aspects of mobility behaviour (e.g. average speed, travelled distances). Focussing on sustainable mobility as the application scenario, we simulate a real-time data stream using a real trajectory dataset collected from 139 users over 3 months in Switzerland.

This paper is structured as follows: First, in Sect. 2 background information is provided starting with a brief review of available methods for surveying human mobility behaviour on the basis of movement trajectory datasets. Then, the focus is shifted to the potential of similar techniques for inducing and analysing changes in mobility behaviour. In the following Sect. 3, our concept is presented and discussed with regards to data preprocessing, completeness assessment, feature extraction and pattern mining, and finally anomaly detection. In Sect. 4, the framework is applied to a test dataset, before the results are discussed and the paper is concluded in Sect. 5.

2   Related Work

In the context of this study, relevant prior work applies one of two distinct perspectives on mobility behaviour and movement data analysis, and is briefly reviewed in this section:

1. Assessing the **present state of mobility behaviour**, i.e. *where*, *when* and *how* a person travels. This is normally achieved by means of GNSS-assisted travel surveys.
2. Aiming to **change existing mobility behaviour** in order to increase its sustainability, e.g. by means of mobile applications which provide both tracking and user feedback functionalities.

2.1   Movement Trajectories for Surveying Human Mobility Behaviour

Before the rise of position tracking technologies, the traditional ways of gaining insights about the mobility behaviour of people were face-to-face interviews, mail-out/mail-back or telephone surveys. Since the late 1990s, however, GNSS-assisted travel surveys emerged as a novel method, and gradually replaced these approaches due to numerous advantages, such as a relatively high accuracy in recording time and position, low cost (especially with modern smartphones), and less problems with regards to trip-misreporting by respondents (Shen and Stopher 2017). Nowadays, exemplary approaches are manifold, and have spread from pilot studies undertaken in the USA (Wagner 1997) to a range of other countries, including Switzerland (Shen and Stopher 2017).

After recording the movements of test persons, the data require extensive processing in order to extract relevant mobility features, in particular places that have been visited for a certain purpose and the travelled routes between these places. With regards to the former category, stay points are typically detected based on various clustering techniques (e.g. Palma et al. 2008), or the movement speed (e.g. Li et al. 2008). With regards to the travelled routes, via map matching, the exact path taken through a road network can be inferred from the tracking points, e.g. by simple point-to-curve snapping (e.g. White et al. 2000) or advanced techniques such as evolutionary algorithms (Quddus and Washington 2015). Apart from the routes, numerous studies have proposed approaches to infer the used traffic mode, for instance based on identifying walking transitions between mode changes (Zheng et al. 2010), analysing a range of movement descriptors (Sester et al. 2012), or the underlying transportation network (Stenneth et al. 2011).

In order to describe a person's mobility behaviour based on trajectory data, these (and other) mobility features need to be further analysed to extract patterns, i.e. observable regularities in movement behaviour such as habits or long-lasting preferences and restrictions. Thus, one can calculate general statistics over certain time intervals, such as the average duration and length of trips, the modal split, or the usual times of travel (Axhausen and Frick 2005), but also more use-case specific aspects such as frequently visited places other than home or the work location (Siła-Nowicka et al. 2015) or the location of regularly performed activities like eating, shopping or physical exercise (e.g. Zheng et al. 2010; Furletti et al. 2013). When being properly interpreted, mobility features and their regular patterns can serve as indicators for higher-level attributes, such as the sustainability of mobility behaviour. In this context, for instance, (Nicolas et al. 2003) formulated a set of potential sustainability indicators which can be extracted from travel survey data. Among others which refer to the aggregate city level, those which could be extracted from trajectory data include the daily number of trips, the structure of trip purposes (e.g. commuting versus leisure), the daily average time budget spent for travelling, the modal split (especially the share of slow mobility, i.e. walking and cycling), the average distance travelled daily, and the average movement speed. Other relevant indicators which have been formulated in the literature include the amount of CO_2 emissions and the degree to which trips are intermodally integrated, i.e. use different traffic modes in combination (World Business Council 2015).

Naturally, the validity of the results computed for mobility features depend to a large degree on the quality of the input trajectory data, in particular the completeness of the recorded movement. Missing trips or even full day gaps will lead to erroneous, in some cases even heavily biased, results (Hecker et al. 2010), however, are a regularly occurring issue in travel surveys (Shen and Stopher 2017). Although this issue is frequently discussed in the literature (e.g. Shen and Stopher 2017; Wolf et al. 2003), only few studies propose solutions, such as evaluating the intrinsic trajectory data quality based on the spatial and temporal resolution (Prelipcean et al. 2015), a statistical approach to detect dependencies between mobility behaviour, socio-demography and missing data (Hecker et al. 2010), or imputation, the process of inferring the missing trips based on observed data using statistical relationships (Polak and Han 1997). Another popular option to improve and ensure the completeness and correctness of the movement data in travel surveys are prompted recall (PR) methods, in which during the tracking phase, respondents are regularly asked to manually validate and complete their recorded movements, for instance at the end of each day (e.g. Bucher et al. 2016).

In traditional travel surveys, the focus is usually put on analysing the status quo of mobility behaviour, since, as (Schlich and Axhausen 2003) argue, there is a general assumption that travel behaviour mainly consists of highly habitual routines, and remains relatively static over time. Thus, in most cases, mobility features are calculated once on the basis of the entire available data in order to assess the present state of transportation system usage (e.g. Schüssler 2008; Kohla and Meschik 2013) rather than analysing its temporal dynamics. Additionally, this snapshot approach is often caused by practical limitations with regards to the available movement data, with durations of the tracking period rarely exceeding two weeks (Shen and Stopher 2017). There are, however, also examples of longitudinal analyses of travel behaviour (e.g. Hanson and Huff 1988; Schlich and Axhausen 2003; Stopher et al. 2013; Gonzalez et al. 2008; Song et al. 2010). These studies were mostly concerned with detecting day-to-day variations, stability measures, and statistical properties of mobility behaviour from movement data of various kinds, such as those obtained with GSM or GPS, or traditional travel survey methods. While GSM data typically covers long durations and large numbers of users, transport surveys and GPS recordings stem from much less persons over the course of merely a few weeks. Gonzalez et al. (2008), for instance, developed an aggregated model of human mobility based on extensive mobile phone data, and found strong inter-personal regularities, but did not distinguish between individual users or temporal changes. Schlich and Axhausen (2003) report on different mobility indicators, and how they can be used to compute similarity measures between mobility behaviour on two different days.

2.2 Inducing Change in Human Mobility Behaviour

Apart from merely monitoring and analysing the status quo of mobility behaviour, other studies have built on similar analytical methods to actively influence users in order to make them travel in a more environmentally sustainable way. For this, mobile applications and a feedback loop were used, with examples including *Ubi-Green* (Froehlich et al. 2009), *PEACOX* (Montini et al. 2015), or *GoEco!* (Bucher et al. 2016). In some cases, apart from merely summarizing the recorded mobility behaviour, the provided feedback also included the proposal of more sustainable travel alternatives. At present, however, most approaches suffer from either short study periods (Hamari et al. 2014), or from basing their feedback and suggestions for more sustainable mobility on a single snapshot, for example data which was recorded during a pre-study or a baseline-tracking phase. This shortcoming hinders the development of long-running applications that continuously monitor mobility behaviour and are thus able to provide feedback based on detected changes of current in comparison to past behavioural patterns.

Thus, a system would be worthwhile with the ability to automatically detect changes in behaviour, which could then, based on established models of behavioural change processes, select actions to be taken to support (in case of increased sustainability) or prevent (in the opposite case) the observed behaviour change. A commonly used psychological conceptualization is the Transtheoretical Model (Prochaska and Velicer, 1997) which separates behaviour change into *precontemplation*, *contemplation*, *preparation*, *action* and *maintenance* phases. Upon detecting a change in mobility, one could for instance infer that a user started contemplating new behavior, and support a transition towards this behavior by supplying her with *information* (e.g. Tulusan et al. 2012; Taniguchi et al. 2003), *rewarding* further good choices (e.g. Ben-Elia and Ettema 2011), *dissuading* unsustainable behavior (e.g. Schade and Schlag 2003), or otherwise engage and motivate her to move to the *preparation* or *action* stage (Weiser et al., 2015). Alternatively, for users without changes in mobility (one could argue they are in a *precontemplation* or *maintenance* phase), a system might foster *self-experience of travel alternatives* (e.g. Abou-Zeid et al. 2012; Bamberg et al. 2003; Bamberg 2006) in order to make them try out new and more sustainable transport options.

Automatically exposing behaviour change is closely related to anomaly detection, the identification of deviations from a certain norm (Chandola et al. 2009). In contrast to filtering out noise, in this case the focus of interest is usually placed on the nature of the abnormalities themselves. In the transportation domain, researchers have been interested in detecting anomalies in large collective mobility datasets (cf. Souto and Liebig 2016; Yang and Liu 2011) for urban traffic applications and emergency management. Another line of research considers (geometrical) pattern matching on trajectory data (e.g. Florescu et al. 2012; Du Mouza et al. 2005), for example by building a higher-order Markov model of a user's transitions from one mobile phone cell to another (Sun et al. 2004). The authors encode the individual patterns in a *mobility trie*, which they in turn use to search for anomalies by computing distances between previous and new, potentially anomalous patterns. They explicitly note on the importance of dynamically updating "normal behaviour", and weighting recent patterns higher than ones which occurred longer ago. However, all these approaches are based on a relatively crude assessment of mobility, which either only considers transitions from one region to another, or aggregate data from many users to get a complete view of the current traffic situation. For detecting individual behaviour change over time, however, a method is needed which works with a continuous stream of non-aggregated movement data on an individual level, and tests multiple dimensions of mobility behaviour for anomalies, by comparing them to the user's past behaviour.

3 Method

In this section, we present a system for detecting mobility behaviour change based on a continuous stream of movement data from individual users. The proposed workflow is illustrated in Fig. 1. We assume that a user's raw movement trajectories, recorded via a smartphone application or a similar device, are constantly streamed to a server, and logged in a database. After a certain time period has passed (we propose one week), the data recorded in this interval are fed into a data processing engine, where they pass through four processing steps: first, the trajectories are preprocessed, i.e. filtered, segmented, annotated with the traffic mode, and matched to the road network. Then, the available data for this time period are tested for completeness in order to evaluate their sufficiency for the following analytical processes. If found insufficiently complete, the data are discarded, if rated appropriate, however, they are fed to the next module, which extracts a range of mobility features and mines for patterns. The results are stored in a database, and provide the input for an anomaly detection sub-process, which identifies behaviour change and triggers an appropriate reaction. As can be seen on the far right of Fig. 1, this may involve sending out notifications to the users or analysts, triggering a response (e.g. encouraging or discouraging the observed behaviour change), logging the occurrence of

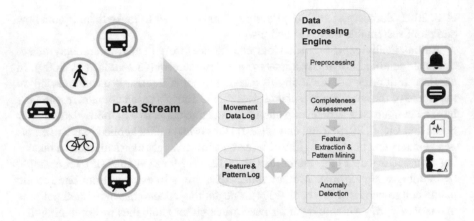

Fig. 1 Workflow

the anomaly, or providing information to an expert for decision support. The exact nature of these system reactions, however, is beyond the scope of this paper. Instead, since our focus is put on the data processing engine, its four sub-modules will be further described in this section.

3.1 Data Preprocessing

As it has been described, movement data are continuously streamed to a server, and logged in a database. In order to evaluate behavioural changes, however, it is necessary to define discrete time intervals (in the following: one week), which will serve as atomic units for later temporal analysis. Thus, after all available data for a full week have been stored in the database, they are fed into the data processing engine (cf. Fig. 1), and further analysed. In a first step, the data need to be preprocessed, which involves the sub-processes noise filtering, stay point detection, segmentation, mode detection, and map matching (Zheng 2015). Please note that whereas exemplary methods for these preprocessing steps are proposed in the following, they could also be replaced by other solutions which are better suited to the respective study aims or data characteristics.

In the beginning, the data are cleaned by removing noisy trackpoints based on a set of filter functions such as a spatial query with a certain study area, or plausibility checks with regards to speed constraints (Zheng 2015). Then, the stay points are detected in the remaining trackpoints, e.g. by means of a clustering technique (Palma et al. 2008). The next preprocessing step detects the traffic mode(s) used, e.g. by computing and analysing various movement descriptors such as the speed or acceleration (Sester et al. 2012). Finally, map matching needs to be performed for all

Fig. 2 The different layers
of movement data
aggregation used in this
study. Note that in contrast to
"home" and "work", the
transition points between
train, *bus* and *tram* are not
considered activities.
Basemap© Mapbox.com

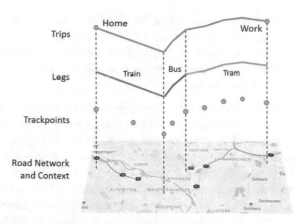

points using one of the available techniques, e.g. evolutionary algorithms (Quddus
and Washington 2015).

After basic preprocessing, it is necessary to structure the movement data into
meaningful units. Inspired by prior approaches (Axhausen and Frick 2005), we propose to distinguish between the following elements: At the most fundamental level,
trajectories (the complete trace of a users movement over a given time frame) are
made up of **trackpoints**. In a first layer of aggregation, trackpoints are grouped into
trip legs based on the used transport mode. Finally, a **trip** consists of one or more
legs, and describes the journey from one 'activity' to another. A **stay point** simply
denotes a location where someone spent longer than a certain time span, and can
qualify as an **activity** if it represents an actual destination of travel (e.g. work, home
or a shop), and not merely a location where a user spent time waiting for a bus or
stuck in a traffic jam. Figure 2 shows an exemplary trip with its constituting elements.

3.2 Data Completeness Assessment

After preprocessing, the available data for the current week are tested for their completeness. As has been discussed in Sect. 2.1, missing trips or other gaps in recording
can have negative effects on downstream analysis processes (e.g. Shen and Stopher
2017; Wolf et al. 2003). In our case, for instance, missing data, if not identified and
filtered previously, might result in misdetections of behaviour changes due to drastically altered values for mobility features. Please note that in this step, we assume the
norm to be continuous tracking over the whole study period, as it is often the case
in related surveys (e.g. Montini et al. 2015; Bucher et al. 2016).

As a first step, we distinguish between different types of recording gaps:

- *Temporal* gaps: the duration with no recorded data between the last recorded time
  stamp of a trip leg or stay point and the first recorded time stamp of the sub-sequent
  trip leg or stay point. The spatial deviance between the position of the last track

point of the former, and the first track point of the latter tripleg or stay point is
smaller than an expected GPS error (e.g. 250 m).

- *Spatio-temporal* gaps: gaps for which the spatial distance between the last track
  point of the former, and the first track point of the latter trip leg or stay point is
  larger than an expected GPS error.

This distinction is motivated by the fact that in the first case, chances are high that
no mobility behaviour has been missed since the user might simply have remained
stationary during the recording gap, whereas in the second case, the user's change
in position proves that movement has certainly taken place but was not recorded.

Both types of gaps can be easily extracted from the database by calculating the
time differences as well as spatial distances between the start and end points of subsequent pairs of trip legs and stay points. The data completeness for the current time
interval can then be evaluated based on two index values:

$$gdur_i = \frac{\sum \Delta g_i}{\Delta t_i}$$

$$gdist_i = \frac{\sum dist(g_i)}{\sum dist(triplegs_i)}$$

where $gdur_i$ is the ratio of the summed durations Δg_i of all temporal and spatiotemporal gaps g_i and the total duration Δt_i of week i. In the second index, $gdist_i$
is the ratio of the summed distances $dist(g_i)$ of all spatio-temporal gaps g_i and the
summed distances $dist(triplegs_i)$ of all trip legs $triplegs_i$ recorded within week i. In
combination, these index values express the temporal extent of recording gaps, as
well as the relative magnitude of missed mobility behaviour. For instance, in a week
in which a user has travelled relatively less compared to others, recording gaps of
similar temporal length can be rated as less critical, since less travelled distance, i.e.
mobility behaviour, might be missing in the data.

3.3   Mobility Feature Extraction and Pattern Mining

After the available data has been confirmed to be of sufficient completeness, selected
mobility features can be extracted. Of course, these will depend to a large degree on
the study aims. As our focus is on sustainability, we compute durations, distances,
speed, and produced CO_2 emissions for each trip leg to serve as basis for computing the indicators listed in Sect. 2.1. Next, in addition to segmenting the movement
trajectories based on their semantics (e.g. trip legs by traffic mode, trips between
activities), as described in Sect. 3.1, we also induce a temporal structure by grouping all movement on a daily basis. Of course, the pre-defined discrete time interval at
which the data is processed (here: one week) provides a further temporal analytical
unit.

Table 1 Units of analysis for deriving mobility features and patterns

Analysis unit	Delimiting factor	Description
Trip leg	Transport mode/vehicle	Mono-modal trip segment between two points without changing mode or vehicle
Trip	Purpose	Trip between two locations for a certain purpose; consists of one or more trip legs
Day	Time	All trips within 24 h; contains one or more complete or incomplete travels (incomplete: beyond temporal delimitation)
Week	Time	All trips within 7 consecutive days

Table 2 Mobility features

Descriptor	Day	Week
Total number of trips		x
Average number of triplegs per trip		x
Total distance travelled		x
Total distance travelled (per trip purpose)		x
Total distance travelled (per traffic mode)		x
Average distance travelled	x	x
Total duration spent travelling		x
Total duration spent travelling (per trip purpose)		x
Total duration spent travelling (per traffic mode)		x
Average duration spent travelling	x	x
Total CO_2 emissions		x
Average travel speed		x
Average travel speed (per traffic mode)		x
Frequently visited places		x

The resulting analytical units for computing mobility features are summarized in Table 1.

For assessing the sustainability of the user's mobility behaviour within the week, we compute a set of indicators (Nicolas et al. 2003; World Business Council 2015) as listed in Table 2.

Whereas the first three indicators can be easily extracted from the preprocessed data, several others require a classification of the stay points and their related trips according to their purpose. Purpose and activity detection can either be achieved by computational methods, e.g. based on visited POI (e.g. Furletti et al. 2013), or by simply asking the users to annotate the data manually in the course of an accompanying PR survey. The total CO_2 emissions produced by travelling depend primarily on the modal split, and can for instance be calculated based on the Mobitool consumption and emission factors (Tuchschmid and Halder 2010), which provide the consumption and emissions of the full life-cycle of a mode of transport per single kilometre travelled in Switzerland.

Finally, although not being directly related to sustainability, the frequently visited places are nevertheless included in the list of mobility features. This is due to the fact that this attribute allows for drastic changes in the personal circumstances to be detected (e.g. moving to a different city). Thus, if other indicators such as the CO_2 emissions change, but the visited places remain unaltered, this could indicate that a user is testing new travel options (e.g. taking the bicycle to work) while her circumstances remain the same. For mining the frequently visited places in a way which allows them to be compared to the results obtained for previous weeks, we choose a clustering approach. Using the DBSCAN algorithm (Sander et al. 1998), we cluster all activities found during the week. Due to the fact that although a user might have visited the same place as in the week before, the recorded activities and their associated point geometries will not correspond spatially, we choose an alternative approach and compute a minimum bounding geometry of the points based on their cluster membership. In order to avoid creating multiple instances of the "same" place in the database, the resulting polygon is tested for overlaps with already existing places in the database. If an overlap is found, no new place instance is created, but rather the id of the overlapping place in the database is extracted and stored in a list of frequently visited places for the current week. If no overlap with already existing places is detected, a new place instance is created in the database, and a new id is assigned. Figure 3 shows an example of activities and the overlapping cluster geometries from different weeks for one user. Since they all overlap, only the first occurrence would be created as an instance and assigned an id. For all the other clusters, only the information that the place has been visited frequently enough to be detected as a cluster would be stored together with its id. After computation, the results for all indicators are stored in a database (see Fig. 1).

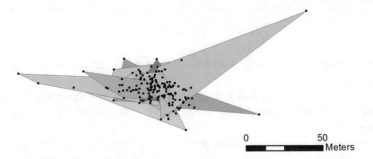

Fig. 3 The activities are shown on top of the overlapping minimum bounding polygons, as derived from the point clusters at different weeks

3.4 Anomaly Detection

At the present stage of the workflow, mobility features and patterns have been detected and stored for the current week. Now, it is possible to load similar data computed for the previous weeks from the database, and assess potential anomalies in mobility behaviour (see Fig. 1). Numerous algorithms available for anomaly detection simply classify individual data points (in our case, aggregations of all mobility features for the current week) as anomalous or normal, without allowing further insight into which feature exactly caused the data point to be classified as anomalous (cf. Chandola et al. 2009). This knowledge, however, is critical for our purposes since merely knowing that an anomaly occurred is not sufficient, but rather the results should allow deeper interpretation of the detected behaviour change. Thus, to decide which system action should be triggered as a reaction, it is critical to explicitly identify the mobility features which have changed, i.e. were detected as anomalous. For instance, an increase in bicycling distance could trigger encouraging feedback, whereas an increase in CO_2 production could lead to a discouraging response. There is work on explaining anomalies in more detail after their detection (e.g. Pevný and Kopp 2014), which could therefore be used in combination with any anomaly detection algorithm. For our purpose, we found this unnecessary and rather detect anomalies for each feature individually.

For each mobility feature f_i (except the *frequently visited places*, which will be explained separately) we compute the mean μ_i and standard deviation σ_i of the n weeks preceding the week currently under investigation, where n is a tunable window size (set to 5 weeks in our tests). Comparing the values computed for the current week, it is now possible to assess if an existing deviation should be considered a normal fluctuation or an anomaly. This is controlled by another parameter λ, i.e. a feature f_i is considered anomalous if $|f_i - \mu_i| > \lambda \cdot \sigma_i$. Accordingly, if the feature re-centred around zero has a deviation larger than what can be expected given previous feature values, it is treated as anomalous. We found a value of $\lambda = 3$ to yield reasonable results.

To compute if a set of *frequently visited places* within a week should be considered anomalous, a similar approach is applied. We encode the presence of a certain place in a given week with a 1, and its absence with a 0. For every place, this results in a list of binary digits, e.g. the sequence $(0, 0, 1, 0, 1)$ encodes a place being visited in weeks 3 and 5, but not in any other week. Using this numerical representation, we can compute if the appearance of an individual place in any week should be considered anomalous or not by using a similar technique as above. However, as this results in every place being an additional mobility feature (which results in frequent cases with large number of anomalous features), we sum the number of anomalous places in every week, and perform another anomaly detection process on the resulting values. For example, a person frequently travelling for work purposes will constantly yield high numbers of anomalous places (i.e. first time visits at new places), a fact which is not particularly useful in terms of behaviour change detection. If, however, this number drops suddenly, and the visited places show a more regular pattern, it signals a behavioural change (which could be due to holidays, a job change, etc.). Summarizing anomalies in the frequently visited places as described allows us to handle them as a single mobility feature, and to report their anomalies for further interpretation by an automated system or an expert.

4   Case Study

We implemented the described method as a Python application (using a PostgreSQL database with the PostGIS extension for all spatial operations), and evaluated it on a large dataset collected over a period of three months, from approximately middle of December 2016–March 2017. 139 people used a smartphone tracking app, which passively recorded all their journeys, inferred a transport mode, and allowed them to change it in case the proposed one was wrong. The dataset consists of 52'370'797 trackpoints, which are divided into 125'759 trip legs and 71'099 trips.

Using these data, we simulated a continuous data stream by feeding data for each week subsequently into the data processing engine. Below, the results for our mobility behaviour change detection process are provided for two exemplary users. Figures 4 and 5 show the detected anomalies for these users per week. The blue dots indicate the number of anomalies for each week, while the yellow dots show the number of anomalies with regards to frequently visited places. Please note that this does not correspond to the total number of places visited by a user, but only to those that were unexpectedly visited or skipped in the respective week. Not surprisingly, the place-related anomalies are relatively more frequent in the first weeks, which is due to the cold start problem, i.e., sparse data making it difficult to assess whether a frequently visited place should represent an anomaly. Weeks which are missing values were filtered out previously, due to insufficient data completeness. For this,

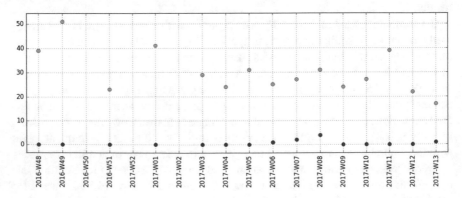

Fig. 4 All (blue) and only place-related (yellow) anomalies for user A of our test sample. In weeks 2016-50, 2016-52, and 2017-02, the data completeness was found insufficient to reliably assess mobility behaviour patterns

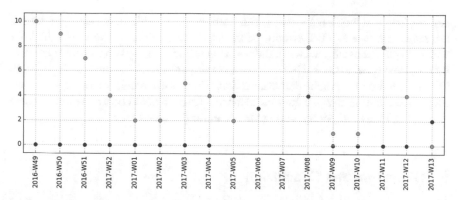

Fig. 5 All (blue) and only place-related (yellow) anomalies for user B of our test sample. In week 2017-07, the data completeness was found insufficient to reliably assess mobility behaviour patterns

we defined threshold values so that data for weeks were only further analysed if their $gdur_i \leq 0.25$ and $gdist_i \leq 0.25$.

The mobility behaviour of user A, whose anomalies are shown in Fig. 4, remains rather constant up until calender week 2017-06, where several anomalies are detected. Whereas in that week, only the average walking speed is noticeably higher compared to preceding weeks, in the following week 2017-07 we detect an increase in the distance ($\mu_d = 7.0$ km $\rightarrow f_d = 33.2$ km) and duration ($\mu_t = 19$ min $\rightarrow f_t = 1$ h 41 min) of travels made by bus. In week 2017-08, one can observe an additional increase in distance and duration of both walking (18.6 km $\rightarrow$ 58.4 km; 2 h 38 min $\rightarrow$ 9 h 43 min) and bicycling (1.9 km $\rightarrow$ 31.2 km; 5 min $\rightarrow$ 1 h 37 min). Due to the fact that in contrast to these anomalies, the frequently visited places still remain largely unchanged compared to the weeks before, we can conclude that this user indeed changed her mobility behaviour by increasingly using slow mobility (walking and bicycling) and public transport. An automated feedback system as described previously could now trigger reinforcing measures for this behaviour, e.g., by providing incentives, and thus assisting the user to transition to a phase where this new mobility behaviour is internalized and does not require further motivation.

The results for user B are shown in Fig. 5. Here, changes in mobility behaviour can be observed between weeks 2017-05 and 2017-08, which in this case, however, originate from increases in the totally travelled distance (e.g., 690 km $\rightarrow$ 1'836 km), the average speed (41.4 km/h $\rightarrow$ 97.1 km/h) the distance covered by car (307 km $\rightarrow$ 1'091 km), bike (1.1 km $\rightarrow$ 14.5 km) and walking (12.3 km $\rightarrow$ 34.1 km), as well as the related durations (plus the duration spent travelling by tram in week 2017-06). Based on the observation of such a general increase in mobility activities (not just one specific mode of transport), and set in combination with the occurrence of several place-related anomalies in weeks 2017-06 and 2017-08, one can interpret this pattern as an exceptional change of behaviour likely caused by altered personal circumstances, e.g., a holiday or business trip, rather than a gradual change of new habit formation. Indeed, when analysing the movement data for this user in more detail, we found several long distance car journeys with destinations outside of Switzerland during the respective weeks. Furthermore, in the user's home Kanton, the weeks 2017-07 and 2017-08 are usually winter holidays. This would also explain the observed data incompleteness in week 2017-07, since the smartphone tracking method deployed in this study relies on a mobile data connection, which is often unavailable when travelling abroad. In this case, an automated system reaction could be to rate the detected changes as likely temporary, and ignore them for the time being.

5 Discussion and Conclusion

In this study, we proposed a framework for continuously mining streams of movement trajectory data of users for detecting mobility behaviour changes. As it has been discussed, after data preprocessing, the completeness of the available movement recordings needs to be assessed in order to avoid misdetections of behavioural anomalies in the later steps of the analysis process. For this purpose, we presented a solution for quantifying recording gaps, hereby distinguishing between purely temporal and spatio-temporal gaps. Furthermore, we calculated a list of mobility features to serve as sustainability indicators, and proposed a method to compute and evaluate frequently visited places. Finally, the anomaly detection process was described which yields detailed results with regards to the exact mobility feature causing the anomaly occurrence. By applying the framework to a simulated stream based on a pre-recorded large-scale trajectory dataset, and evaluating the plausibility of the results obtained for two exemplary users, we could demonstrate its functionality and practical value.

In our view, this work provides a first step towards the development of personalized, automated mobility support systems which provide adaptive intervention strategies for gradually changing people's mobility behaviour towards a higher sustainability. The proposed framework, however, is not restricted to this application domain, but could be applied for other purposes as well, e.g. for general monitoring of mobility behaviour and computing descriptive statistics, or for detecting anomalies in the movements of animals or even automated vehicles or drones. A practical advantage of our approach worth mentioning is the fact that whereas the derived mobility feature values are stored for every week (feature and pattern log in Fig. 1), the actual movement data (movement data log in Fig. 1) can be deleted immediately after processing. This not only reduces the resources necessary for data storage, but also addresses privacy concerns, since the most sensitive data are deleted regularly.

There are, however, still some limitations to our approach. Thus, although the most sensitive movement data can be deleted after analysis, there still remain concerns with regards to location privacy. With mobile devices constantly gaining in computation and storage capabilities, however, a potential solution could be to shift critical parts of the analytical process to the client, and simply transmit the computed index values to the server for anomaly detection. Moreover, the list of used sustainability indicators is not exhaustive, and more complex values, e.g. incorporating car occupancy, would increase the realism with which sustainability is quantified in our study. These restrictions, however, largely depend on the quality and level of detail of the available data. Furthermore, in the exemplary application of our system, we could clearly observe problems for the first iterations due to the cold start problem, which is a usual challenge for user profiling and sequence mining applications. The usefulness of our system would therefore be reduced to a certain degree in the first phase of application. In addition, it would certainly be worthwhile to include more detailed mobility features, e.g. the usual times of travel, distinguish between the weekend and working days, or incorporate contextual information (e.g. the weather) for better results. However, special care needs to be taken for correlating features (e.g., distance and duration), as they would be flagged as anomalous in the same weeks, thus leading to a wrong assessment of behaviour change. At the same time, it can be expected that an increase in the number of features could complicate their semantic interpretation. Decision support, e.g. in the form of automated feature classification could therefore be worthwhile. Finally, due to the fact that at the current stage of this study, we have no access to ground truth data with regards to the behavioural anomalies (e.g. in the form of user interviews), a systematic evaluation of the proposed method must be regarded as future work.

Apart from testing and evaluating the model with a subset of users who can provide additional information with regards to their mobility behaviour, it is planned to refine the list of mobility features and develop a prototype of an expert system capable of interpreting the detected behavioural changes. It would also be interesting to apply a semantic perspective to the interpretation of place-related anomalies, e.g. by incorporating POI from additional data sources to assess the type of places visited.

Acknowledgements This research was supported by the Swiss National Science Foundation (SNF) within NRP 71 "Managing energy consumption" and by the Commission for Technology and Innovation (CTI) within the Swiss Competence Center for Energy Research (SCCER) Mobility.

References

Abou-Zeid M, Witter R, Bierlaire M, Kaufmann V, Ben-Akiva M (2012) Happiness and travel mode switching: findings from a Swiss public transportation experiment. Transport Policy 19(1):93–104

Andrienko G, Andrienko N, Fuchs G (2016) Understanding movement data quality. J Loc Based Serv 10(1):31–46

Axhausen KW, Frick M (2005) Nutzungen—Strukturen—Verkehr

Bamberg S (2006) Is a residential relocation a good opportunity to change peoples travel behavior? results from a theory-driven intervention study. Env Behav 38(6):820–840

Bamberg S, Rölle D, Weber C (2003) Does habitual car use not lead to more resistance to change of travel mode? Transportation 30(1):97–108

Banister D (2008) The sustainable mobility paradigm. Transport policy 15(2):73–80

Ben-Elia E, Ettema D (2011) Changing commuters behavior using rewards: a study of rush-hour avoidance. Trans Res Part F Traffic Psychol Behav

Boulouchos K, Cellina F, Ciari F, Ciari F, Cox B, Georges G, Hirschberg S, Hoppe M, Jonietz D, Kannan R, Kovacz N, Küng L, Michl T, Raubal M, Rudel R, Schenler W (2017) Towards an energy efficient and climate compatible future swiss transportation system. SCCER mobility working paper

Brunauer R, Rehrl K (2016) Big data in der mobilität–FCD modellregion salzburg. In: Big Data, pp 235–267. Springer

Bucher D, Cellina F, Mangili F, Raubal M, Rudel R, Rizzoli RE, Elabed O (2016) Exploiting fitness apps for sustainable mobility-challenges deploying the Goeco! app. ICT for sustainability (ICT4S)

Bundesamt fuer Umwelt (BAFU), Treibhausgasemissionen der Schweiz 1990–2014

Chandola V, Banerjee A, Kumar V (2009) Anomaly detection: a survey. ACM Comput Surv (CSUR) 41(3):15

Du Mouza C, Rigaux P, Scholl M (2005) Efficient evaluation of parameterized pattern queries. In: Proceedings of the 14th ACM international conference on information and knowledge management, pp 728–735. ACM

Feng Z, Zhu Y (2016) A survey on trajectory data mining: techniques and applications. IEEE Access 4:2056–2067

Florescu S, Körner C, Mock M, May M (2012) Efficient mobility pattern stream matching on mobile devices. In: Proceedings of the ubiquitous data mining workshop (UDM 2012), pp 23–27

Froehlich J, Dillahunt T, Klasnja P, Mankoff J, Consolvo S, Harrison B, Landay JA (2009) Ubigreen: investigating a mobile tool for tracking and supporting green transportation habits. In: Proceedings of the sigchi conference on human factors in computing systems, pp 1043–1052. ACM

Furletti B, Cintia P, Renso C, Spinsanti L (2013) Inferring human activities from GPS tracks. In: Proceedings of the 2nd ACM SIGKDD international workshop on urban computing—13. Association for Computing Machinery (ACM)

Gonzalez MC, Hidalgo CA, Barabasi A-L (2008) Understanding individual human mobility patterns. Nature 453(7196):779–782

Hamari J, Koivisto J, Pakkanen T (2014) Do persuasive technologies persuade?-a review of empirical studies. In: International conference on persuasive technology, pp 118–136. Springer

Hanson S, Huff OJ (1988) Systematic variability in repetitious travel. Transportation 15(1):111–135

Hecker D, Stange H, Korner C, May M (2010) Sample bias due to missing data in mobility surveys. In: 2010 IEEE International conference on data mining workshops, Dec, pp 241–248

Kohla B, Meschik M (2013) Comparing trip diaries with gps tracking: Results of a comprehensive austrian study. In: Transport survey methods: best practice for decision making, pp 305–320. Emerald Group Publishing Limited

Li Q, Zheng Y, Xie X, Chen Y, Liu W, Ma W-Y (2008) Mining user similarity based on location history. In: Proceedings of the 16th ACM SIGSPATIAL international conference on advances in geographic information systems, p 34. ACM

Montini L, Prost S, Schrammel J, Rieser-Schüssler N, Axhausen KW (2015) Comparison of travel diaries generated from smartphone data and dedicated GPS devices. Trans Res Proc 11:227–241

Nicolas J-P, Pochet P, Poimboeuf H (2003) Towards sustainable mobility indicators: application to the lyons conurbation. Transport Policy 10(3):197–208

Palma AT, Bogorny V, Kuijpers B, Alvares LO (2008) A clustering-based approach for discovering interesting places in trajectories. In: Proceedings of the 2008 ACM symposium on Applied computing, pp 863–868. ACM

Pevnỳ T, Kopp M (2014) Explaining anomalies with sapling random forests. In: Information technologies—applications and theory workshops, posters, and tutorials (ITAT 2014)

Polak J, Han X (1997) Iterative imputation based methods for unit and item non-response in travel surveys. In: 8th meeting of the international association of travel behaviour research. Austin, Texas

Prelipcean AC, Gidofalvi G, Susilo YO (2015) Comparative framework for activity-travel diary collection systems. In: 2015 International conference on, models and technologies for intelligent transportation systems (MT-ITS), pp. 251–258. IEEE

Prochaska JO, Velicer WF (1997) The transtheoretical model of health behavior change. Am J Health Promotion 12(1):38–48

Quddus M, Washington S (2015) Shortest path and vehicle trajectory aided map-matching for low frequency gps data. Trans Res Part C Em Technol 55:328–339

Sander J, Ester M, Kriegel H-P, Xu X (1998) Density-based clustering in spatial databases: the algorithm gdbscan and its applications. Data Mining Knowl Discovery 2(2):169–194

Schade J, Schlag B (2003) Acceptability of urban transport pricing strategies. Trans Res Part F Traffic Psychol Behav 6(1):45–61

Schlich R, Axhausen KW (2003) Habitual travel behaviour: evidence from a six-week travel diary. Transportation 30(1):13–36

Schüssler N (2008) Processing GPS raw data without additional information

Sester M, Feuerhake U, Kuntzsch C, Zhang L (2012) Revealing underlying structure and behaviour from movement data. KI—Künstliche Intelligenz 26(3):223–231

Shen L, Stopher PR (2017) Review of GPS travel survey and GPS data- processing methods. Trans Rev 1–19

Siła-Nowicka K, Vandrol J, Oshan T, Long JA, Demšar U, Fotheringham AS (2015) Analysis of human mobility patterns from GPS trajectories and contextual information. Int J Geograph Inf Sci 30(5):881–906

Song C, Qu Z, Blumm N, Barabási A-L (2010) Limits of predictability in human mobility. Science 327(5968):1018–1021

Souto G, Liebig T (2016) On event detection from spatial time series for urban traffic applications. In: Solving large scale learning tasks. Challenges and algorithms, pp 221–233. Springer

Stenneth L, Wolfson O, Yu PS, Xu B (2011) Transportation mode detection using mobile phones and GIS information. In: Proceedings of the 19th ACM SIGSPATIAL international conference on advances in geographic information systems—GIS 11. Association for Computing Machinery (ACM)

Stopher PR, Moutou CJ, Liu W (2013) Sustainability of voluntary travel behaviour change initiatives: a 5-year study

Sun B, Yu F, Wu K, Leung V (2004) Mobility-based anomaly detection in cellular mobile networks. In: Proceedings of the 3rd ACM workshop on Wireless security, pp 61–69. ACM

Taaffe EJ (1996) Geography of transportation. Morton O'Kelly, New Jersey, USA

Taniguchi A, Hara F, Takano S, Kagaya S, Fujii S (2003) Psychological and behavioral effects of travel feedback program for travel behavior modification. Trans Res Record J Trans Res Board 1839:182–190

Tuchschmid M, Halder M (2010) Mobitool-grundlagenbericht: Hintergrund, methodik & emissionsfaktoren. Tuchschmid und M, Halder im Auftrag von SBB, Swisscom, BKW und ÖBU

Tulusan J, Steggers H, Fleisch E, Staake T (2012) Supporting eco-driving with eco-feedback technologies: recommendations targeted at improving corporate car drivers' intrinsic motivation to drive more sustainable. In: 14th ACM international conference on ubiquitous computing (UbiComp), p 18. Ubicomp

Wagner DP (1997) Lexington area travel data collection test: GPS for personal travel surveys. Final report, office of highway policy information and office of technology applications. Federal Highway Administration, Battelle Transport Division, Columbus, pp 1–92

Weiser P, Bucher D, Cellina F, De Luca V (2015) A taxonomy of motivational affordances for meaningful gamified and persuasive technologies. In: Proceedings of the 3rd international conference on ICT for sustainability (ICT4S), ser. Adv Comput Sci Res 22, pp 271–280. Atlantis Press, Paris

White CE, Bernstein D, Kornhauser AL (2000) Some map matching algorithms for personal navigation assistants. Trans Res Part C Em Technol 8(1):91–108

Wolf J, Loechl M, Thompson M, Arce C (2003) Trip rate analysis in GPS-enhanced personal travel surveys. In: Transport survey quality and innovation. Emerald Group Publishing Limited, pp 483–498

World Business Council for Sustainable Development (WBCSD) (2015) Methodology and indicator calculation method for sustainable urban mobility. WBCSD, Geneva, Switzerland

Yang S, Liu W (2011) Anomaly detection on collective moving patterns: a hidden Markov model based solution. In: Internet of things (iThings/CPSCom), 2011 international conference on and 4th international conference on cyber, physical and social computing, pp 291–296. IEEE

Zheng VW, Cao B, Zheng Y, Xie X, Yang Q (2010) Collaborative filtering meets mobile recommendation: a user-centered approach. In: Proceedings of the twenty-fourth AAAI conference on artificial intelligence, ser. AAAI'10, pp 236–241. AAAI Press

Zheng Y (2015) Trajectory data mining. TIST 6(3):1–41

Zheng Y, Chen Y, Li Q, Xie X, Ma W-Y (2010) Understanding transportation modes based on GPS data for web applications. ACM Trans Web 4(1):1:1–1:36

An Overall Framework for Personalised Landmark Selection

Eva Nuhn and Sabine Timpf

Abstract This paper proposes a multidimensional model for the selection of personalized landmarks. The model is based on an existing landmark salience model, which was designed to be open to adaptations regarding individual user preferences. The conventional model is based solely on landmark dimensions (i.e. visual, semantic and structural dimension). We add an additional personal dimension to account for different familiarities and interests. Further, we add an environmental dimension to accommodate different routing situations and a descriptive dimension to consider the brevity of a landmark description. In this paper we identify the attributes of the dimensions of the multidimensional model and investigate methods for calculating the salience of the attributes. The applicability and usefulness of the (still evolving) model is shown with three different case studies.

1 Introduction

Awareness has been increasing that people with different backgrounds and preferences prefer different landmarks (Hamburger and Röser 2014; Quesnot and Roche 2015). The latter study showed that people familiar with an environment clearly preferred local semantic landmarks, while people unfamiliar with an environment preferred landmarks with salient visual and structural characteristics. It is also known that the level of interest can enhance memory for some information (McGillivray et al. 2015). Obviously, it is a challenging task to find the best landmark based on spatial knowledge and individual interests of a traveler.

The term landmark exhibits many different meanings. The most fundamental one is that of an object or structure that serves as external point of reference (Lynch 1960). Thus, a landmark has an outstanding visual characteristic, a unique importance

E. Nuhn (✉) · S. Timpf
Geoinformatics Group, University of Augsburg, Alter Postweg 118, 86159
Augsburg, Germany
e-mail: eva.nuhn@geo.uni-augsburg.de

S. Timpf
e-mail: sabine.timpf@geo.uni-augsburg.de

© Springer International Publishing AG 2018
P. Kiefer et al. (eds.), *Progress in Location Based Services 2018*, Lecture Notes
in Geoinformation and Cartography, https://doi.org/10.1007/978-3-319-71470-7_12

or meaning or is in a central location (Sorrows and Hirtle 1999). Landmark salience is additionally affected by the perspective of the observer, the surrounding environment and the objects contained therein (Caduff and Timpf 2008). For the purpose of this study, we are defining the term landmark as "any outstanding urban structure". We do not restrict our work to buildings and treat also other urban structures (e.g. water wheels, information panels or dust bins). We focus on three-dimensional local landmarks for pedestrians at decision points.

There is a large number of possible landmarks, which can be included in route instructions in different situations and for different travelers. Different travelers would find different landmarks to be most useful in a given situation (Götze and Boye 2016). Humans choose landmarks based on several criteria, such as the mode of travel, the desired route characteristics (Lovelace et al. 1999) but also using personal dimensions. Several studies have proposed landmark salience models. These models are either typically *landmark identification* or *landmark integration* models (Richter and Winter 2014). *Landmark identification* models are based on landmark dimensions and identify landmarks' salience based on the well-established visual, semantic and structural dimensions by Sorrows and Hirtle (1999). The degree to which each of these dimensions influences the total measure of landmark salience is determined using weights for each dimension. How these weights should be chosen to adapt to the mode of travel or individual user preferences has not yet been studied extensively. *Landmark integration* models by contrast are based on environmental dimensions. They detect route-dependent landmarks according to attributes such as uniqueness in a given environment, position along a route or visibility from the route. None of the models investigated so far include personal preferences or knowledge that influence the process of landmark integration.

The contribution of this paper is a multidimensional model that helps to select personalized landmarks. The goal is to extend an existing landmark salience model by including a so called personal dimension of landmarks. Specifically, we take the existing landmark salience model by Raubal and Winter (2002) and add personal attributes. Furthermore, we add an environmental dimension to account for different routing situations and a descriptive dimension to consider the brevity of a landmark description. The result of the model is a measure of the personal landmark salience of a landmark candidate for a specific person. The measure can then be integrated in the generation of a route (Nuhn and Timpf 2016). This paper tackles the challenges of designing such a multidimensional model, while the integration of the results in routing algorithms is treated elsewhere.

Section 2 gives an overview of related work, focusing on existing landmark salience models based on landmark, descriptive, environmental as well as personal dimensions. Section 3 introduces the multidimensional model for personalized landmarks. In Sect. 4 we present example case studies to demonstrate the proposed model. The final section concludes and identifies future work.

2 Related Work

In this section existing work regarding landmark salience models based on landmark, descriptive, environmental and personal dimensions is reviewed.

Landmark Dimensions

In landmark research *landmark identification* is done considering several dimensions of landmarks. A classification was presented by Sorrows and Hirtle (1999) and modified by Raubal and Winter (2002). The framework defines three landmark dimensions: the visual, the semantic and the structural dimension. There are many approaches based on these landmark dimensions to assess the salience of objects for route instructions. One very fundamental approach was proposed by Raubal and Winter (2002). They suggested measures to formally specify the salience of buildings (see Sect. 3.3.1). Nothegger et al. (2004) further extended and implemented the approach on façades and showed that the model is applicable to assessing landmark salience. Elias (2003) was the first to propose data mining methods: She used existing spatial databases instead of manual collection methods and thus focused on buildings as landmark candidates.

There are other studies addressing the lack of available data sources. Newer approaches are based on VGI (Volunteered Geographic Information) and crowdsourcing initiatives. For example Kattenbeck (2016) proposed an empirically validated model and approach for a survey-based assessment of object salience. The model incorporates the results of prior studies on features that are important for salience. After testing the model with a large-scale in-situ experiment it turned out that route related features as well as visual aspects are the most important influences for the prediction of the overall salience of a feature. Another approach used Open Street Map (OSM) data as source and implemented tagging OSM objects as potential landmarks (Wolfensberger and Richter 2015). They implemented a mobile application, which enables user-generated collection of landmarks. Other approaches used OSM data to automatically identify landmarks. Nuhn et al. (2012) proposed a landmark index based on attributes of the landmark dimension to automatically extract landmarks from OSM. These approaches can be used to provide methods for real-world crowd sourcing scenarios, which are important for mobile pedestrian navigation systems. However, all approaches have in common that they only consider attributes contributing to the landmark dimension.

Descriptive Dimensions

The brevity of a landmark description relates to the number of words or terms needed to refer to it in route instructions (Burnett et al. 2001). The description of a landmark should be as precise as possible. According to Burnett et al. (2001) a good landmark requires a minimum of additional information to be usable in route instructions. A detailed description of an object can prevent confusion with other objects but the complexity of the description should be minimized to reduce the cognitive load (Elias 2003). Too much information has an adverse effect on efficient wayfinding (Schneider and Taylor 1999). Objects with lengthy descriptions require the wayfinder to process several different information elements (Burnett et al. 2001). The length of the landmark description can vary depending on the perspective and the familiarity of the traveler with the environment. A description can be coarse such as "the church", but it can also be refined in various ways (e.g. "the church with the red façade and the two steeples") (Tenbrink and Winter 2009).

Environmental Dimensions

Approaches that focus on environmental dimensions are known as *landmark integration* approaches (Richter and Winter 2014). Here the focus is on environmental attributes (e.g. distance to the decision point, visibility from the route or uniqueness in the neighborhood of the route). The *advance visibility* of an object informs if the object can be clearly seen from the route in all conditions (Burnett et al. 2001). Winter (2003) introduced advance visibility into the basic model of Raubal and Winter (2002). He investigated the identifiability of an object along the route, taking into account that a geographic feature that is visible early on along a route is more suitable as a landmark than a feature that is spotted at the very last moment. Klippel and Winter (2005) also integrated landmarks in route instructions with regard to a specific route. Besides advance visibility they considered the configuration of the street network as well as the route along the network. An approach to integrate landmark information directly into the routing algorithm was proposed by Elias and Sester (2006) using a modified Dijkstra algorithm to calculate an optimal route based on landmark quality. Weights were adapted according to the permanence, visibility, usefulness of location, uniqueness and brevity of the landmark description. In a similar fashion the Landmark-Spider-Algorithm from Caduff and Timpf (2005) calculates the clearest route in terms of spatial references and uses selected landmarks to give route instructions. The model selects landmarks based on distance and orientation of the traveler with respect to the landmark and salience of the objects. Another approach which uses types of landmarks tackles the incorporation of landmarks in computer-generated route instructions (Duckham et al. 2010). Here a weighting system is proposed that is based on expected average properties of the types of landmarks (e.g. ubiquity, length of description, permanence…). Those objects are determined that are best suited to describe how to follow a given route.

Personal Dimensions

The landmark salience of an object is not only dependent on landmark or environmental attributes but also on personal dimensions. Different travelers would find different landmarks to be the most useful ones in a given situation (Götze and Boye 2016). The landmarkness of an object is dependent on mobility, gender, age, education or hometown of the traveler (Winter et al. 2012). There is only little work that deals with the idea that salience is not the same for every person. Burnett et al. (2001) were the first who showed that travelers familiar with an environment choose other landmarks than people unfamiliar with an environment. More recent studies confirmed their findings and showed that familiar buildings are more easily recognized than unfamiliar ones (Hamburger and Röser 2014). Based on these results Quesnot and Roche (2015) assumed that travelers who know the area by heart prefer different landmarks than travelers unfamiliar with an environment. They confirmed this assumption and showed that persons that are familiar with a specific environment prefer landmarks with personal significance.

Balaban et al. (2014) showed that emotions may have an influence on landmark selection as well. They showed that negatively laden landmarks are remembered better than positively laden or neutral ones. In addition, Palmiero and Piccardi (2017) showed that both positive and negative emotional landmarks equally enhance the ability to learn a path, and thus influence the acquisition of spatial knowledge. Furthermore, they found that positive emotional landmarks improved the reproduction of a path on the map as compared to negative or neutral emotional landmarks. The investigation of emotions and landmarks is also a personalization, which however neglects other personal dimensions. Götze and Boye (2016) model every landmark that a person refers to in route instructions as a vector of features. Then an individual mathematical model of salience is derived for every person. Currently this approach is restricted to landmark dimensions, since the feature vectors only include spatial attributes (distance and angle to a landmark as well as name and type extracted from OSM data).

An approach to adapt the model by Raubal and Winter (2002) to different contexts was proposed by Winter et al. (2005) by modeling the weights of the salience measures. In addition they investigated the proposed method in a thorough human subject test and found evidence that the variation of the context changes the selection of the landmarks. However, their work focused on weights based on different contexts (here, the time of the day). Apart from gender differences in weighting landmarks by day and by night no other attributes of the personal dimension were investigated. Although the familiarity with the environment was collected from test persons on a simple binary scale, this attribute was not further evaluated. In our work we focus on additional fundamental attributes of the personal dimension to provide help for the automatic selection of personalized landmarks.

3 The Multidimensional Model for Personalized Landmarks

In this section we introduce our multidimensional model for personalized landmarks. In a first step the dimensions of the multidimensional model are discussed. Then the saliences of all attributes are calculated. In a final step the overall salience of a landmark is calculated using the model from Raubal and Winter (2002) and compared to our extended multidimensional model.

3.1 Dimensions of the Multidimensional Model

In this section we identify, investigate and discuss the attributes of the dimensions of the multidimensional model.

3.1.1 Landmark Dimensions

We follow the preceding definitions of Sorrows and Hirtle (1999) and Raubal and Winter (2002) for the landmark dimensions.

Visual Dimension

Our multidimensional model includes four attributes for the visual dimension. One of them is the *surface structure*. Buildings are visually salient if they have e.g., bay windows or balconies. Other objects are salient if they are not shaped uniformly (e.g. a water wheel with its blades is salient). An object with a differently shaped roof than all the others within an environment (e.g. a street light with a peaked "roof") has a salient *surface area*, which is another visual attribute. An object can also be outstanding because of the visual attribute *height* (e.g. a city gate is higher than all the other objects around it). Another attribute of the visual dimension is *color*. For example, multicolored recycling bins in a street with houses with no outstanding coloring can attract the traveler's attention.

Semantic Dimension

We calculate semantic salience by taking into account the *cultural* and *historical importance*. Culturally important objects are for example museums, sports centers or cinemas. Objects with historical importance are city walls or historic buildings. In addition we investigate if *explicit marks* are available, because objects showing explicit marks specify their semantics to the traveler (Raubal and Winter 2002) and are therefore easy to identify.

Structural Dimension

Following Raubal and Winter (2002) we focus on local landmarks for wayfinding, thus we include only local structural elements. The *number of adjacent routes* gives information if the object is located at a street intersection. Such objects are more important for route instructions than objects not connected to a street intersection. The *number of adjacent objects* shows if the object is freestanding or not. Freestanding objects (e.g. a city light) are more salient than objects that are part of an assembly (e.g. terraced houses).

3.1.2 Descriptive Dimension

The descriptive dimension has not been considered in the work of Raubal and Winter (2002). We propose to use *explicit marks* and *number of words* as attributes. An object with an *explicit mark* can be explicitly named within route instructions. Further, the traveler can easily identify the intended object. Thus, an explicit mark is very valuable and can be directly used in route instructions. Furthermore, the *number of words* is an important attribute for the descriptive dimension. It can be assumed that the reference to a "long elongated blue building" needs more working memory than the simple reference to the "casino" (Schneider and Taylor 1999).

3.1.3 Environmental Dimension

There are several studies (see Sect. 2) proposing several environmental attributes. Based on that our multidimensional model includes *advance visibility, orientation, distance* and *uniqueness. Advance visibility* for a person approaching a decision point is a cognitively relevant factor for the determination of landmarks (Winter 2003). To consider the *orientation* of an object to the traveler the geographical space is divided into sections (i.e. in front, beside and behind). The sections are dependent on the traveler's heading which corresponds to the orientation of the route segment leading to the decision point. Objects close to a decision point are useful for navigation purposes (Waller et al. 2000). Thus, we consider the *distance* to the decision point as attribute of the environmental dimension. Landmarks which are not *unique* can be mistaken with other objects within the environment. Therefore we investigate the neighboring street intersections if there are similar misleading objects.

3.1.4 Personal Dimensions

In a former work we identified personal dimensions to consider in determining personalized landmarks (Nuhn and Timpf 2017). Based on that we include the personal dimensions *prior spatial knowledge, personal interests* and *personal background* in our multidimensional model.

Prior Spatial Knowledge
The prior spatial knowledge of a traveler seems to be the most important dimension to consider. It is commonly divided into three distinct types: landmark knowledge, route knowledge and survey knowledge (Siegel and White 1975). In Nuhn and Timpf (2017) we introduced four attributes to consider the prior spatial knowledge of a traveler: no knowledge, landmark knowledge, route knowledge and survey knowledge. While traveling through the environment people notice various objects and encode images in a database. Thus, people are able to recall the objects they have seen and to remember the names of certain buildings and locations (Thorndyke 1980). These *landmark knowledge* landmarks can be used within route instructions in order to link already known elements with new ones along the route (Nuhn and Timpf 2017). *Route knowledge* is gained when a traveler is exposed to a route. This also includes the knowledge of objects along the route. These objects can be divided in two groups: objects that were part of previous route instructions and objects that were not yet used for navigating. Route instructions, for a route segment part of route knowledge, can be coarser, i.e., merely enriched with additional landmarks (Tenbrink and Winter 2009). *Survey knowledge* is defined as the result of the mental integration of two or more routes (Herrmann et al. 1998). This is in contrast to route knowledge, which is related to a single route. If the traveler has never been to the environment and has never seen a map or photos then he has *no prior spatial knowledge* at all.

Personal Interests

Travelers must look around in order to perceive things. But looking by itself is not enough (Rensink et al. 1997). A traveler whose mind wanders during route following may often miss important objects, even if these are highly salient. The key factor for perceiving things is attention, which is dependent on the degree of interest (Rensink et al. 1997). Banerjee et al. (2015) confirmed that the voluntary focus of attention on environmental inputs is influenced by an observer's level of interest in an object. There are two types of interests: individual and situational interest (Hidi and Renninger 2006). Individual interests refer to an ongoing relation of a person to a particular content (Hidi and Renninger 2006). Situational interest describes interest that is caused by certain conditions and/or concrete features in the environment (Renniger and Su 2012). In this work we consider individual interests. It represents personality-specific orientation and provide important categories for action goals in a situation where a person is free to do as one pleases (Krapp et al. 2017). There are many different possible interests for a pedestrian in an urban environment. For example, a traveler, who is passionate about soccer but bored by historical monuments, will obviously be more attentive to soccer related things than urban features such as city walls or statues.

Personal Background

The personal background is a common name for attributes describing the traveler's experience outside of a specific domain (Brusilovsky and Millán 2007), in our case navigation and wayfinding. It gives information about the personal characteristics of a traveler and includes geographic data as well as data describing the traveler's characteristics (Kobsa et al. 2001). The *country of residence* is considered, because travelers not living in the country of the environment they need to navigate may be used to environments or objects shaped differently (Kattenbeck 2016). For example, if a Dutchman refers to recycling bins he maybe thinks of a tube-like object set into the ground (see Fig. 1, left) whereas a Frenchmen would search for a completely different object (see Fig. 1, upper right). The second geographically related attribute is the *cultural background* of the traveler. Travelers, who grew up in another country may be used to completely different environments and objects. For example, someone who grew up in a small village in Africa, where the next bigger city is several kilometers away, has a different background compared to somebody who grew up in the middle of a modern central European city. There are also attributes important for the multidimensional model concerning the traveler's characteristics. This includes the *education* of the traveler. It was revealed that users' knowledge in a domain varies considerably according to their background and job (Berry and de Rosis 1991). Concerning navigation and wayfinding, the education of a traveler can influence the way visual and structural dimensions are perceived (Kattenbeck 2016). Further attributes concerning travelers' characteristics are *gender* and *age* of the traveler. The incorporation of these attributes into this first proposal of a multidimensional model would require deeper analysis of their influence on the overall salience of a landmark, which is beyond the current scope of this paper. Nevertheless, we mentioned these attributes for the sake of completeness.

Fig. 1 Recycling bins in the Netherlands (left), France (upper right) and Germany (bottom right)

3.2 Calculating Salience

In this section the salience of the landmark attributes defined above is calculated. Methods for the calculation of salience for the attributes of all dimensions are investigated.

3.2.1 Salience of the Attributes of the Landmark Dimensions

In this section the salience for the attributes of each landmark dimension is calculated. We assign salience values to each attribute. If for a landmark candidate all attributes of a landmark dimension are salient it gets a 100% salience. For example, if an object meets all requirements of the structural dimension, it gets 100% for structural salience. The conditions that must be fulfilled in order to assign a percentage of a salience value to the attributes is shown in Table 1 and explained below.

Surface area and *color* are considered salient if their value is different from all others in a local environment. For the definition of this local environment a buffer of 100 m is chosen in this work. The *surface structure* is salient if the object has an outstanding surface (see Sect. 3.1.1). The assessment if the attribute value of *height* of an object is significantly different from mean characteristics within the buffer is done by hypothesis testing (see Raubal and Winter (2002) for details). As soon as the model is complete, a sensitivity analysis to identify the importance of the individual attributes of the visual dimension will be carried out. This will enable us to give different weights to different attributes. However, for the current study, we assume that each of the attributes of the visual dimension has the same effect on the overall salience of an object. Therefore, we assign a salience value of 25% if the attribute is salient. Zero percent means the attribute is not salient.

The attributes of the semantic dimension are salient if their attribute values are "True". Because the availability of explicit marks is of a higher value than cultural or historical importance it gets a salience of 50%. The other two attributes get a salience of 25%.

Table 1 Rules for the computation of landmark, descriptive and environmental saliences

Dimension	Attribute	Salient	Salience (Attribute)	Salience (Dimension)
Visual	Surface Structure λ	If $\lambda = $ True	$s_\lambda \in \{0, 25\%\}$	$s_{\text{vis}}[\%] = s_\lambda + s_\phi + s_\mu + s_\gamma$
	Height ϕ	See text below	$s_\phi \in \{0, 25\%\}$	
	Surface Area μ		$s_\mu \in \{0, 25\%\}$	
	Colour γ		$s_\gamma \in \{0, 25\%\}$	
Semantic	Cultural importance ϵ	If True	$s_\epsilon \in \{0, 25\%\}$	$s_{\text{sem}}[\%] = s_\epsilon + s_\iota + s_\xi$
	Historical importance ι		$s_\iota \in \{0, 25\%\}$	
	Explicit marks ξ		$s_\xi \in \{0, 50\%\}$	
Structural	Number of adjacent routes η	If $\eta > 1$	$s_\eta \in \{0, 50\%\}$	$s_{\text{str}}[\%] = s_\eta + s_\theta$
	Number of adjacent objects θ	If $\theta = 0$	$s_\theta \in \{0, 50\%\}$	
Descriptive	Explicit marks D_e	If True	$s_{\text{De}} \in \{0, 100\%\}$	$s_{\text{desc}}[\%] = \max(s_{\text{De}}, s_{\text{Dn}})$
	Number of words D_n	dependent on the number	$s_{\text{Dn}} \in \{0, 50\%, 75\%, 100\%\}$	
Environmental	Advance visibility E_v	If visible	$s_{\text{Ev}} \in \{0, 25\%\}$	$s_{\text{env}}[\%] = s_{\text{Ev}} + s_{\text{Eo}} + s_{\text{Ed}} + s_{\text{Eu}}$
	Orientation E_o	If "in front" OR "beside"	$s_{\text{Eo}} \in \{0, 25\%\}$	
	Distance E_d	If $E_d = \min(D_{e1}, \ldots D_{ei})$	$s_{\text{Ed}} \in \{0, 25\%\}$	
	Uniqueness E_u	If True	$s_{\text{Eu}} \in \{0, 25\%\}$	

Concerning the structural attributes, the number of adjacent routes is salient if there is more than one route next to the object. Freestanding objects, where the number of adjacent objects is zero, are also significant. Similar to the case of visual attributes we assume that each of the attributes has the same effect on the overall salience and therefore assign a salience value of 25%.

3.2.2 Salience of the Attributes of the Descriptive Dimension

A landmark that can be described with an explicit mark is a valuable navigation aid and can be directly used in route instructions. Therefore such a landmark gets a salience for the explicit mark of 100% (see Table 1). If there is no explicit mark available the number of words is investigated. A landmark, which can be described with a single word is easy to remember for the traveler and therefore gets a salience for number of words of 100%. Descriptions with two terms are still easy to memorize but are more complicated than a one-word description. Therefore, such a landmark only gets a salience of 75% for number of words. Landmarks with descriptions including three words get a 50% salience. Landmarks with descriptions of more than three words get no salience for the attribute number of words, because they get too long.

3.2.3 Salience of the Attributes of the Environmental Dimension

One attribute of the environmental dimension is *advance visibility* of a landmark (see Table 1). The implementation of Winter (2003) approach for visibility analysis would require deeper analysis of our data, which is beyond the scope of this paper. Therefore we use as a first step a simple line of sight analysis (see Fig. 2). It is investigated if the line of sight from the street intersection before the decision point intersects another object. If that is not the case then the attribute *advance visibility* is salient for this object.

Landmarks which are located next to or in front of the route get a salience of 25%. Landmarks at the back of the traveler are not as good as landmarks at the front or next to the route. Therefore, such a landmark gets no salience for *orientation*. Landmarks are useful navigation aids if they are close to the next decision point. Therefore, the object with the smallest distance to the decision point is assigned a salience of 25%. The other objects do not get any salience for this attribute.

The landmark is *unique* if there is no other misleading object within the environment of the route. Thus, neighboring street intersections are investigated. If there are no similar objects in one of these environments, a salience of 25% is assigned. For the environmental attributes the same applies as for the attributes of the visual and the structural dimension. We assume the same effect of the environmental attributes on the overall salience and apply a salience of 25% if it is salient. If it is not salient, the salience is zero.

Fig. 2 Example advance visibility

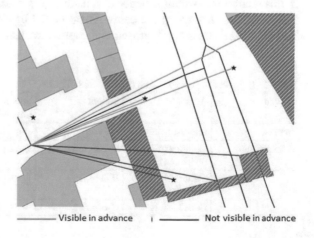

——— Visible in advance ┆ ——— Not visible in advance

3.2.4 Salience of the Attributes of the Personal Dimensions

Landmarks should be selected according to the interests, background and prior spatial knowledge of the traveler. Each landmark belongs to a number of areas of interests. For example a city gate could belong to the areas of interests *historical monuments* and *architecture*. If an interest of the traveler matches one of the areas of interests that the landmark belongs to, then a significance value of 100% is assigned to the landmark (see Table 2).

For the attributes of the background dimension *country of residence, cultural background* and *education* are considered. The first two attributes are salient if the traveler has grown up or rather lives in the environment that she has to navigate. *Country of residence* and *cultural background* are attributes that are connected to each other. This means: if one attribute holds true the possibility is high that the other attribute also holds true. In order to avoid a higher weighting of these attributes compared to *education* only a salience value of 25% is assigned.

For education the same approach applies as for the interests. Hence, each landmark belongs to one or more educations. For example, measuring points are salient objects for surveyors. If the education of the traveler matches one education to which the landmark belongs, a significance value of 50% is assigned to the landmark. Which is as high as the salience values for *Country of residence* and for *cultural background* together.

The prior spatial knowledge is a dimension that influences most of the other dimensions and their attributes. For that reason the prior spatial knowledge is considered using weights within the multidimensional model. How the weighting is done is investigated in Sect. 3.3.2.

3.3 Overall Salience

In this section the overall salience of a landmark is determined. First, we calculate the overall salience using the conventional model by Raubal and Winter (2002). Secondly, we discuss our approach of the multidimensional model.

Table 2 Rules for the computation of personal saliences

Dimension	Attribute	Salient	Salience (Attribute)	Salience (Dimension)
Interest	Interest I	If $I = I_{LM}$	$s_i \in \{0, 100\%\}$	$s_I[\%] = s_i$
Background	Country of residence C	If C = True	$s_C \in \{0, 25\%\}$	$s_{PB}[\%] = s_C + s_B + s_E$
	Cultural background B	If B = True	$s_B \in \{0, 25\%\}$	
	Education E	If $E = E_{LM}$	$s_E \in \{0, 50\%\}$	

3.3.1 Conventional Model by Raubal and Winter (2002)

The approach from Raubal and Winter (2002) is also based on the well-established visual, semantic and structural dimensions. They include different attributes, which differ slightly from ours. However, we use our attributes (see Sect. 3.1.1), to make the models comparable.

Raubal and Winter (2002) determine in a first step values for each attribute. Then, it is investigated whether an attribute value is significantly different from the others. This is done using hypothesis testing. The significance value is set to 1 if there is a significant difference, i.e. the attribute is salient. Otherwise, the significance value is zero. We consider the attribute values as salient if they fulfill the conditions defined in Table 1 (column salient). For the salience values for the attributes we use also 1 and zero for the conventional model. Note, that this is a difference to our multidimensional model (see Sect. 3.3.2), where we use the salience values defined in Table 1 (column Salience (Attribute)).

Next, the significance values are grouped for visual, semantic and structural dimensions (see Eqs. 1–3). The total measure of landmark salience for each building is determined by adding up the grouped significance values (see Eq. 4). Raubal and Winter (2002) mentioned that the weights in this total measure can be used for an adaptation to the context or individual user preferences, but did not discuss this any further.

$$s_{vis} = (s_\lambda + s_\phi + s_\mu + s_\gamma)/4 \tag{1}$$

$$s_{sem} = (s_\epsilon + s_\iota + s_\xi)/3 \tag{2}$$

$$s_{str} = (s_\eta + s_\theta)/2 \tag{3}$$

$$s_{convM} = w_{vis} * s_{vis} + w_{sem} * s_{sem} + w_{str} * s_{str} \tag{4}$$

3.3.2 Multidimensional Model

In this section we add the additional dimensions which we defined above to the conventional model of Raubal and Winter (2002) (see Sect. 3.3.1). Analog to their model the values for each attribute for each dimension are determined. Then, it is investigated whether an attribute value is salient. This is done according to the rules for the computation of saliences in Tables 1 and 2 (column salient). The attribute saliences are assigned according to Tables 1 and 2 (column Salience (Attribute)).

Then the significance values are grouped for visual, semantic and structural dimensions (see Table 1, column Salience (Dimension)). The same is executed for the environmental dimension (see Table 1, column Salience (Dimension)) and the background dimension of the personal dimensions (see Table 2, column Salience (Dimension)). To determine the overall salience value for the descriptive dimension the higher salience value of the attributes number of words and explicit marks is chosen. The interest dimension of the personal dimensions consists of only one attribute, therefore no further processing is needed.

Next, the overall salience value of an object is determined according to Eq. 5. The values for the weights are shown in Table 3. The weights represent the consideration of the prior spatial knowledge of the traveler within the multidimensional model. Travelers not familiar with an environment use landmarks which are highly visual salient (Quesnot and Roche 2015), therefore visual salience is weighted with a factor of 3. The semantic salience is zero because it is more appropriate for people familiar with an environment (Quesnot and Roche 2015). Structural salient landmarks should be used if there are no visual outstanding landmarks (Quesnot and Roche 2015), therefore structural salience is not as important as visual salience. For people not familiar with the environment explicit marks are not important, because it tells them nothing. Therefore, only the number of words is considered within the model. Because the environmental dimension, the interest and the background dimensions are important dimensions for selecting personalized route dependent landmarks they are weighted twice for all types of prior spatial knowledge (see Table 3).

$$
\begin{aligned}
s_{multidimM} = (w_{vis} * s_{vis} &+ w_{sem} * s_{sem} + w_{str} * s_{str} \\
&+ w_{desc} * s_{desc} + w_{De} * s_{De} + w_{Dn} * s_{Dn} \\
&+ w_{env} * s_{env} \\
&+ w_I * s_I + w_{PB} * s_{PB})/100
\end{aligned}
\tag{5}
$$

A traveler with route knowledge is familiar with the route and therefore prefers landmarks that have a special meaning to him. Therefore, the semantic salience is weighted with a factor of 3. In this case the landmarks should also be describable by an explicit mark, therefore explicit marks are considered within the multidimensional model for route knowledge. Similarly to the other cases, the environmental and the personal dimensions are weighted with a factor of 2.

A traveler with survey knowledge should be familiar with the area. Nevertheless, it can be assumed that not all of the available landmarks are familiar because of their semantic salience but also because of their visual salience. Therefore, the visual as well as the semantic salience are weighted with a factor of 3. Structural salience

Table 3 Weights

Weights	No	Route	Survey
w_{vis}	3	0	3
w_{sem}	0	3	3
w_{str}	1	0	1
w_{desc}	0	0	1
w_{De}	0	1	0
w_{Dn}	1	0	0
w_{env}	2	2	2
w_I	2	2	2
w_{PB}	2	2	2

is also considered. In this case it is important if the object has a high descriptive salience no matter if this is because of a low number of words or an explicit mark.

4   Case Studies

4.1   Provenance of Data

The modeling of the landmarks in the multidimensional model requires a number of data sources. According to the defined attributes, visual, semantic, structural and descriptive data are required. In our case studies we used OSM data. The data not available from OSM (e.g. color or description) were collected through a field survey. The height of the buildings was extracted from a official 3D city model (block model). The height of the other objects was estimated manually for this preliminary study. For this paper we assume that the attribute values for the personal dimensions are available. In Sect. 4.3 possible methods to acquire the attribute values of the personal dimensions are discussed.

4.2   Personalized Landmarks—Examples

This section demonstrates the applicability and usefulness of the multidimensional model using three different case studies. In Fig. 3 an example decision point with 8 potential landmark candidates is shown. There are five buildings and three other objects. Within Table 4 the saliences for all the dimensions except the personal dimensions of the landmark candidates are listed.

In Table 5 the overall saliences based on Eq. 4, for the conventional model are presented. The results identify the recycling bins at the decision point as the most suitable landmark. Although the semantic salience of this landmark candidate is low, it has a high visual and structural salience (see Table 4). The next salient landmarks according to the conventional model are the streetlight and the casino. The streetlight is salient because it is freestanding. Whereas, the casino gets a high overall salience because of its semantics.

In the following sections we demonstrate the overall saliences based on Eq. 5 for different case studies for a traveler with no knowledge, survey and route knowledge. Note that, if there are landmarks part of landmark knowledge or part of previous route instructions available, theses landmarks should be used for route instructions. Then, no further investigations are needed. Therefore, in these example case studies we assume that neither landmark nor route knowledge (used in previous route instructions) landmarks are available.

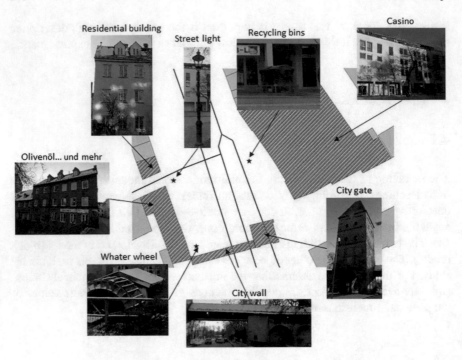

Fig. 3 Example decision point with landmark candidates

Table 4 Examples of object saliences (in percent)

Object	Visual	Semantic	Structural	Descrip.	Number of words	Explicit marks	Environm.
Olivenöl... und mehr	0	50	50	100	50	100	25
Residential building	0	0	50	75	75	0	25
Streetlight	25	0	100	75	75	0	50
Recycling bins	75	0	100	75	75	0	50
Casino	25	75	50	100	100	100	75
City gate	50	25	50	75	75	0	50
City wall	50	25	50	75	75	0	50
Water wheel	75	25	50	75	75	0	25

Table 5 Object saliences for a traveler with unknown interest and education

Object	No	Route	Survey	Conventional model
Oliven... und mehr	2.5	4	4.5	1.67
Residential building	2.75	1.5	2.75	1
Streetlight	4.5	2	4.5	2.25
Recycling bins	**6**	2	6	**2.75**
Casino	4.75	**5.75**	7	2.25
City gate	4.75	2.75	5.5	1.83
City wall	4.75	2.75	5.5	1.83
Water wheel	5	2.25	5.75	2.08

4.2.1 Traveler with Unknown Interest and Education

In Table 5 the overall saliences based on Eq. 5 are presented for a traveler with no, survey and route knowledge. These values are based on the attribute values of a traveler with unknown interest and education (Table 6). It is assumed that the traveler lives and has grown up within the country of the environment to navigate.

If we assume a traveler with no knowledge, the recycling bins would be the most suitable landmark (see Table 5), which is in line with the conventional model. This landmark shows a 75% visual salience, which is weighted with a factor of 3 (see Table 3). The water wheel also carries a visual salience of 75% but the environmental salience, which is weighted twice, is lower. The water wheel is located at the back of the traveler and has no advance visibility at all. Whereas the recycling bins are located in front of the traveler and are visible in advance. Also the structural salience

Table 6 Interest and background saliences for a traveler with unknown interest and education

Object	Interest	Country of residence	Cultural background	Education	Background
Olivenöl... und mehr	0	25	25	0	50
Residential building	0	25	25	0	50
Street light	0	25	25	0	50
Recycling bins	0	25	25	0	50
Casino	0	25	25	0	50
City gate	0	25	25	0	50
City wall	0	25	25	0	50
Water wheel	0	25	25	0	50

of the recycling bins is high, because they are freestanding objects. With regard to the number of words, which are considered to determine landmarks salience (see Table 3), only the casino is describable with a single word. The others need at least two words. Nevertheless, the recycling bins are the most salient landmark at the decision point, because the saliences of the other dimensions are high.

For someone with route knowledge the casino is a suitable landmark. The semantic salience is weighted with a factor of 3, while the visual and structural characteristics of a landmark candidate are not considered (see Table 3). For the determination of the salience of the landmark candidates in route knowledge areas it is considered if an object shows explicit marks. The casino is one of two landmarks (beside "Olivenöl... und mehr") which shows explicit marks (see Table 4). Furthermore, the environmental salience is high, because the casino is visible early on while approaching the decision point, its orientation is "in front" and it is unique within the environment.

The best landmark for someone with survey knowledge would be the casino as well (see Table 5). To determine the landmarks in survey knowledge areas the visual and the semantic salience are weighted with a factor of 3 (see Table 3). Because the casino shows explicit marks it has a high semantic and a high descriptive salience as well.

4.2.2 Traveler with Interest in Historical Monuments

In this case study we assume a traveler who is a professor of cultural history and very interested in historical monuments. He also lives and has grown up in the country of the environment to navigate. In Table 7 the interest and background saliences for this case are shown. The city gate and the city wall get a salience for the interest and the background of 100%. The saliences of the landmark candidates dependent on these attribute values are shown in Table 8.

In this case, the most salient landmark for a traveler independent of his prior spatial knowledge are the city wall and the city gate. Because of the additional factor for the interest and the background their saliences are exceeding the saliences of the recycling bins (for no knowledge) and the casino (for survey and route knowledge). For a traveler with route knowledge the casino stays a good choice because it is highly semantic.

4.2.3 Traveler with Different Cultural Background

In this last case study we assume a traveler who has not grown up and does not live within the environment to navigate. This could be for example a tourist who is just for a few days within the city. In Table 9 the interest and background saliences for this case are shown. Recycling bins are often shaped differently in different countries (see Fig. 1). They can differ in form, size and color, therefore they get a value of 0 for *country of residence* and *cultural background*. Because the other buildings/objects

Table 7 Interest and background saliences for a traveler with interest in historical monuments

Object	Interest	Country of residence	Cultural background	Education	Background
Olivenöl... und mehr	0	25	25	0	50
Residential building	0	25	25	0	50
Streetlight	0	25	25	0	50
Recycling bins	0	25	25	0	50
Casino	0	25	25	0	50
City gate	100	25	25	50	100
City wall	100	25	25	50	100
Water wheel	0	25	25	0	50

Table 8 Object saliences for a traveler with interest in historical monuments

Object	No	Route	Survey	Conventional model
Oliven... und mehr	2.5	4	4.5	1.67
Residential building	2.75	1.5	2.75	1
Streetlight	4.5	2	4.5	2.25
Recycling bins	6	2	6	**2.75**
Casino	4.75	**5.75**	7	2.25
City gate	**7.75**	**5.75**	**8.5**	1.83
City wall	**7.75**	**5.75**	**8.5**	1.83
Water wheel	5	2.25	5.75	2.08

are the same in their appearance in different countries they get a salience of 25% for country of residence and for cultural background. The best landmarks for someone with no knowledge are the recycling bins or the water wheel (see Table 10). The water wheel has a low environmental salience, because the only attribute, which is salient of the environmental dimension, is the uniqueness. However, it has a high visual salience. In addition, it is an object that is shaped more or less the same in different countries. That makes the water wheel a good choice for someone with no knowledge for this case study.

The recycling bin is still a valuable landmark for someone with no prior spatial knowledge, although it has no salience for the personal background. But as already mentioned in Sect. 4.2.1, its visual, structural and environmental salience is high. For future work it is necessary to consider if landmarks with no *country of residence* or no *cultural background* salience should be excluded from the potential landmarks for the traveler.

Table 9 Interest and background saliences for a traveler with different cultural background

Object	Interest	Country of residence	Cultural background	Education	Background
Olivenöl... und mehr	0	25	25	0	50
Residential building	0	25	25	0	50
Street light	0	25	25	0	50
Recycling bins	0	0	0	0	0
Casino	0	25	25	0	50
City gate	0	25	25	0	50
City wall	0	25	25	0	50
Water wheel	0	25	25	0	50

Table 10 Object saliences for a traveler with a different cultural background

Object	No	Survey	Route	Conventional model
Oliven... und mehr	2.5	4.5	4	1.67
Residential building	2.75	2.75	1.5	1
Street light	4.5	4.5	2	2.25
Recycling bins	**5**	5	1	**2.75**
Casino	4.75	**7**	**5.75**	2.25
City gate	4.75	5.5	2.75	1.83
City wall	4.75	5.5	2.75	1.83
Water wheel	**5**	5.75	2.25	2.08

For someone with survey knowledge or route knowledge the casino is still (as in the two other case studies) a suitable landmark. The casino can be recognized by its explicit marks and is located in a normal building which has the same appearance in different countries.

4.3 Discussing Data Collection Methods for the Personal Dimension

In Nuhn and Timpf (2017) we discussed first methods to acquire the attribute values of the personal dimensions. A possible method to capture the prior spatial knowledge of the traveler is to store already navigated routes. Also, landmarks that were already used for navigation could be stored as landmarks part of *landmark knowledge*. The attribute values for the personal background must be provided explicitly because it is nearly impossible to deduce them by sensors or by simply watching the traveler. The personal interests of a traveler can be learned with the help of a learning system (see also Richter 2017) because entering all the values explicitly would be too exhausting and time consuming for a traveler.

There are still attributes of the personal dimensions missing (for example, gender and age of the traveler) which we have to include in our model. As soon as the model is complete a sensitivity analysis to check the models logic and robustness will be carried out. The identification of the importance of the individual attributes enables to estimate the effort which must be invested in data acquisition for different attributes. If the sensitivity analysis indicates that the model includes a number of attributes to which the model is insensitive, then we can maybe exclude these attributes from our multidimensional model to minimize the acquisition effort.

5 Conclusion and Future Work

This paper proposes a multidimensional model for landmarks that incorporates landmark, descriptive, environmental and personal dimensions. The dimensions of the model and their attributes were defined and debated. Further, methods for the calculation of salience for the attributes of all dimensions were investigated. Finally, the dimensions were integrated in a multidimensional model to calculate the overall salience. We showed that varying attribute values for the attributes for the personal dimension changed the most salient landmark in our case studies. In this paper, weights were chosen based on consideration. This provides a good framework for an empirical study in a real usage context to fine-tune the current approach.

In this paper, first ideas on how to consider the traveler's interests were proposed. In this work we considered the interest of a person to a particular content. In future work we will also consider interest that is caused by certain conditions such as the goal of wayfinding. Further attributes concerning travelers' background are gender and age of the traveler. In this paper we neglected these attributes because their incorporation in the multidimensional model would require deeper analysis. In future work we will investigate their influence on the overall salience of an object.

In this paper we provided an example how to consider a traveler with different cultural background. But there are also people with multi-cultural background. For example someone who grew up in a small village in Africa and then moved to a central European city is used to differently shaped objects with the same meaning. So we have to investigate the question after which period of time such a person is familiar enough with the city that he is also able to use country-specific objects for navigation.

References

Balaban CZ, Röser F, Hamburger K (2014) The effect of emotions and emotionally laden landmarks on wayfinding. In: Proceedings of the 36th annual conference of the cognitive science society, pp 1880–1885

Banerjee S, Frey HP, Molholm S, Foxe JJ (2015) Interests shape how adolescents pay attention: the interaction of motivation and top-down attentional processes in biasing sensory activations to anticipated events. Eur J Neurosci 41(6):818–834

Berry DC, de Rosis F (1991) Designing an adaptive interface for epiaim. In: Mario S, Arie H, Marius F, Jan T (eds) AIME 91, Springer, pp 306–316

Brusilovsky P, Millán E (2007) User models for adaptive hypermedia and adaptive educational systems. In: Peter B, Alfred K, Wolfgang N (eds) The adaptive web. Springer, pp 3–53

Burnett G, Smith D, May A (2001) Supporting the navigation task: characteristics of 'good' landmarks. In: Hanson MA (ed) Contemporary ergonomics 2001, Taylor & Francis, pp 441–446

Caduff D, Timpf S (2005) The landmark spider: Representing landmark knowledge for wayfinding tasks. In: Barkowsky T, Freksa C, Hegarty M, Lowe R (eds) AAAI spring symposium: reasoning with mental and external diagrams: computational modeling and spatial assistance. AAAI Press, pp 30–35

Caduff D, Timpf S (2008) On the assessment of landmark salience for human navigation. Cog Process 9(4):249–267

Duckham M, Winter S, Robinson M (2010) Including landmarks in routing instructions. J Locat Based Serv 4(1):28–52

Elias B (2003) Extracting landmarks with data mining methods. In: Kuhn W, Worboys MF, Timpf S (eds) Spatial information theory. COSIT 2003. Lecture notes in computer science, vol 2825. Springer, Berlin, pp 375–389

Elias B, Sester M (2006) Incorporating landmarks with quality measures in routing procedures. In: Raubal M, Miller HJ, Frank A, Goodchild MF (eds) Geographic information science. GIScience 2006. Lecture notes in computer science, vol 4197. Springer, Berlin, pp 65–80

Götze J, Boye J (2016) Learning landmark salience models from users route instructions. J Locat Based Serv 10(1):47–63

Hamburger K, Röser F (2014) The role of landmark modality and familiarity in human wayfinding. Swiss J Psychol 73(4):205–213

Herrmann T, Schweizer K, Janzen G, Katz S (1998) Routen- und überblickswissen - konzeptuelle überlegungen. Kognitionswissenschaft 7(4):145–165

Hidi S, Renninger KA (2006) The four-phase model of interest development. Educ Psychol 41(2):111–127

Kattenbeck M (2016) Empirically measuring salience of objects for use in pedestrian navigation. PhD thesis, University of Regensburg, Chair for information science

Klippel A, Winter S (2005) Structural salience of landmarks for route directions. In: Cohn AG, Mark DM (eds) Spatial information theory. COSIT 2005. Lecture notes in computer science, vol 3693. Springer, Berlin, pp 347–362

Kobsa A, Koenemann J, Pohl W (2001) Personalised hypermedia presentation techniques for improving online customer relationships. Knowl Eng Rev 16(2):111–155

Krapp A, Hidi S, Renninger A (2017) Interest, learning and developement. In: Ann R, Suzanne H, Andreas K (eds) The role of interest in learning and development. Psychology Press, pp 3–26

Lovelace KL, Hegarty M, Montello DR (1999) Elements of good route directions in familiar and unfamiliar environments. In: Freksa C, Mark DM (eds) Spatial information theory. COSIT 1999. Lecture notes in computer science, vol 1661. Springer, Berlin, pp 65–82

Lynch K (1960) The image of the city. MIT press

McGillivray S, Murayama K, Castel AD (2015) Thirst for knowledge: the effects of curiosity and interest on memory in younger and older adults. Psychol Aging 30(4):835

Nothegger C, Winter S, Raubal M (2004) Selection of salient features for route directions. Spat Cogn Comput 4(2):113–136

Nuhn E, Timpf S (2016) A multidimensional model for personalized landmarks. In: International conference on location based services. Austria, Research Group Cartography, Vienna University of Technology, Vienna, pp 4–6

Nuhn E, Timpf S (2017) Personal dimensions of landmarks. In: Bregt A, Sarjakoski T, van Lammeren R, Rip F (eds) Societal Geo-innovation: selected papers of the 20th AGILE conference on geographic information science. Springer International Publishing, pp 129–143

Nuhn E, Reinhardt W, Haske B (2012) Generation of landmarks from 3D city models and osm data. In: Gensel J, Josselin D, Vandenbroucke D (eds) Proceedings of the AGILE 2012 international conference on geographic information science, pp 365–369

Palmiero M, Piccardi L (2017) The role of emotional landmarks on topographical memory. Front Psychol 8:763

Quesnot T, Roche S (2015) Quantifying the significance of semantic landmarks in familiar and unfamiliar environments. In: Fabrikant SI, Raubal M, Michela B, Davies C, Freundschuh S, Bell S (eds) Spatial information theory. COSIT 2015. Lecture notes in computer science, vol 9368. Springer, pp 468–489

Raubal M, Winter S (2002) Enriching wayfinding instructions with local landmarks. In: Egenhofer MJ, Mark DM (eds) Geographic information science. GIScience 2002. Lecture notes in computer science, vol 2478. Springer, Berlin, pp 243–259

Renniger AK, Su S (2012) Interest and its development. In: Ryan RM (ed) The Oxford handbook of human motivation, OUP USA, pp 167–190

Rensink RA, O'Regan JK, Clark JJ (1997) To see or not to see: the need for attention to perceive changes in scenes. Psychol Sci 8(5):368–373

Richter KF (2017) Identifying landmark candidates beyond toy examples. KI 31(2):135–139

Richter KF, Winter S (2014) Landmarks—GIScience for intelligent services. Springer International Publishing

Schneider LF, Taylor HA (1999) How do you get there from here? mental representations of route descriptions. Appl Cogn Psychol 13(5):415–441

Siegel AW, White SH (1975) The development of spatial representations of large-scale environments. Adv Child Dev Behav 10:9–55

Sorrows ME, Hirtle SC (1999) The nature of landmarks for real and electronic spaces. In: Freksa C, Mark DM (eds) Spatial information theory. COSIT 1999. Lecture notes in computer science, vol 1661. Springer, Berlin, pp 37–50

Tenbrink T, Winter S (2009) Variable granularity in route directions. Spat Cogn Comput 9(1):64–93

Thorndyke PW (1980) Spatial cognition and reasoning. In: Harvey JH (ed) Cognition, social behavior, and the environment, Rand Corporation, pp 137–149

Waller D, Loomis JM, Golledge RG, Beall AC (2000) Place learning in humans: the role of distance and direction information. Spat Cogn Comput 2(4):333–354

Winter S (2003) Route adaptive selection of salient features. In: Kuhn W, Worboys MF, Timpf S (eds) Spatial information theory. COSIT 2003. Lecture notes in computer science, vol 2825. Springer, Berlin, pp 349–361

Winter S, Raubal M, Nothegger C (2005) Focalizing measures of salience for wayfinding. In: Meng L, Zipf A, Tumasch R (eds) Map-based mobile services theories, methodsand implementations. Springer, pp 125–139

Winter S, Janowicz K, Richter KF, Vasardani M (2012) Knowledge acquisition about places. SIGSPATIAL Spec 4(3):20–21

Wolfensberger M, Richter KF (2015) A mobile application for a user-generated collection of landmarks. In: Gensel J, Tomko M (eds) Web and wireless geographical information systems. W2GIS 2015. Lecture notes in computer science, vol 9080. Springer International Publishing, pp 3–19

Part IV
Location Based Social Media and Citizen Participation

"Thanks for Your Input. We Will Get Back to You Shortly." How to Design Automated Feedback in Location-Based Citizen Participation Systems

Andreas Sackl, Sarah-Kristin Thiel, Peter Fröhlich and Manfred Tscheligi

Abstract Location-based citizen participation systems have so far mostly been characterized by mediated human-to-human communication between citizens, authorities and other stakeholders. However, in the near future we will see more automatized feedback elements, which inform citizens about the expectable financial or legal implications of their requests. We conducted an experiment to provide research-driven guidance for interaction design in this application context. Thirty participants submitted tree planting proposals with an experimental prototype that varied along the dimensions immediacy, implicitness, and precision. They rated the different forms of provided automatic feedback with regard to satisfaction, and they ranked them in a subsequent card sorting trial. The results show that users have considerably high expectations towards the immediacy and precision of automated feedback, regardless of the inherently higher responsiveness compared to human-operated participation systems. With regard to interaction design, results indicate that the automatically processed information should be made available as early and as possible to users.

1 Introduction

The research field "smart cities" investigates various aspects of modern urban systems to provide profound solutions for actual and upcoming issues and challenges like sustainable energy generation and consumption, mobility concepts. The integration of citizen in urban development processes is one of the key challenges in this

A. Sackl (✉) · S.-K. Thiel · P. Fröhlich · M. Tscheligi
AIT, Giefinggasse 2, Vienna, Austria
e-mail: andreas.sackl@ait.ac.at

S.-K. Thiel
e-mail: sarah.kristin.thiel@gmail.com

P. Fröhlich
e-mail: peter.froehlich@ait.ac.at

M. Tscheligi
e-mail: manfred.tscheligi@ait.ac.at

© Springer International Publishing AG 2018
P. Kiefer et al. (eds.), *Progress in Location Based Services 2018*, Lecture Notes
in Geoinformation and Cartography, https://doi.org/10.1007/978-3-319-71470-7_13

257

research context. For this type of active citizen participation, location-based services (LBS) need to be designed and implemented in a way that the resulting user experience is high enough to encourage citizens to actively use the system. So, in this paper we want to address user-centric design issues and how these aspects should be implemented in LBS-based citizen participation systems.

Citizen participation has become a central aspect of modern societies, and there is an increasing amount of interactive computing systems that help innovate the way people discuss, contribute and influence public decisions (Conroy and Evans-Cowley 2006). However, as the recent history on eParticipation shows, the roll out of specialized platforms rarely scales and typically does not reach a large amount of users (Prieto-Martín et al. 2012). Studies have shown that citizens' satisfaction with participation technologies is, amongst other factors (e.g., user-friendliness of the application, trust in politics), determined by authorities' responsiveness to citizens (Kweit and Kweit 2004; Parasuraman et al. 2005; Webler and Tuler 2000; Harding et al. 2015). Also, receiving meaningful feedback from authorities helps increase citizens' internal political efficacy (i.e., their subjective belief that they understand community issues) (Kim and Lee 2012). These requirements of timely and meaningful feedback are clearly not sufficiently met in current digital participation services.

Although the paradigm of such services is slowly changing from one-way to more interactive participation forms (Conroy and Evans-Cowley 2006; Lukensmeyer and Torres 2008), there still seem to be significant barriers for responsiveness (Thiel et al. 2016). A central problem often mentioned by administrative staff who have to deal with citizens' initially posted contributions is that these are often perceived as "naive" in terms of their administrative, legal or economic implications. In that respect officials wish for "better qualified" complaints and proposals (Bohøj et al. 2011). For example, suggestions for the location of a new bus stop may not take into account certain traffic regulations or road construction constraints. In such cases, much effort is needed to provide feedback to citizens on "basic issues", without actually gaining significant benefits for their urban planning work.

In order to provide solutions to this problem space, automatic feedback technology has been proposed that may enable citizens to probe and refine their ideas, which in turn should provide urban planners and city authorities with validated, 'useful' input (Poplin 2012; Vogt and Fröhlich 2016). In order to provide meaningful feedback that allows for a higher level of participation (i.e. from consolidation to cooperation; see Tambouris et al. 2007), communication with authorities should go beyond currently available solutions of "bots" that compile databases to help with automatizing customer services, tax return process (Karsten and West 2016), or voting procedures (Phoneia 2016). Rather, it should offer answers and ideally also comment to citizens' requests. For instance, when proposing the development of a new park at a certain location, the feedback should give an indication of whether that is in principle possible.

2 Related Work

In recent years e-participation platforms have been ported to mobile devices. With their manifold features and sensors (e.g. gyroscope, GPS), devices such as smartphones allow to augment citizens' input with valuable information making them even more meaningful for representatives (Fröhlich et al. 2011). Considering the wide penetration of mobile devices [cite stats], making use of this technology is anticipated to broaden the scope of involved citizens and potentially also encourage those previously eager to participate. With mobile technology facilitating in-situ location-based participation (i.e., collecting input directly from citizens on-site, Korn 2013), this participation method further mitigates traditional participation barriers (i.e., spatial and temporal).

Albeit existing mobile participation services including affordances such as location-awareness (Schröder 2015), taking pictures and even augmented reality (Allen et al. 2011), it has been stated that available applications do not exploit the potential of pervasive technology such as mobile devices by far (Desouza and Bhagwatwar 2012). Previous research in the field of e-participation mostly focused on exploring novel interaction techniques (Valkanova et al. 2014; Steinberger et al. 2014) as well as the integration of open data. While employing novel technology arguably attracts curiosity, we see relevance in addressing prevailing challenges of participation first. The one addressed with the study presented in this paper is associated with unmet expectations of e-participation. By capitalizing on open data, it is aimed to make content produced in participation platforms more relevant for both citizens and city officials as well as improving the responsiveness by providing automatic feedback. With the latter it is further envisioned to relieve city officials.

3 Goals and Hypotheses

The location-based technology that could enable such levels of an informed dialog between the city and their inhabitants is still at a research stage (West 2004). Apart from questions related to the feasibility of semantic processing of open data, the interaction design space has so far not been explored. The main goal of the user study presented in this paper is to evaluate how automated feedback has to be implemented in a location-based, participatory application to match the needs of the users. Our main focus of interest is on the user experience in terms of satisfaction. The following section discusses the related hypotheses.

3.1 Immediacy (H1)

A critical success factor associated with personalized feedback is the time-lag between the complaint/request and the governmental answer (Kearns et al. 2002; West 2004). While automatic feedback is in itself significantly faster than the response by a human administrator, studies on system response time (SRT) in interactive systems (e.g. Kohlisch and Kuhmann 1997; Szameitat et al. 2009; Rhodes and Wolf 1999) indicate that users may even be sensitive to small delays of one or more seconds. Hence, we assume that *fast feedback* (=low delay between sending proposal and receiving feedback) is crucial in our context (Hypothesis H1.1).

We also hypothesize that in correspondence to classical studies on system response time (Kohlisch and Kuhmann 1997), users in this application context are also more tolerant regarding higher response delays, if *additional information* (e.g. "data is transmitted to the server") is provided while the data is processed at the server (Hypothesis H1.2).

3.2 Precision (H2)

A crucial aspect with regard to the feasibility of location-based, automatic feedback is the level of detail and accuracy that must be provided by the system. We posit that learning about costs of various proposals such as the planting of a tree will render citizens more sensible about the complexity of its implementation and value. We assume that *precise information about costs* are preferred compared to providing a range of costs or price probabilities (Hypothesis H2). However, we also expect that, in case of unavailability of precise data, the provision of less definite information are viable, such as a price range or probabilities.

3.3 Implicitness (H3)

Apart from the above discussed issues, guidance is also needed on how and at what point in the participation process to display information in the interface. Usability research and practice have shown that providing implicit feedback, such as using mouse-over effects (cf. Dix 2009; Mace et al. 1998), should be implemented, i.e., available information can be displayed without changing the screen state. Hence, instead of the classical way of subsequently providing the feedback *after* the user's suggestion, users may even want information on possible options *before* posting their suggestion. In this sense, our hypothesis is that citizens wish for a *highlighting of constraints* related to their proposals, e.g. when placing objects on a map (Hypothesis H3).

3.4 Social Awareness (H4)

Citizen participation is a process of social exchange, and it has been demonstrated that knowledge about what the community is thinking is an important feature in such platforms. For example, it has been suggested that particularly for siting problems combining individually developed "idea maps" can support the identification of physically and socially robust solutions (Carver and Openshaw 1996). Simão adds, that the indication of other users' contributions provides a strong sense of the public's feelings (Simão et al. 2009). Moreover, social awareness is a condition of collective reflection, which enables citizens to broaden their knowledge and understanding of processes and specific roles in urban government (Gordon and Baldwin-Philippi 2014). Thus, also when developing automated feedback functions it is important to consider ways to embed social awareness in the interaction. Based on the above mentioned findings from standard participation platforms, we assume that information about existing proposals should be communicated *before the user submits* a new proposal instead of providing this information after the submission (Hypothesis H4).

4 Method

To test the above stated hypotheses, we conducted an experimental user study with 30 participants (16 males, 14 females) in an enriched laboratory setting. For the test participant selection, that was done by a specialized market research company, volunteer sampling was employed while taking care to achieve a balanced sample. Test persons were compensated for their participation. The mean age was 36.9 years, 10 (33%) participants were between 18 and 30 years old, 11 (36%) participants were between 31 and 45 years old and 9 (31%) participants were older than 45 years. Two of the participants (6.7%) had completed only the compulsory school. Seven persons (23.3%) owned a degree from a professional school or a apprenticeship. Five participants (16.7%) had a grammar school qualification. Six participants (20%) had either a vocational school or college degree; 33.3% of the study sample (ten persons) owned a university degree. All study participants were experienced with smartphone usage, and one third had used a digital participation platform before. Only one third of this user group received some kind of a feedback while interacting with this system. In terms of age and education level, our sample matches typical users of mobile e-participation platforms (Åström and Karlsson 2016).

We focused on the concrete user task of proposing tree planting positions by means of a map-based mobile participation app, as this had been identified by city officials to be both relevant and representative for the exploration of the idea of automatic feedback in citizen participation. Participants used a clickable HTML5/JS-prototype on a smartphone to place a tree symbol on a 2D map to create a proposal about a new tree, which should be planted at the selected position. Then, the submitted proposal was processed on a server and feedback was transmitted to the smart-

Fig. 1 Test user receives feedback on the clickable prototype & Google Street view is displayed in the background as context simulation to enhance involvement

phone and displayed, see Fig. 1 for an example. Figure 1 also shows that the user was standing in front of a large projection of Google Street View. This context simulation was meant to enhance the immersivity of the laboratory setup (please see Busch et al. 2014 for a detailed discussion about immersivity in laboratory settings).

In the experimental part of the study, there were four test blocks, which were corresponding to the four investigated issues (see Fig. 2, left side). These blocks and the respective study alternatives within the block were presented to the participants in random order. Each condition was complemented by a short question about satisfaction ("How satisfied were you with the feedback related to the specific aspect?") with answering options ranging from "not satisfied" (=1) to "very satisfied" (=5). Furthermore, after each condition, participants were interviewed about the currently evaluated aspect to get further qualitative feedback.

In the sub-test related to immediacy, the conditions to test H1.1 were realized by presenting a default feedback page with different delays (5, 10, 30, and 60 s). To validate H1.2 there were two further alternatives which provided additional feedback during the loading phase (e.g. "Request is sent to the server", "Data is being analyzed"). For the precision sub-test (H2), information was given either as the precise costs ("6000€"), a range of costs ("between 4000€ and 8000€") or a probability ("6000€ (80% accuracy)"). In the implicitness sub-test (H3), one alternative included feedback *after* a submission and two further alternatives presented the respective information *before* the submission: always visible or only visible when hovering over the respective area. For the social awareness sub-test (H4), information about others' opinions were either provided *before* or *after* submission of the proposal.

After this experimental part, participants were asked to complete an adapted card sorting exercise, to gain a direct comparative view on user preference of the provided alternatives. The participants were asked to define their "perfect" interface for imme-

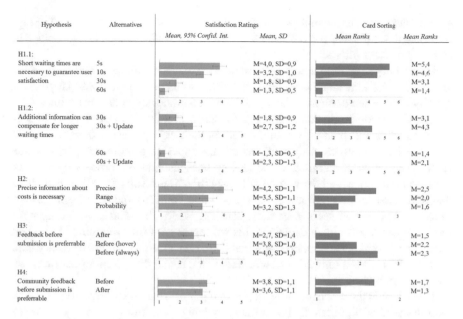

Hypothesis	Alternatives	Satisfaction Ratings		Card Sorting	
		Mean, 95% Confid. Int.	Mean, SD	Mean Ranks	Mean Ranks
H1.1: Short waiting times are necessary to guarantee user satisfaction	5s		M=4,0, SD=0,9		M=5,4
	10s		M=3,2, SD=1,0		M=4,6
	30s		M=1,8, SD=0,9		M=3,1
	60s		M=1,3, SD=0,5		M=1,4
H1.2: Additional information can compensate for longer waiting times	30s		M=1,8, SD=0,9		M=3,1
	30s + Update		M=2,7, SD=1,2		M=4,3
	60s		M=1,3, SD=0,5		M=1,4
	60s + Update		M=2,3, SD=1,3		M=2,1
H2: Precise information about costs is necessary	Precise		M=4,2, SD=1,1		M=2,5
	Range		M=3,5, SD=1,1		M=2,0
	Probability		M=3,2, SD=1,3		M=1,6
H3: Feedback before submission is preferrable	After		M=2,7, SD=1,4		M=1,5
	Before (hover)		M=3,8, SD=1,0		M=2,2
	Before (always)		M=4,0, SD=1,0		M=2,3
H4: Community feedback before submission is preferrable	Before		M=3,8, SD=1,1		M=1,7
	After		M=3,6, SD=1,1		M=1,3

Fig. 2 Overview of Results. The middle column provides mean satisfaction ratings and 95% confidence intervals from the experimental part; 1 stands for "not satisfied" and 5 for "very satisfied". In the right column "Card Sorting", the means of the inverted ranks from the card sorting part are displayed; "1" represents the least preferred option and "5" stands for the most preferred option

diate feedback in the context of location-based citizen participation. All alternatives were printed out on cards and the users laid them in the order of their preference. Also the card-sort task was complemented with a short interview to better interpret and weight participants' responses.

5 Results

In Fig. 2, bar charts and descriptive statistics, the satisfaction ratings from the experiment and the preference ranks from the card sorting activity are shown, grouped by the four sub-tests, their related hypotheses and experimental alternatives. For each of the above mentioned experimental alternatives, we calculated Kendall rank correlations between the mean satisfaction ratings and card sorting rank values that were derived from the experimental and the card sorting part, respectively. These correlations were significant ($p < 0.05$), except for the alternatives of H4. Thus, participants mostly provided consistent feedback about the satisfaction of the aspects (evaluated via the clickable prototype in the experimental part) and the individually selected interfaces (via card sorting). In order to derive evidence on the pairwise statistical differences between the experimental alternatives, we decided to calculate Wilcoxon signed-ranks tests, as normal distribution of our data could not be assumed. The significance threshold was $p < 0.5$, which was Bonferroni-adjusted in each test block to avoid alpha-error inflation.

Immediacy: The immediacy sub-test (see the related results grouped under H1.1 and H1.2 in Fig. 2) resulted in significant differences for all comparisons. Satisfaction already diminished at short delays (H1.1) and continuously decreased with longer delays. Correspondingly, many participants said that waiting times of up to 10 seconds would be acceptable for them, based on their experiences with other mobile apps. In their responses, only few participants appeared to consider performance feasibility aspects of automated feedback systems, such as the processing of open data, and thus conceded 30 s to be still tolerable. The results for H1.2 also show that additional information about data processing compensated for longer waiting times to some extent: users were more satisfied even if they had to wait longer for the feedback. A participant explained this willingness to wait by arguing "[...] when I see that something is happening and that there is an effort to get the necessary data then it's okay to wait longer". Another participant added that promptness surpasses additional information.

Precision: In the precision sub-test (H2 in Fig. 2), displaying precise pricing information was rated significantly better than providing a range or probability. Comparing the preference for range and cost, no statistically significant difference was found. Some participants however stated that displaying a range of estimated costs, rather than the precise amount, would be more realistic and honest. Also it was stated that if inaccuracies cannot be avoided, the term "ca." could be used instead of displaying a range or probabilities, because among others "you need to calculate the value to understand what the probability means". Participants further stressed the importance of communicating that the provided value might not be exact and might vary to some degree.

Implicitness: With regard to implicitness of feedback (confer H3 in Fig. 2), test participants significantly preferred getting information about alternatives (i.e. where it is generally possible or not to plant a tree) before an actual proposal was made, as opposed to receiving this feedback afterwards. Participants highlighted that they do not want to "waste time by hazarding guesses" of where a tree might be plantable. In addition, many participants mentioned that an important feature would be to get information about the reasons why a certain tree cannot be planted in a certain area before submission is sent. There was no significant difference between the two approaches for offering feedback before the submission is transmitted, i.e., displayed when hovering vs. always displayed. Participants preferring the hovering approach highlighted its dynamic and playful interaction and better map visibility, while those favoring the persistent visibility liked to see all information without further need to act. The downside of the map becoming too cluttered was also uttered several times.

Social awareness: As regards social awareness (confer H4 in Fig. 2), results on when community opinion should be disclosed to users was not consistent among participants. Some stated they would like to make their own decision and thus would not want to see the other users, while others participants saw the aspect of getting influenced as a positive feature. One of this positive aspects of being able to see other users' proposals beforehand was based on the assumption that "other users propose the planting of trees in regions they are familiar with", hence increasing the meaningfulness of the suggestion. Related to that argument is the statement of another user who stated that s/he would not mind seeing the suggestions after submission in case s/he is familiar with the area; otherwise s/he would prefer to see them beforehand. The statistical analysis of the satisfaction scores revealed no significant differences, but the comparison preference ranks from the card sorting resulted in a significantly higher preference for displaying the community opinion *before*, rather than *after* proposal submission.

Relevance of feedback: In order to verify the assumed necessity of providing (automatic) feedback in context of public participation, we further asked participants to indicate how they perceive the impact of feedback. This was done after the experimental part of the study in the form of a short questionnaire using 5-point Likert scales. More than half of all participants fully agreed to the statement that it is essential for participation services to provide feedback ($M = 4.43$, $SD = 0.82$). Only one participant indicated to not entirely agree. Aiming to explore the potential impact of feedback, we further assessed its influence on motivation and trust. Regarding motivation, 67% agreed to the statement that being provided feedback would increase their motivation to actively engage in participatory processes. Hence, feedback can be considered as a highly contributing factor to promote public participation. Participants however were more skeptic about feedback's impact on their trust in institutions such as city administration ($M = 3.70$, $SD = 1.40$).

6 Conclusions

Referring back to our hypotheses, we can say that H1.1 was confirmed: user satisfaction is decreased if longer waiting times are experienced. Our observation, that already 10 s are regarded as a minimum quality threshold by the majority of users, points to an important requirement that designers should seriously consider in the conception of future location-based automated feedback features. The benefit of time savings compared to standard participation setups, where people often wait for days or weeks to receive feedback, obviously are overriden by expectations evoked by "fast" mobile apps and Web services. Our finding that longer waiting times can be compensated by displaying additional information about the feedback process confirms research hypothesis H1.2. We assume, that by using more advanced forms of progress feedback than we had in this experiment, expectations could be managed even to a better extent.

Also hypothesis H2 was verified, that is, precise information about costs of the submitted proposal should be communicated as often as possible. This implies even more demanding requirements on automated feedback technology for digital participation. However, our qualitative data also suggests that there remains a certain tolerance, i.e., some participants appreciated that authorities and companies are not always in the position to provide definite figures (e.g. due to liability concerns, insufficient data availability, etc.). Two of such alternatives to enhance "fuzziness" of information have been tested, but no clear preference between the price range and the probabilities could be found. We also discussed further suggestions with the test participants, such as using disclaimers like "ca.". Follow-up studies should seek to get more conclusive insight into the optimal trade-off between information precision and real-world feasibility in various contexts.

With regard to implicitness, we could also confirm hypothesis H3, as the presentation of available options before proposal submission led to higher user satisfaction than afterwards. Enabling social awareness is, not surprisingly, also highly important for the design of automated feedback in location-based participation systems. With regard to the question of when to present community opinions, parts of our data (the ranking results from the card sorting) support hypothesis H4: the community opinion should be provided before the proposal submission, rather than after the proposal submission.

7   Implications for Location-Based Participation Services

If we consider the study results as how to reflect on a conceptual level how "a dialog with the city" should be realized, by means of location-based automated feedback interfaces, two general statements could be made: First, in order to comply with the identified severe performance and precision requirements with the current technological state, tasks should probably not be much more complex than the tree planting application tested within this study. These should encompass well prepared use cases with detailed and purpose-structured data in the background, in order to deliver fast and precise results. Second, as a reaction to our findings regarding implicitness and social awareness (H3 and H4), realizing a dialog with the city shall not be literally or idealistically envisioned as an interactive conversation, in the sense that users submit proposals, which are then iterated by citizens, systems and authorities. Rather, information should best possibly reduce interaction steps while still providing all relevant information. As we have found in our study, this is especially a challenge for mobile applications, where limited screen real estate may not allow for also providing all necessary rationale for decisions or constraints on the screen.

Regarding implications for location-based participation on a more general level, our study confirmed the importance of providing feedback as it might promote engagement and lead to more trust in the population. Yet, in order to achieve this, the feedback needs to be designed in a way that citizens perceive it as meaningful. Having investigated the design of specific aspects, this work lies a foundation for the development of more effective and efficient location-based participation services. Further studies are recommended to explore related design options for other contexts in the field of digital participation. In that respect, studies should also be conducted in real-life settings over a longer period of time in order to evaluate the presumed benefit for city administrations as well as dynamics arising from a large number of users interacting with the service.

References

Allen M, Regenbrecht H, Abbott M (2011) Smart-phone augmented reality for public participation in urban planning. In: Proceedings of the 23rd Australian computer-human interaction conference. ACM, pp 11–20

Åström J, Karlsson M (2016) Will e-participation bring critical citizens back in? In: International conference on electronic participation. Springer, pp 83–93

Bohøj M, Borchorst NG, Bødker S, Korn M, Zander PO (2011) Public deliberation in municipal planning: supporting action and reflection with mobile technology. In: Proceedings of the 5th International conference on communities and technologies. ACM, pp 88–97

Busch M, Lorenz M, Tscheligi M, Hochleitner C, Schulz T (2014) Being there for real: presence in real and virtual environments and its relation to usability. In: Proceedings of the 8th Nordic conference on human-computer interaction: fun, fast, foundational (NordiCHI '14). ACM, New York, NY, USA, pp 117–126

Carver S, Openshaw S (1996) Using GIS to explore the technical and social aspects of site selection for radioactive waste disposal facilities. Accessed 14 September 2016 from http://eprints.whiterose.ac.uk/5043/1/96-18.pdf

Conroy MM, Evans-Cowley J (2006) E-participation in planning: an analysis of cities adopting on-line citizen participation tools. Environ Plann C Govern Policy 24(3):371–384

Desouza KC, Bhagwatwar A (2012) Citizen apps to solve complex urban problems. J Urban Technol 19(3):107–136

Dix A (2009) Human-computer interaction. Springer

Fröhlich P, Oulasvirta A, Baldauf M, Nurminen A (2011) On the move, wirelessly connected to the world. Commun ACM 54(1):132–138

Gordon E, Baldwin-Philippi J (2014) Civic learning through civic gaming: community planit and the development of trust and reflective participation. Int J Commun 8(2014):759–786

Harding M, Knowles B, Davies N, Rouncefield M (2015) HCI, civic engagement & trust. In: Proceedings of the 33rd annual ACM conference on human factors in computing systems. ACM, pp 2833–2842

Karsten J, West DM (2016) Streamlining government services with bots (07 June 2016). Accessed 12 Sept 2016 from https://www.brookings.edu/blog/techtank/2016/06/07/streamlining-government-services-with-bots/

Kearns I, Bend J, Stern B (2002) E-participation in local government. Institute for Public Policy Research

Kim S, Lee J (2012) E-participation, transparency, and trust in local government. Public Adm Rev 72(6):819–828

Kohlisch O, Kuhmann W (1997) System response time and readiness for task execution the optimum duration of inter-task delays. Ergonomics 40(3):265–280

Korn M (2013) Situating engagement: ubiquitous infrastructures for in-situ civic engagement. PhD Dissertation. Aarhus University, Science and Technology, Institute for DatalogiDepartment of Computer Science

Kweit MG, Kweit RW (2004) Citizen participation and citizen evaluation in disaster recovery. Am Rev Public Adm 34(4):354–373

Lukensmeyer CJ, Torres LH (2008) Citizensourcing: citizen participation in a networked nation. Civic Engagem Netw Soc 2008:207–233

Mace RL, Story MF, Mueller JL (1998) The universal design file: designing for people of all ages and abilities. NC State University

Parasuraman A, Zeithaml VA, Malhotra A (2005) ES-QUAL a multiple-item scale for assessing electronic service quality. J Servi Res 7(3):213–233

Phoneia Technology & Entertainment (2016) Politibot, the first bot Telegram to follow the elections 26J (10 July 2016). Accessed 12 Sept 2016 from http://phoneia.com/politibot-the-first-bot-telegram-to-follow-the-elections-26j/

Poplin A (2012) Playful public participation in urban planning: a case study for online serious games. Comput Environ Urban Syst 36(3):195–206

Prieto-Martín P, de Marcos L, Martínez JJ (2012) A critical analysis of EU-funded eParticipation. In: Empowering open and collaborative governance. Springer, pp 241–262

Rhodes DL, Wolf W (1999) Overhead effects in real-time preemptive schedules. In: Proceedings of the international workshop on HW/SW codesign, pp 193–197

Schröder C (2015) Through space and time: using mobile apps for urban participation. In: Conference for e-democracy and open governement, p 133

Simão A, Densham PJ, Haklay MM (2009) Web-based GIS for collaborative planning and public participation: an application to the strategic planning of wind farm sites. J Environ Manage 90(6):2027–2040

Steinberger F, Foth F, Alt F (2014) Vote with your feet: local community polling on urban screens. In: Proceedings of the international symposium on pervasive displays. ACM, p 44

Szameitat AJ, Rummel J, Szameitat DP (2009) Behavioral and emotional consequences of brief delays in human-computer interaction. Int J Hum Comput Stud 67(7):561–570

Tambouris E, Liotas N, Tarabanis K (2007) A framework for assessing eParticipation projects and tools. In: 40th annual Hawaii international conference on system sciences HICSS 2007. IEEE, p 90

Thiel S-K, Fröhlich P, Sackl A (2016) Experiences from a living lab trialling a mobile participation platform. In: Real Corp'16: 21st international conference on urban planning and regional development in the information society geomultimedia, pp 263–272

Valkanova N, Walter R, Moere AV, Müller J (2014) MyPosition: sparking civic discourse by a public interactive poll visualization. In: Proceedings of the 17th ACM conference on computer supported cooperative work & social computing. ACM, pp 1323–1332

Vogt M, Fröhlich P (2016) Understanding cities and citizens: developing novel participatory development methods and public service concepts. In: Proceedings of 21st international conference on urban planning, regional development and information society. RealCORP, pp 991–995

Webler T, Tuler S (2000) Fairness and competence in citizen participation theoretical reflections from a case study. Adm Soc 32(5):566–595

West MD (2004) E-government and the transformation of service delivery and citizen attitudes. Public Adm Rev 64(1):15–27

Captcha Your Location Proof—A Novel Method for Passive Location Proofs in Adversarial Environments

Dominik Bucher, David Rudi and René Buffat

Abstract A large number of online rating and review platforms allow users to exchange their experiences with products and locations. These platforms need to implement appropriate mechanisms to counter malicious content, such as contributions which aim at either wrongly accrediting or discrediting some product or location. For ratings and reviews of locations, the aim of such a mechanism is to ensure that a user actually was at said location, and did not simply post a review from another, arbitrary location. Existing solutions usually require a costly infrastructure, need proof witnesses to be co-located with users, or suggest schemes such as users taking pictures of themselves at the location of interest. This paper introduces a method for location proofs based on visual features and image recognition, which is cheap to implement yet provides a high degree of security and tamper-resistance without placing a large burden on the user.

1 Introduction

In recent years the impact of online ratings and reviews on the decisions people make has steadily risen, and has thus also moved into the focus of research. Among others, the empirical analysis by Ye et al. (2011) showed that a 10% increase in the online rating of hotels led to a boost of online bookings by more than 5%, while Anderson et al. (2012) found that an extra half star rating causes restaurant sales to go up by 19%. As a consequence, the number of corresponding platforms has risen and ratings and reviews have become an important factor for business owners.

D. Bucher (✉) · D. Rudi · R. Buffat
Institute of Cartography and Geoinformation, ETH Zurich, Stefano-Franscini-Platz 5, 8093 Zurich, Switzerland
e-mail: dobucher@ethz.ch

D. Rudi
e-mail: davidrudi@ethz.ch

R. Buffat
e-mail: rbuffat@ethz.ch

© Springer International Publishing AG 2018
P. Kiefer et al. (eds.), *Progress in Location Based Services 2018*, Lecture Notes in Geoinformation and Cartography, https://doi.org/10.1007/978-3-319-71470-7_14

Generally, online ratings and reviews (in the following only referred to as reviews) can be separated into those that refer to a product and those that refer to a point of interest (POI), such as a hotel or a restaurant. This work focuses on the latter. While some platforms are specialized on certain domains, such as restaurants or hotels (as in the case of TripAdvisor[1]), general purpose platforms such as Yelp[2] or Google Maps[3] also exist. Using these platforms, it is possible to review any POI from a train station to a national park.

As online reviews directly influence the revenue of businesses, the incentive to create fake reviews is high. Not surprisingly, Hu et al. (2011) were able to detect manipulations in online reviews on both Amazon[4] and Barnes & Nobles.[5] Currently the major platforms do not implement any sophisticated measures to prevent the creation of fake reviews, other than that reviews need to be written by humans. This makes it easy to create malicious reviews, such as fake positive reviews for the own business or fake negative reviews for the competitors. Thanks to crowdsourcing platforms such as Mechanical Turk,[6] companies that offer the creation of fake reviews have access to a large human workforce. As a consequence, the state of the art systems are not able to efficiently prevent the creation of fake reviews.

Hence, in order to prevent fake reviews, it is not only necessary to check if a user is human, but also to verify that she visited the location she wishes to review. Such a "proof of location" or "location proof" can for example be achieved by letting the user solve a location based challenge. To do so, the location based challenge needs to fulfill the following properties:

- The challenge should only be solvable if the user is present at the correct location.
- It should not be possible to use a solved challenge a second time.
- A solution to the challenge should only work for one particular location.
- Business owners or other entities with positive or negative intentions with regard to the location of the challenge should not have any influence on the challenge.
- In order to ensure scalability there should be no need to locally install additional hardware or to require other users or entities to be present at the same time.

The first three properties are obviously required in order to ensure that a user is at a specific location. For example, if the challenge would simply consist of a secret code that is attached to the wall of a restaurant, this key could easily be distributed to other users. Furthermore, the business owner can choose whom he wants to show the key, thus avoiding potential bad reviews.

In this work, we propose a location based challenge using photographs. A user must take a picture of a location with her smartphone camera in order to prove that

---

[1]https://www.tripadvisor.com.

[2]https://www.yelp.com.

[3]https://www.maps.google.com.

[4]https://www.amazon.com.

[5]https://www.barnesandnoble.com.

[6]https://www.mturk.com/mturk/welcome.

she actually is at that location. Specifically, the user has to take a picture that overlaps by 50% with the right part of an existing picture. The verification is achieved by matching the photo to the previous photos. As nowadays nearly everybody has a smartphone with a decent camera, no additional infrastructure is required. Matching of photos is a well-studied problem. State-of-the art algorithm can automatically match multiple images to create panoramic images (Brown and Lowe 2007). These algorithms work by first extracting and matching scale-invariant feature transform (SIFT) features (Lowe 2004) from the different images. Thus, these algorithms inherently must decide if two images can be matched or not. Even when having a large database of existing photos, it is unlikely that an existing photo, not yet known to the system, matches a previously taken photo.

This paper is structured as follows. The next section discusses the relevant literature. We then present our method followed by an analysis of the adversarial model, as well as a simulation thereof. Finally, we discuss the location proof and draw our conclusions.

2 Related Work

A large body of work treats location proofs and secure location claims, as proving that a device or person is at a claimed position is an important step in many tasks. Location-based access control and authentication (Sastry et al. 2003; Francillon et al. 2011), interaction with online location-based services (Zhu and Cao 2011; Javali et al. 2015; Khan et al. 2014), or people-centric sensing (with smartphones) (Talasila et al. 2013) are examples where adversarial users might want to fake their locations in order to gain access, get additional benefits from services, or simply disturb the system. Our motivating examples are *online reviews*, where users can post their experiences with a service or at a location to a central system, making it available for other people which might be thinking about using the same service, or visiting the same location. In such settings, it is not uncommon for the different entities to cheat by posting fake reviews about their own service or a competitors' one (Mayzlin et al. 2014). Mayzlin et al. (2014) propose a methodology for detecting review manipulations, and find examples of both positive as well as negative manipulation on different review platforms, in particular when a competing service is located closely to the one being reviewed. Optimally, users would have to prove that they a) actually were or currently are at the location they are reviewing, and b) that they are not owning the business, nor being otherwise closely affiliated with it.

In general, such proofs require an active component, i.e., a device that has both computational and communicative powers, and a specialized communication protocol between the device and a user's smartphone. For example, Sastry et al. (2003) present the Echo protocol, which allows a set of *verifiers V* to verify that a *prover p* is in a certain *region of interest R*. Echo requires the verifier and prover to be able to communicate both using radio frequency as well as sound – in its simplest form the prover echos a request by the verifier using ultrasound, and the time required for this action bounds the maximal distance the prover is from the verifier. Ultimately, such *time-of-flight* or *time-difference-of-arrival* based approaches are very common for location proofs (Brands and Chaum 1994; Waters and Felten 2003), but require specialized devices at each location, which makes them an unfavorable choice for review platforms. In addition, to bound the round trip time (from verifier to prover and back), the prover has to be able to compute a response within a short time frame, requiring a device capable of doing so. This also implies that the active device is in control or possession of the reviewee, which is usually not given for online reviews, where the central platform has no direct connection to the individual services and locations. In the past, different communication technologies, such as Wi-Fi (Waters and Felten 2003; Luo and Hengartner 2010; Saroiu and Wolman 2009; Sastry et al. 2003; Javali et al. 2016), Bluetooth, (Mengjun et al. 2016; Wang et al. 2016; Zhu and Cao 2011) or RFID (Gao et al. 2012) were used for *time-of-flight* location proofs.

Another line of research concerns location verification utilizing third-party witnesses. Khan et al. (2014) require a spatio-temporally co-located entity to be present when the verifier tries to verify the provers location claim. Witnesses register with a location authority, and are used to generate location proofs (which are sent to the prover), which in the end can be presented to another party requiring location verification. Their protocol is resistant to malicious verifiers, provers, and witnesses, but requires the presence of even more active devices. Similar methods requiring witnesses are described in literature (Mengjun et al. 2016; Wang et al. 2016). While finding a witness might be possible for online reviews, it again is difficult for the review system to control location authorities at every location.

In our work, we propose a weaker form of location verification, inspired by so-called Captchas (Von Ahn et al. 2003). In essence, "a Captcha is a cryptographic protocol whose underlying hardness assumption is based on an [artificial intelligence (AI)] problem" (Von Ahn et al. 2003, p. 296). Commonly, Captchas are known from registration websites, where they appear as distorted images, in which a user has to recognize letters or numbers in order to verify his or her *human* nature. Saroiu and Wolman (2009) present an approach inspired by Captchas, which has the intention of proving that a person (i.e., not a device) is at a certain location. For this, they require that a picture of the person at the location is being sent to the verifier. The proof is made stronger by making the person hold up a paper, on which a certain requested message by the verifier is written. This prevents a malicious user from simply reusing the same picture over and over again. Our approach differs from the approach in Saroiu and Wolman (2009) in that we do not require the person to be present in the picture (i.e., we verify the location of the device), and that the repeated use of the same picture is prevented by requesting a picture at a random (but well-defined) location. It also differs from conventional Captchas, because its underlying hardness assumption is based on location properties, i.e., only someone present at the location is able to easily solve the problem.

3 Method

This section introduces the "Captcha your Location Proof" (CLP) method. The CLP method works with little to no infrastructural overhead, and has a similar mechanism to that known from Captchas. In particular, CLP uses the smartphones' capability of taking pictures for proving a client's location. The section starts with presenting the stakeholders, i.e., the natural persons or organizations involved in the execution of the proof. Then, the actual procedure describing both the underlying data structure and the communication between a client and a location based service provider employing the location proof method is explained in detail.

3.1 Stakeholders

The CLP approach involves three different stakeholders, the location based service provider, the location owner and the client.

Location Based Service Provider. The *location based service provider* (called: provider) is an organization or a group which wants to use CLP to ensure that clients of its (crowd sourced) services receive authentic information about locations. In particular, the provider acts as a trusted entity throughout this process. The provider typically hosts CLP within a web application, such as an online review site.

Location Owner. The *location owner* is an organization or a group which manages a particular location of interest, such as a restaurant or a park. The location owner does not have to set up any infrastructure for CLP to work.

Client. The *client* (or also *user*) is both the consumer and producer of information regarding the locations made available through the provider.

3.2 Procedure

The Data Structure. For the CLP approach we focus on the client as a producer of information, i.e., someone who wants to add some kind of information (e.g., a rating, a description, etc.) to a particular location she previously visited (e.g., a restaurant, a park, etc.). The provider (technically represented by an appropriate IT infrastructure) requires an *image* of the location from the client as a proof of presence. We define the combination of these three data items (i.e., information, location and image) together with the client as a *contribution* (cf. Fig. 1).

The images of contributions of different users are required to overlap partially, i.e., new images need to partly depict scenes of already existing images. In our data structure, this means that each image contains a reference to its successor, forming a directed tree $\mathcal{I}_t$. A successor of an image i_l is an image i_k that overlaps with 50% of the right half of i_l (cf. Fig. 4). Images taken by different clients from the same

Fig. 1 The *simplified* class diagram for the data structure underlying the CLP approach

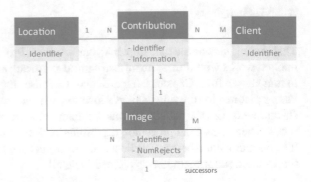

Fig. 2 An UML sequence diagram depicting the "The Contribution" step of the CLP procedure

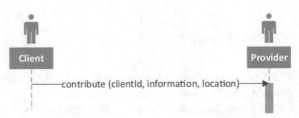

perspective constitute the level l in the image tree $\mathscr{I}_t$. In particular, all images $i_{l,1}, \ldots, i_{l,n}$ at level l of the image tree have all images $i_{k,1}, \ldots, i_{k,m}$ of level k in the image tree as successors. Consequently, connecting the images of a path through the image tree would result in a big panoramic image representation of a location. We also store the number of times an image was rejected or accepted as a valid representation of a location.

Every now and then we ask clients to start a new image tree $\mathscr{I}_t$ for a particular location. Therefore, a location refers to one or more image trees $[\mathscr{I}_0, \ldots, \mathscr{I}_n]$ represented by their respective root image, and t indicates the tth tree in that list of trees.

The Contribution. The CLP approach starts with a *contribute* message sent from the client to the provider. The message contains the client's ID, the information she wants to contribute and the location the contribution is for (see also Fig. 2). For example, Alice wants to create a contribution for a restaurant she visited in Tokyo (see Fig. 3). She therefore sends the following (simplified) message to her provider: "(id: alice2018, location: (Tasty Edamame, Tokyo, Japan), information: ('5 stars', 'amazing sushi place'))".

The Challenge. After receiving the client's contribution request, the provider creates a temporary "Contribution" instance and returns its ID (for an appropriate communication tracking) together with a challenge (see also Fig. 5). That is, the provider challenges the user to prove she actually is at a location by taking a picture of it. More precisely, this is done by choosing from one of the following challenges, i.e., either to:

- *append* a picture to a known image tree of the location, i.e., taking a new picture that partially overlaps with an existing one in that tree, or to

Fig. 3 The location the
"Alice" wants to create a
contribution for

- *prove* an existing picture, i.e., taking a new picture that completely overlaps with a known picture in a known image tree, or to take a completely
- *new* picture, which is not compared to any other picture and constitutes the root for a new image tree.

Each of these challenges is chosen with a different probability. Formally speaking, the provider chooses the challenge $c \in \{append, prove, new\}$, where $p_{append} \in [0, 1], p_{prove} \in [0, 1]$ and $p_{new} \in [0, 1]$, with $p_{append} + p_{prove} + p_{new} = 1$.

To reduce the complexity, we assume that images can only be appended to the right of an existing image, i.e., there is at most one image per tree a user can append to, namely the rightmost one. We denote the list of all images for a location, i.e., the concatenation of all images of all image trees $\mathscr{I}_t$ as $I = [\mathscr{I}_0, \ldots, \mathscr{I}_n]$. We further define an image $i \in I$ by $I[idx]$, where $idx = (idx_t, idx_l, idx_i)$ references a single tree by idx_t, a particular level by idx_l, and an image on this level by idx_i.

The image for an *append* challenge is defined as $i_c = I[idx]$, with $idx_l = |\mathscr{I}_t| - 1$ for a randomly chosen image tree $\mathscr{I}_t$ ($idx_t = t$; idx_i is randomly chosen from all the available images in the tree at that level). For the *prove* challenge the image is defined as $i_c = I[idx]$, where $idx_l < |\mathscr{I}_t| - 1$ for a chosen tree, i.e., the image is not the rightmost image in this image tree. Finally, to create a *new* image tree $\mathscr{I}_{n+1}$ the provider asks the client for an image and accepts whatever the client responds.

Note that, even though the provider actually chooses out of three different challenges, from the client's perspective there is no actual difference between *append* and *prove*, since the provider always sends some image i_c and marks 50% to the right of it for matching or asks her to take a completely new picture. If we assume that each user can only make one contribution for any location, the request is rejected if such a contribution has already been made.

In our example from Alice's perspective the provider responds to her depending on whether she is the first or *n*th contributor for the location '(Tasty Sushi, Tokyo, Japan)':

(a)   **(b)**

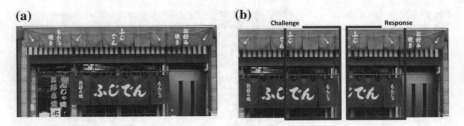

Fig. 4 **a** The picture Alice could take given the *new* challenge. **b** After receiving the *append* challenge with the picture to the left i_c, Alice could take the picture to the right i_r to match the right hand side (red boundary) of it

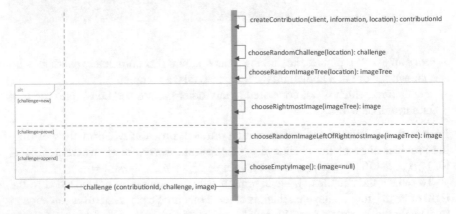

Fig. 5 An UML sequence diagram depicting the "The Challenge" step of the CLP procedure

- *In case of the first contribution* Alice receives the response (contributionId: 42, challenge: 'new', image: null) and is thus asked to take a picture of the "Tasty Sushi" (see Fig. 4a).
- *In case of the nth contribution* Alice either receives the response (contributionId: 42, challenge: 'new', image: null) and is again asked to take a picture of the "Tasty Sushi" (see Fig. 4a), or she receives the response (contributionId: 42, challenge: 'append', image: i_c) and is asked to append to the right of an existing picture (see Fig. 4b).

The Response. After receiving the challenge, the client responds with either one of the following (see also Fig. 6):

- *img*, i.e., taking the picture, or
- *reject*, i.e., not taking the picture (except if we had a *new* challenge)

Rejecting to send an image could be either because the client cannot or does not want to respond to the challenge.

In case of Alice this means that she could either:

- *Take a picture* and respond with (contributionId: 42, image: i_r) (see also Fig. 4b), or

Fig. 6 An UML sequence diagram depicting the "The Response" step of the CLP procedure

- *Reject i_c* and respond with (contributionId: 42, image: null)

  The Reaction. The provider's reaction to the the client's response depends on the previously chosen challenge (see also Fig. 7):

- if the provider chooses an *append* challenge with the image i_c

  - and the client responded with *img* and the image i_r: the provider checks whether there is a 50% overlap between i_c and i_r. If not, the client's response is rejected and she is asked to submit a new image $i_{r,new}$. Otherwise, i_r is added to $\mathscr{I}_t$ at level $|\mathscr{I}_t|$ and is defined as successor to all images at level $|\mathscr{I}_t| - 1$. The provider then responds with an "OK" message to the client.
  - and the client responded with *reject*: the provider increments the "NumRejects" counter of i_c and goes back to the **The Challenge** step and reruns the process from there on.

- if the provider chose a *prove* challenge with the image i_c

  - and the client responded with *img* and the image i_r: the provider checks for a 50% overlap between i_c and i_r, as well as for a 100% overlap with all images stored at the same level as i_c. If there was already no 50% overlap, the client's response is rejected and she is asked to submit a new image $i_{r,new}$. If however the 100% comparison fails, the provider accepts the contribution anyway, but increases the "NumRejects" counter for the images that do not match i_r. i_r is then added to $\mathscr{I}_t$ at the level of $i_c + 1$ and is defined as successor to all images at the level of i_c. The provider responds with an "OK" message to the client.
  - and the client responded with *reject*: the provider increments the "NumRejects" counter of i_c and all images at the same level in $\mathscr{I}_t$ that match it, and goes back to the **The Challenge** step and reruns the process from there on.

- if the provider chose a *new* challenge

  - and the client responded with *img* and the image i_r: the provider creates a new image tree $\mathscr{I}_{n+1}$ with i_r as the root image and n being the number of image trees for the location. The provider responds with an "OK" message to the client.

At any point in time, the client can cancel the procedure; this is not treated separately here.

In case of Alice this means that she could either:

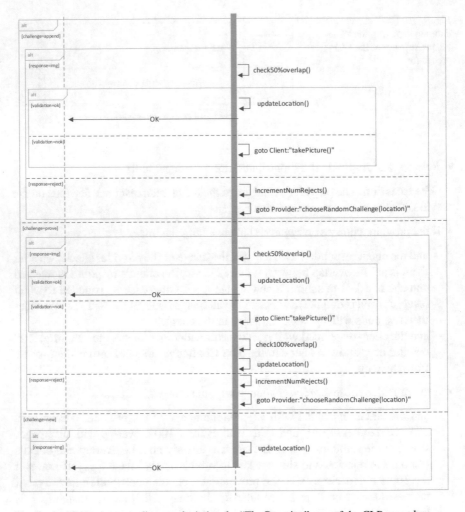

Fig. 7 An UML sequence diagram depicting the "The Reaction" step of the CLP procedure

- *Take a picture* and respond with (contributionId: 42, image: i_r), or
- *Reject i_c* and respond with (contributionId: 42, image: null)

3.3 Adversarial Models

3.3.1 Assumptions

We assume that the service provider has a crowd sourced location based service. In particular, the service provider wishes to ensure that its clients' contributions are authentic, i.e., that the client actually was at the location she made the contribution for. The communication between the clients and the server is secure, and the server (and thus also the service provider) is a trusted entity. We assume that the location owner wants the information which is made available to the clients by the service provider to be of a positive nature.

Furthermore, the clients, to prove the authenticity of their contribution, use a mobile phone to take a picture at the locations they wish to contribute information about. The picture along with their current (GPS) position, as well as a set of any other information (e.g., a rating) is then sent to the service provider. Moreover, the clients trust the service provider to ensure that the location based information they receive from other clients is authentic. The image processing is assumed to be able to detect if two images are similar or completely distinct. In addition, we assume the challenge images to be watermarked, i.e., it is not possible for an adversary to simply reuse these images at a later point in time.

Finally, we we will assume scenarios, where adversarial contributions appear in "bursts", i.e., ignoring a location's actual quality, adversarial contributions will either occur at the beginning or the end of a contribution "history". This is based on the observations from Hu et al. (2011), i.e., an adversary might either create novel contributions for a location that has none, or try to negate whatever contributions already exists. In either case the contributions aim at either discrediting a location or boosting its reputation. To realize this, they have to occur within a certain clustered timeframe. This holds true even if adversaries would try to make the contributions with temporary displacements to avoid attracting attention.

3.3.2 Threats

Threats can originate from both clients and location owners. Reasons to manipulate contributions could be that the client is a competitor of the location owner and wishes to discredit the location, or that she is employed by the location owner to add positive information regarding the location. The location owner could also act as a client herself.

Generally, an adversary (either a client or location owner) can easily spoof her GPS location. That step would by itself however not necessarily break the CLP approach. In particular, we identified the following additional actions that are necessary to pose a threat:

- **Malicious first contribution**. An adversary (client or location owner) can contribute with an initial image which is not of the location L she claims to be at. For example, she can take a picture of any other location L' and have that image uploaded. Afterwards the adversary can add further images and information from the fake initial location L'.
- **Similar looking locations**. An adversary might try to make a contribution from a location L' for the location L by trying to find an image that she believes might be accepted by the system. That is, after receiving the challenge for a segment of

an image of L, she searches an image database and finds one that she believes has similar enough features.

- **Catalog of real images**. An adversary could take many pictures of the actual location L and store those in a data storage. Afterwards the adversary could create contributions and use the image database for looking up an appropriate proof that complies with the CLP approach.
- **Systematically boosting fake contributions**. Adversaries could systematically reject all images they know are real and try to find images fitting to fake images.
- **Creating a fake image tree**. Adversaries could systematically reject all images until they receive the *new* challenge and afterwards continue rejecting any image that is not within that image tree and only append to that tree.
- **Accepting only new challenges**. Adversaries could systematically reject all challenges until they receive the *new* challenge and create only new image trees with fake images.
- **Man-in-the-middle**. A known attack against the Captcha system was an application that relayed Captchas to third users, which were incentivized by various rewards to return the solved problem to the original user. With our system, such man-in-the-middle attacks are much harder as somebody needs to be present at the actual location.

The next section will evaluate these threats using a simulation approach, and discuss countermeasures for the threats which CLP cannot prevent.

4   Technical Evaluation

In this section, we will discuss the advantages and disadvantages, as well as the presence of adversaries in the CLP approach in more depth. Recall that upon receiving an image, a user has two options to respond: $\{img, reject\}$, where the first response means sending an *image i* (either appended to the *challenge image* i_c with an overlap of 50%, or a completely new one), and the second means to *reject* the challenge image, thus discrediting it, and receiving a new challenge (this second response is not available for the *new* challenge).

4.1   Honest and Adversarial User Responses

For each challenge, we have to evaluate the possible answers of a user u_h (honest), and a user u_a (adversarial). Note that we do not make any statements on the honesty of the contribution of a user, but only on its location—someone who is present at a certain location can still send reviews discrediting a competitor. The user has to be at the location, though, and cannot simply outsource the task to some company operating from another location.

The *new* challenge. If a user receives a *new* challenge, there is only one possible response, namely to send a new image i, which will be stored in the CLP system as part of a user's contribution. As we do not know if u is honest or adversarial, the system simply stores the image without any further processing.

The *append* challenge. When appending to an image i_c, honest and adversarial users react differently. In the following, we assume that u_h always is correctly following protocol, while u_a chooses either the potentially best outcome for herself, or the worst impact on the system (to destabilize it). As such, user u_h will perform one of the following two actions:

- The user finds the scene depicted by the challenge image i_c at the location, and is able to append as requested. She will send the appended image i, which will be stored as part of a contribution associated with the given location.
- The user is not able to find the scene depicted by i_c, and thus has to *reject* the challenge. The challenge image i_c is now discredited, i.e., the CLP system increases the rejection count of image i_c. This will later be important, when we assess which users to trust, and which users to flag as adversarial.

A dishonest user u_a will react completely different upon receiving an *append* challenge. Let us first assume the challenge image i_c is honest, i.e., the scene depicted by it can be found at the location. In this case, the adversarial user has two options:

- She can reject the honest image i_c, thus discrediting it, and receive a new challenge. The CLP provider will store that u_a discredited the image, and continue as usual.
- Append a dishonest image by taking the overlapping half of i_c, generating some arbitrary image for the other half, and sending it back to the CLP provider. Again, the CLP system does not know this is an adversarial image yet, so it will store the the image as part of the user's contribution.

Choosing the first option, u_a builds up distrust towards honest users, which can be an advantage, as long as her adversarial identity is not revealed. With the second option, u_a builds up trust, as long as the contribution is not discredited by honest users. In case the challenge image i_c itself is dishonest, u_a would have the same choices again. However, rejecting a dishonest image would discredit the adversarial owner of that image, and result in uncovering the owner's mischievous doings. As such, it would always benefit honest users, and the trustworthiness of the system. We can thus assume that in order to get the maximal benefit for herself, a malicious user always appends to a dishonest image.

We introduce the probabilities p_r (reject) and $p_a = 1 - p_r$ (append) with which the response of an adversarial user is chosen. As discussed above, $p_a = 1$ for a dishonest challenge image, as the adversary will always append. We will evaluate different values of p_r and p_a for a honest challenge image below, but for example, if the probability to reject is $p_r = 1$, this means the adversarial user will reject until she either has to *append* to a dishonest image or gets a challenge for a *new* image (the *prove* challenge will also always be rejected, as from a user perspective, *append* and *prove* look equal).

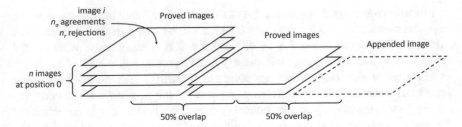

Fig. 8 Schematic representation of the images in the CLP system. Each image can stem from a honest or from an adversarial user, and has a certain number of agreements (successfully appended) and rejections associated with it (number of times the image was rejected during an *append* or *prove* challenge). Whenever a *new* challenge is responded with a new image, another such a "list of images" is started

The *prove* challenge. From a user's perspective, the *prove* challenge cannot be distinguished from the *append* challenge. As such, the responses are the same, but since the CLP system has additional knowledge this results in a different procedure on the system.

Namely, the system will check for any appended image, if it corresponds to any of the other images taken at this position. This results in a number of agreeing images, and a number of disagreeing ones. In the worst case, all adversaries work together, i.e., they send images that agree with each other, but not with the images of the honest users. On the other hand, all images of honest users always agree with each other, per our definition of honesty. Figure 8 shows the images associated with a given tree of images, displayed as a list of stacks of proved images. Such a list is started whenever an image is entering the system as response to a *new* challenge. The images in this list can stem from honest and adversarial users, and thus create a number of *trust votes* for each other. For example, in a stack with 5 honest images, and 3 adversarial ones, each of the honest images has 4 votes of trust, and 3 votes of distrust, while each of the adversarial images has 5 votes of distrust, and 2 votes of trust. The votes are simply computed by applying image processing techniques to all images, and measuring how many of the features of two images agree with each other. Note that these images would all overlap each other, for clarity they were drawn separated in Fig. 8.

In addition to these trust votes, each image can be appended to or be rejected in response to *append* and *prove* challenges, which is counted for each image in variables n_a (number of agreements) and n_r (number of rejections).

4.2 Assessing Adversarial Users

We are ultimately interested in determining which users are adversaries, and which users can be trusted. As a system, our initial trust towards each user is the same, but given user votes, we can determine which users we can trust more, and as a result, which users' reviews we have to remove from the system again. We can do this using the knowledge gained from rejected images, as well as from proved images that do not match.

In the following analysis, we restrict ourselves to saying that every user can at most prove her location once (i.e., each user is unique and does not appear for multiple images nor multiple locations). We reason that security improves if users are allowed to prove their location several times and at multiple locations, as honest users can build up trust, which can be weighted more when wanting to detect adversaries. Insofar, the situation considered here is a worst-case scenario which should improve in real-world systems.

As described above, we have two sets of votes for each image, one from rejections, and one from comparison to other proved images. The rejection votes make statements about users thinking that a particular image is adversarial (i.e., it will contain an arbitrary number of votes, depending on the number of times the image was chosen to be part of a challenge). The prove votes make a similar statement implicitly, and thus always contains a number of votes corresponding to all other proved images at the same location.

We can now count the positive votes in a set of prove images I_p (which should depict the same location) for any given image i (and thus the user who posted it) as:

$$v_{i,p} = n_a + agree(i, I_p)$$

where n_a is the number of users who successfully appended to this image and $agree(i, I_p)$ is the number of images of the same location that agree (i.e., have enough matching features for the image processing to recognize them as the same location) with this image. The number of negative votes is computed as:

$$v_{i,n} = n_r + (|I_p| - agree(i, I_p))$$

where n_r is the number of users who rejected to append to this image and $|I_p| - agree(i, I_p)$ is the number of images at the same place which do not agree with this image. We now trust an image (and thus the user who posted it) if $v_{i,p} > v_{i,n}$. This also means that we trust the location of the user to be genuine. Note that with this scheme, we can discover adversarial users at any later point in time, and eventually remove their mischievous reviews. We thus have to recompute trust whenever a new user contribution enters the system, and assess which users we trust, and whom we have to classify as adversarial.

4.3 Simulating CLP Challenges and User Responses

To evaluate the influence of different probabilities and parameters on the number of adversaries CLP is able to identify, we now present a simulation of the approach. The simulation model exactly follows the above described method, i.e., honest and adversarial users want to add a contribution to a system, and are challenged to either provide a *new* image, *append* to an existing one, or *prove* an image. While the *append* and *prove* challenges look equal from a user's point of view, the system handles their responses differently. Within the simulation, we model a restricted view of the world for the CLP system, as well as for the clients themselves. As the simulator knows everything (in particular, which users are honest, and which ones are adversarial), we can easily compute how many users were correctly classified as adversarial by the system, and how many honest users are wrongly accused of being dishonest. In an optimal system, no honest user would be treated as an dishonest, and no adversary would be treated as honest.

The simulation routine is shown in Algorithm 1. The functions *generateUser* (L1) and *generateChallenge* (L4) simply generate either *honest* or *adversarial* users, and *new*, *append* or *prove* challenges, according to the rules defined above. *insertImage* (L10) increases the agreement count, and inserts the image at the correct position in the right image tree. The core function *simulateStep* first generates a random user and a random challenge (L13–15), and then applies the logic described above to the image collection, depending on the type of challenge. If a user rejects a certain challenge image, the *goto* (L22/26) statements cause the simulator to generate a new challenge, and restart the procedure. Finally, we count the number of distrusted honest users $h_{distr.}$ as the percentage of all honest users, for whom the majority vote yielded that they should not be trusted, and $a_{trust.}$ as the percentage of all adversaries, for whom the vote yielded that they should be considered honest.

We ran the simulation for a range of values for the parameters p_{new}, p_{prov}, p_r and p_{adv}, in order to determine the best values for p_{new} and p_{prov} (p_{app} can be calculated from them), given different adversary strategies. Table 1 shows the best values (when minimizing a_{trust}) for p_{new} and p_{prov} for given probabilities of adversaries, and different adversarial strategies. *beg.* and *end* are two adversary strategies, where there either is a burst of adversaries in the beginning ($p_{adv} = 0.7$ for the first 25 contributions, and 0.1 for the rest) or in the end ($p_{adv} = 0.7$ for the last 25 contributions, otherwise 0.1). For each parameter combination, we ran the simulations five times for 100 contributions each, and measured the average of the final percentages for distrusted honest users and trusted adversaries.

Figure 9 gives exemplary outputs for three simulations. On the left, a scenario with $p_{adv} = 0.2$ is shown, while in the middle the adversaries are clustered in the

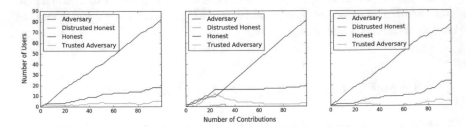

Fig. 9 Output of three simulation runs, where honest and adversarial persons use CLP to upload contributions to a review site

Algorithm 1: The simulator function, which is executed iteratively. Each iteration adds another user contribution, either from an honest or from an adversarial user. $\mathbf{I} = \{I_0, \ldots, I_n\}$ corresponds to the image trees associated with a certain location, and is initially empty. The function *random()* generates a random number $r \in [0, 1]$, the function *randomImage($\cdot$)* selects a random image from all images in the trees passed to it, and the probabilities p_{adv}, p_{new}, p_{app}, and p_r describe the percentage of adversarial users, *new* challenges, *append* challenges, and the reject strategy of adversarial users. We assume *insertImage($\cdot$)* knows how to *append* a new image i_r to the image tree, given a predecessor i_c. A and R are accept and reject counters for all images, and are initially 0 for every image. Finally, *filter($\cdot$,f)* only returns the images which fulfill the predicate function f.

1	**Function** *generateUser ()*
2	**if** *random()* $< p_{adv}$ **then** **return** *AdversarialUser*
3	**else** **return** *HonestUser*
4	**Function** *generateChallenge (**I**)*
5	**if** $|\mathbf{I}| = 0$ **then** **return** *NewChallenge*
6	$r \leftarrow$ random()
7	**if** $r < p_{new}$ **then** **return** *NewChallenge*
8	**else if** $r < p_{new} + p_{prov}$ **then** **return** *ProveChallenge(randomImage($\mathbf{I}$))*
9	**else** **return** *AppendChallenge(randomImage($\mathbf{I}$))*
10	**Function** *insertImage (i_c, i_r)*
11	$A[i_c] \leftarrow A[i_c] + 1$
12	append($\mathbf{I}$, i_c, i_r)
13	**Function** *simulateStep ()*
14	$u \leftarrow$ generateUser()
15	$c \leftarrow$ generateChallenge($\mathbf{I}$)
16	**switch** c **do**
17	**case** *NewChallenge* **do**
18	$\mathbf{I} \leftarrow \mathbf{I} \cup \{I(i_r)\}$ ▷ *Where $I(i_r)$ starts a new image tree.*
19	**case** *AppendChallenge(i_c) or ProveChallenge(i_c)* **do**
20	**if** *typeof* $u = HonestUser$ **then**
21	**if** *typeofowner* $i_c = HonestUser$ **then** insertImage(i_c, i_r)
22	**else** $R[i_c] \leftarrow R[i_c] + 1$; **go to** 15
23	**else**
24	**if** *typeofowner* $i_c = DishonestUser$ **then** insertImage(i_c, i_r)
25	**else**
26	**if** *random()* $< p_r$ **then** $R[i_c] \leftarrow R[i_c] + 1$; **go to** 15
27	**else** insertImage(i_c, i_r)
28	$h_{distrusted} \leftarrow$ |filter($\mathbf{I}$, $i \mapsto$ **typeofowner** $i = HonestUser$ and $v_{i,n} > v_{i,p}$)|
29	$a_{trusted} \leftarrow$ |filter($\mathbf{I}$, $i \mapsto$ **typeofowner** $i = DishonestUser$ and $v_{i,n} < v_{i,p}$)|

Table 1 Different scenarios, and the percentage of identified adversaries. $h_{distr.}$ is the percentage of honest users that are distrusted, $a_{trust.}$ is the percentage of adversaries that are wrongly trusted

p_{adv}	p_r	p_{new}	p_{prov}	$h_{distr.}$ [%]	$a_{trust.}$ [%]
0.1	0.2	0.05	0.35	0.0	1.7
	0.5	0.05	0.70	0.1	8.1
	0.8	0.05	0.05	3.3	9.5
0.2	0.1	0.05	0.25	0.0	4.8
	0.5	0.05	0.50	0.1	7.8
	0.8	0.05	0.25	1.1	16.0
0.4	0.1	0.05	0.50	6.7	27.4
	0.5	0.05	0.80	2.4	27.8
	0.8	0.05	0.80	7.6	42.4
0.8	0.1	0.10	0.35	15.3	87.5
	0.5	0.05	0.50	11.3	87.6
	0.8	0.05	0.30	14.1	89.4
beg.	0.1	0.05	0.20	0.4	14.2
	0.5	0.05	0.20	0.1	26.5
	0.8	0.05	0.75	0.4	33.1
end	0.1	0.05	0.85	0.4	13.6
	0.5	0.05	0.75	0.8	16.2
	0.8	0.05	0.40	0.7	24.4

beginning, and on the right in the end (as described in the previous paragraph). $p_{new} = 0.1, p_{prov} = 0.3$ and $p_r = 0.2$ were constant.

It can be seen from both Table 1 and Fig. 9 that CLP usually is able to identify a large number of adversaries. The best strategies for adversaries are to either simply suppress the honest users (large p_{adv}), or to choose a large p_r, i.e., simply reject everything until the are allowed to send a new image. However, we argue that such users could be identified (for example, by always sending a *append* images with many trust votes), and thus it is not a good choice for adversaries to choose p_r very large. In reality, we would hopefully also always see a substantial number of honest users (otherwise, the place would not be of interest), i.e., p_{adv} cannot get too large. Finally, it is interesting to see that p_{new} should be low. This is because *new* challenges are the primary means for adversaries to hide their malicious intent. In reality, we cannot chose p_{new} too small, as honest users must have a chance to eventually submit a genuine *new* picture (otherwise, they would always have to reject).

4.4   Countermeasures

As the previous section argued, CLP is inherently able to detect a substantial share of adversarial users. Attacks generally become more difficult with a growing number of images, because the number of possible segments that we could challenge the user with is equal to the number of images for the location. Thus, even when an attacker finds a similar location, assuming a high enough number of images exist, finding a match becomes more difficult. In particular, since we have the 100% check it is possible that the system chooses the same image multiple times countering the possibility of a similar looking image.

Another easy way to handle malicious contributions is to allow clients to request a reset of the contributions, i.e., removing all contributions for a location. Graduations of such a solution are possible too, e.g., removing only the last n contributions. In practice, this could be implemented by adding a trust or gamification (cf. Weiser et al. 2015) layer on top of CLP, which would allow "power users" to manually assess all images at a location in return for points or other game elements.

Additionally, it is possible to adapt the size and position of image segments. For example instead of choosing 50% to the right of an image, we could extract two snippets, one of size 200 × 200 pixels and one with 400 × 400 pixels at an randomly chosen position of the "stitched up" image and ask a client to take a picture containing the smaller segment. Afterwards we could extract a segment of 400 × 400 pixels from the response image and match that against the challenge image of same size.

Allowing users to make contributions for multiple locations would allow them to build up trust, which can be used to spot malicious users. This could be further enhanced by using time geography (Miller 2005) to assess whether a person could have traveled between two locations within a given time frame.

5   Discussion

In this work, we presented a method for posing a location based challenge that allows a location based service provider to verify if a user was as at a particular location. Such a location proof is particularly important in crowd-sourced scenarios where a service provider wishes to ensure a certain degree of authenticity of the collected information.

The underlying idea is simple: We ask a user to take a picture of the location she wants to make a contribution for. That picture needs to partially match an existing picture contributed by some other user. Should the image match, we store it in the system and compare it to an ever-increasing set of images depicting the same scene, which allows assessing whom we can trust, and whom we should classify as adversarial. The advantage of such a principle is that it can be applied in large scale, as nowadays nearly everybody has a smartphone with a decent camera and fast image processing algorithms exist. We thus avoid any additional infrastructure or contexts that are difficult to achieve, which current state of the art methods for location verification usually require (additional local infrastructure or co-located witnesses).

We thoroughly analyzed our method for possible attack scenarios and found that while we cannot ensure the authenticity of the location of every user, in many realistic scenarios CLP can detect the majority of attackers. That is, our method is not "bullet-proof" (like a cryptographic method) but makes malicious contributions in a real world scenario considerably more expensive (similar to Captchas, where the challenges are more difficult for automated systems, CLP challenges are more difficult for people absent from a certain location). Moreover, we were able to quantify how many attackers we can identify under which circumstances (in particular for which given probabilities). While our analysis is based on a simulation, the same technology could be used in a real (prototypical) system if supplied with data from a computer vision subsystem and integrated into a web application. As part of our work on location based need matching systems (Bucher et al. 2017), we are working on different location proof system implementations.

Nonetheless, the proposed method has certain limitations. The location challenge requires an honest user to be present at the location. This is a drawback regarding the convenience for the user. From a practical point of view the location of the photo of a location challenge needs to be identifiable by the user. This requires that this location is within the line-of-sight of the user. For normal locations such as restaurants this will not be a problem. However, POIs vary greatly in sizes, e.g., for a national park spanning multiple square kilometers finding the right location will not be feasible. This can be circumvented by also considering the GPS location of the user and only using images that are close to that GPS location. An additional challenge is changing POIs, for example when a restaurant is renovated or a fair moves to another location. However, one can argue that for these POIs the accuracy of an old review also potentially decreases very fast. Additionally, the countermeasure of resetting the contributions could be exploited here as well.

For practical deployment, gradual introductions of CLP are possible (as the hard requirement to take a matching picture could discourage people from contributing to a crowd-sourced service). For example, when using an "add pictures to your review" functionality, users could be asked to try to append to existing images. This could be transformed into a "trust score", which still allows anyone to contribute, but gives a higher weight to honest users. The CLP system can also be used to remove fake reviews long after they have been posted. For example, somebody could have created a large corpus of photos in advance and used them to create fake reviews. If one review was identified as a fake review, due to the graph structure of the matching photos, all connecting reviews can be found and removed from the system. Generally, even though our approach poses a comparably weak authentication proof, extending it with the many countermeasures we introduced and even combining it with other verification mechanisms can make it completely infeasible for adversaries to attack.

6 Conclusion and Future Work

This paper introduced the novel "Captcha your Location Proof" (CLP) approach, which in contrast to earlier research in the field of Location Proofs does neither rely on any additional infrastructure, nor on the client being in the picture. CLP rather relies on the principle that as long as there are more honest than dishonest clients, most of the adversarial ones can be detected and removed. Moreso, providers of location based services employing CLP can be sure that they are providing their clients with authentic contributions or reviews for locations or POIs. Furthermore, from an honest client's perspective the CLP method consists only of declaring the wish to create a review for a POI and taking a picture thereof. Together with the high availability of cameras in smartphones or even regular mobile phones, the simplicity of CLP makes it especially attractive for location based service providers such as review platforms.

Like all location proof approaches, CLP both strengths and weaknesses. We analyzed them in detail in our description of the adversarial model. In particular, we worked out a clear differentiation between honest and dishonest clients, as well as the possible threats they could pose. We then conducted a simulation of a possible real world application with predefined probability measures to demonstrate the behavior and practicability of CLP in practice. Finally, we presented a set of countermeasures that help increase the reliability of CLP and efficiently counter the threats. For future research it might be interesting to realize CLP with a real infrastructure and conduct user studies to both test the stability, practicability, as well as the user acceptance of the system in a real world environment.

Nonetheless, our simulation showed that the CLP approach as it was presented in this paper poses a scalable, easy to implement, easy to use, user friendly solution for location proofs in an adverserial environment.

Acknowledgements This research was supported by the Swiss National Science Foundation (SNF) within NRP 71 "Managing energy consumption" and by the Commission for Technology and Innovation (CTI) within the Swiss Competence Center for Energy Research (SCCER) Mobility and FURIES (Future Swiss Electrical Infrastructure).

References

Anderson M, Magruder J (2012) Learning from the crowd: regression discontinuity estimates of the effects of an online review database*. Econ J 122(563):957–989. http://dx.doi.org/10.1111/j.1468-0297.2012.02512.x

Brands S, Chaum D (1994) Distance-bounding protocols. Lect Notes Comput Sci 765:344–359

Brown M, Lowe DG (2007) Automatic panoramic image stitching using invariant features. Int J Comput Vis 74(1):59–73. https://doi.org/10.1007/s11263-006-0002-3

Bucher D, Scheider S, Raubal M (2017) A model and framework for matching complementary spatio-temporal needs. In: Proceedings of the 25th ACM SIGSPATIAL international conference on advances in geographic information systems. ACM

Francillon A, Danev B, Capkun S (2011) Relay attacks on passive keyless entry and start systems in modern cars. In: Proceedings of the 18th annual network and distributed system security symposium. the internet society. Citeseer

Gao H, Lewis RM, Li Q (2012) Location proof via passive RFID tags. Springer, Berlin, Heidelberg, pp 500–511

Hu N, Liu L, Sambamurthy V (2011) Fraud detection in online consumer reviews. Decis Support Syst 50(3):614–626. On quantitative methods for detection of financial fraud. http://www.sciencedirect.com/science/article/pii/S0167923610001363

Javali C, Revadigar G, Hu W, Jha S (2015) Poster: were you in the cafe yesterday?: location proof generation and verification for mobile users. In: Proceedings of the 13th ACM conference on embedded networked sensor systems. ACM, pp 429–430

Javali C, Revadigar G, Rasmussen KB, Hu W, Jha S (2016) I am Alice, I was in wonderland: secure location proof generation and verification protocol. In: 2016 IEEE 41st conference on local computer networks (LCN), Nov 2016, pp 477–485

Khan R, Zawoad S, Haque MM, Hasan R (2014) Who, when, and where? Location proof assertion for mobile devices. In: IFIP annual conference on data and applications security and privacy. Springer, pp 146–162

Lowe DG (2004) Distinctive image features from scale-invariant keypoints. Int J Comput Vis 60(2):91–110. https://doi.org/10.1023/B:VISI.0000029664.99615.94

Luo W, Hengartner U (2010) Veriplace: a privacy-aware location proof architecture. In: Proceedings of the 18th SIGSPATIAL international conference on advances in geographic information systems, GIS '10. ACM, New York, NY, USA, pp 23–32. http://doi.acm.org/10.1145/1869790.1869797

Mayzlin D, Dover Y, Chevalier J (2014) Promotional reviews: an empirical investigation of online review manipulation. Am Econ Rev 104(8):2421–2455

Mengjun L, Shubo L, Rui Z, Yongkai L, Jun W, Hui C (2016) Privacy-preserving distributed location proof generating system. China Commun 13(3):203–218

Miller HJ (2005) A measurement theory for time geography. Geogr Anal 37(1):17–45

Saroiu S, Wolman A (2009) Enabling new mobile applications with location proofs. In: Proceedings of the 10th workshop on mobile computing systems and applications, HotMobile '09. ACM, New York, NY, USA, pp 3:1–3:6. http://doi.acm.org/10.1145/1514411.1514414

Sastry N, Shankar U, Wagner D (2003) Secure verification of location claims. In: Proceedings of the 2nd ACM workshop on wireless security, WiSe '03. ACM, New York, NY, USA, pp 1–10. http://doi.acm.org/10.1145/941311.941313

Talasila M, Curtmola R, Borcea C (2013) Improving location reliability in crowd sensed data with minimal efforts. In: 2013 6th Joint IFIP wireless and mobile networking conference (WMNC). IEEE, pp 1–8

Von Ahn L, Blum M, Hopper NJ, Langford J (2003) Captcha: using hard AI problems for security. In: International conference on the theory and applications of cryptographic techniques. Springer, pp 294–311

Wang X, Pande A, Zhu J, Mohapatra P (2016) Stamp: enabling privacy-preserving location proofs for mobile users. IEEE/ACM Trans Netw 24(6):3276–3289

Waters B, Felten E (2003) Proving the location of tamper-resistant devices. Technical Report

Waters B, Felten E (2003) Secure, private proofs of location. Technical report

Weiser P, Bucher D, Cellina F, De Luca V (2015) A taxonomy of motivational affordances for meaningful gamified and persuasive technologies. In: Proceedings of the 3rd international conference on ICT for sustainability (ICT4S). Advances in computer science research, vol 22. Atlantis Press, Paris, pp 271–280

Ye Q, Law R, Gu B, Chen W (2011) The influence of user-generated content on traveler behavior: an empirical investigation on the effects of e-word-of-mouth to hotel online bookings. Comput Hum Behav 27(2):634–639. Web 2.0 in travel and tourism: empowering and changing the role of travelers. http://www.sciencedirect.com/science/article/pii/S0747563210000907

Zhu Z, Cao G (2011) Applaus: a privacy-preserving location proof updating system for location-based services. In: 2011 Proceedings IEEE INFOCOM, Apr 2011, pp 1889–1897

Data Quality of Points of Interest in Selected Mapping and Social Media Platforms

Hartwig H. Hochmair, Levente Juhász and Sreten Cvetojevic

Abstract A variety of location based services, including navigation, geo-gaming, advertising, and vacation planning, rely on Point of Interest (POI) data. Mapping platforms and social media apps oftentimes host their own geo-datasets which leads to a plethora of data sources from which POIs can be extracted. Therefore it is crucial for an analyst to understand the nature of the data that are available on the different platforms, their purpose, their characteristics, and their data quality. This study extracts POIs for seven urban regions from seven mapping and social media platforms (Facebook, Foursquare, Google, Instagram, OSM, Twitter, and Yelp). It analyzes the POI data quality regarding coverage, point density, content classification, and positioning accuracy, and also examines the spatial relationship (e.g. segregation) between POIs from different platforms.

Keywords VGI · Crowd-sourcing · Point of interest · Data quality

1 Introduction

Location based services (LBS) play an important part in our everyday life. For many tasks LBS use an inventory of points of interest (POI), also often called places, venues, or businesses, which can originate from commercial (e.g. Google) or crowd-sourced platforms (e.g. OpenStreetMap (OSM)). POIs can be used to geo-reference social activities, such as posting a picture on Instagram, sending a geo-tagged tweet, or checking into a Foursquare/Swarm location (Rösler and Liebig

H. H. Hochmair (✉) · L. Juhász · S. Cvetojevic
Geomatics Program, University of Florida, Davie, FL 33314, USA
e-mail: hhhochmair@ufl.edu

L. Juhász
e-mail: levente.juhasz@ufl.edu

S. Cvetojevic
e-mail: scvetojevic@ufl.edu

© Springer International Publishing AG 2018
P. Kiefer et al. (eds.), *Progress in Location Based Services 2018*, Lecture Notes in Geoinformation and Cartography, https://doi.org/10.1007/978-3-319-71470-7_15

2013). POIs are also relevant for navigation solutions, e.g. when providing reference points in travel directions (Nothegger et al. 2004; Duckham et al. 2010) or for suggesting venues along a route as part of personalized route recommendations (Lim et al. 2015). Due to the importance of POIs in geo-applications including LBS a solid understanding of their nature and data quality is essential to determine their fitness for purpose. Through successful conflation of POIs from different sources, attributes could be complemented, the number of objects increased, and the data quality of POIs improved (Hastings 2008). Successful conflation necessitates, however, to handle the challenging problem of data integration and achieving data interoperability. It requires also an understanding of the nature of POIs provided on the different platforms in order to identify promising candidates for conflation and integration in the first place. This study will tackle the latter task by a joint comparison of various quality aspects of POIs from seven commercial and Volunteered Geographic Information (VGI) crowd-sourcing platforms. This joint analysis is the novel aspect of this contribution, which builds on quality measures (e.g. richness in POI categories) that have been applied in other similar studies before.

From among the various data quality elements that are commonly used to determine how well a geo-spatial dataset meets its specified criteria, this research will closer examine relative completeness (abundance, categorization) and positional accuracy, as well as location bias by analyzing attraction and repulsion between marked point sets from different data sources through Cross-K functions. Most analyses in this study do not use ground truth data since perfect knowledge about POI locations in the different cities is difficult to obtain. Instead it applies comparative measures.

Current studies of POI quality assessment focus primarily on single data sources. They often use intrinsic quality measures (Barron et al. 2014; Gröchenig et al. 2014) or compare the data source in question to a proprietary or governmental reference data set (Senaratne et al. 2017). Especially OSM received considerable attention in these aspects (Jackson et al. 2013; Fan et al. 2014). Using a set of OSM POIs, Mülligann et al. (2011) use geo-ontologies to determine the plausibility of a POI type within a given neighborhood, which can be used for tag recommendation, data cleaning, and coverage recommendation. Several studies address also quality and conflation aspects of multiple POI sources. For example, a comparison of POIs from proprietary (TomTom, NAVTEQ, ESRI), governmental (TIGER/Line, USGS GNIS) and crowd-sourced (OSM) data sources finds that categorization schemes in the different platforms change over time, and that no single data source outperforms another in all aspects (Hochmair and Zielstra 2013). To address the challenges of integrating heterogeneous POI data sets from different sources, McKenzie et al. (2014) developed a weighted multi-attribute method which matches POIs from different sources and applies a variety of similarity measures, such as the Levenshtein distance on feature names, or category alignment based on WordNet. Similarly, Li et al. (2016) allocate Entropy based weights to POI attributes (e.g. distance between objects, name and sound similarity, category similarity) to improve POI matching from different sources.

The digital divide determines in which geographic regions users have the technical and economic means to participate in VGI and social media activities (Heipke 2010), which has a direct effect on VGI data quality. Furthermore, in OSM data coverage is affected by data imports, including a 2009 import of POIs from GNIS (Hochmair and Zielstra 2013), and a 2007 import of TIGER/Line road data in the US (Zielstra et al. 2013). Such an import will affect not only data coverage but also the range of OSM categories found in local datasets.

2 Data Collection

Data for the study were collected from sub-areas of the following seven cities: Albuquerque (New Mexico), Cairns (Australia), Gainesville (Florida), London (England), Nairobi (Kenya), Qingdao (China), and Salzburg (Austria). POI data with geographic coordinates were downloaded through Application Programming Interfaces (APIs) from seven selected data sources (Facebook, Foursquare, Google, Instagram, OSM, Twitter, and Yelp) and inserted into a PostgreSQL database. Twitter places at the different hierarchical levels (POI, neighborhood, city, administration) were extracted from worldwide tweets posted between 20 September 2016 and 20 October 2016 rather than from the Twitter REST API because of faster data access. The location of all Twitter place types except for POIs are defined through a bounding box. Tweets themselves were downloaded in JavaScript Object Notation (JSON) format through the Twitter Streaming API using the Tweepy python library.

For data download of other sources, requests were made in a Python environment using existing API wrappers, where available (Instagram, Facebook, Yelp, Foursquare), or using custom solutions (OSM, Google). The typical approach was to search places within a given radius around a center point (Facebook, Instagram, Google) or within a rectangular area (Foursquare, Yelp), which was moved along in a grid like pattern to cover the area to be analyzed. Since APIs often limit the returned data volume, locally refined rectangles or circles were inserted to ensure the capture of all POIs within an area whenever this threshold was met. An illustration of this refinement process for Yelp data retrieval is provided in (Juhász and Hochmair 2017).

For OSM, a different approach was chosen since large areas can be queried via the OverpassAPI. The query extracted all nodes with names, all ways with names (except for waterways and routes), all ways that are bridges and have names, and all relations (an ordered lists of nodes or ways) with a name and type = multipolygon tag. Since in OSM certain map features, such as parks and buildings, are often mapped as ways or relations, these were represented by their centroids in the final dataset. A set of working code examples that illustrate the different methods and libraries to use APIs for selected VGI and social media services, including Twitter, Instagram, Foursquare, and OSM, are provided in the literature (Juhász et al. 2016).

Technical details about the geo-tagging process in Twitter and Instagram, and its effect on positioning accuracy, are discussed in another study (Cvetojevic et al. 2016). Instagram positions were obtained in October 2015 where the API could still be used without going through an approval process, which changed effective June 1, 2016.[1] All other place data, except for those from tweets, were download in May 2017.

In several data sources many POIs were stacked on top of each other at the exact same point location. Such a case is plausible if various parties reside at a single building. Examples are hospitals (with doctor's offices as individual POIs geo-coded at the same geographic location), or commercial buildings that host several business offices. In many other cases, however, the stacking of POIs appears incorrect, especially if there are no major buildings in the vicinity. In some cases, stacked POIs aggregate places (e.g. businesses, plazas, parks) from across a whole city district. Several possible explanations can be found, such as (1) different locations being aggregated to a single point location by the platform, or (2) users uploading information of different POIs from one physical location (e.g. after a wireless network connection became available), and being unaware of the app attaching that same location to all POIs. Large POI stacks often contain POIs with made up place names (e.g., My Bed, Smoker's Paradise, Hell, Mi Casa, LETS OPEN Our BIBLE), people's names or unlikely business names (e.g. Herstyle, Happy Healthy Life) with no actual business found nearby. These kind of stacked POIs could be the outcome of location spoofing, i.e., the intentional falsifying of one's locational information (Zhao and Sui 2017). To avoid massive, incorrect POI stacks biasing subsequent cluster and density analyses, all point locations with 15 or more stacked POIs were manually reviewed for plausibility. If no building or market plaza of appropriate size was found at the posted position, the stacked POIs were removed from further analysis. The POI stack size of 15 is arbitrarily chosen. Although this stack size does not capture small clusters, it reduces the workload for manual cluster checking to a manageable amount.

Table 1 lists the size of identified POI clusters in descending order that were removed before further data analysis. Most removed clusters are found in Facebook and Instagram, which suggests that user-added places on these platforms undergo only little review and quality control. Incorrect clusters occur in all analyzed cities, but mostly in Nairobi, which is possibly indicative of poor positioning accuracy in that city. In fact, one point had 1195 different Facebook places stacked on top of each other.

All seven analyzed platforms operate worldwide, however, with some differences in data coverage between countries. Yelp and Twitter POIs, for example, are available in only five of the seven analyzed cities.

---

[1]http://developers.instagram.com/post/133424514006/instagram-platform-update.

Table 1 Size of removed clusters

Platform	City	Sizes of removed clusters
Facebook	Albuquerque	16
	Cairns	445, 246, 118, 37, 25, 24
	Gainesville	279, 54, 37, 21, 20
	London	–
	Nairobi	1195, 813, 245, 121, 64, 37, 37, 31, 26, 24, 23, 20, 18, 18, 17
	Qingdao	357
	Salzburg	113
Instagram	Albuquerque	105, 37, 28, 28, 21, 18
	Cairns	33, 20, 19
	Gainesville	58, 34, 30, 29, 28, 18
	London	33, 24, 23, 17
	Nairobi	33, 33, 33, 33, 33, 31, 26, 26, 20, 20, 18, 18, 17, 17, 16
	Qingdao	33, 33
	Salzburg	33, 31
Google	Nairobi	15, 15

3 POI Classification

3.1 Platform Comparison

Facebook, Google, OSM, Yelp, and Foursquare provide a categorization of their POIs into different hierarchical levels. New places added by users need to follow the provided POI categorization for these platforms. As opposed to this, the content of Twitter and Instagram POIs can only be characterized by their name since these platforms do not provide POI categories. For an application there can be differences in categories between POIs shown on a Web map and POIs downloaded through an API. The latter categorization is typically limited in detail. One example is a youth hostel that in the Google Places API Web Service is classified as type "lodging" whereas on Google Maps (in the Web browser) the same feature is classified as a more detailed "2-star hotel". Another example is "school" (API) versus "high school" (browser map).

The list of Google place types cannot be directly retrieved from the API. Alternative methods include extraction of place types from downloaded POIs (similar to how it was done for Twitter), or using classifications from third parties, such as Blumenthals.[2] That Website lists Google place categories for different language-country combinations. POI category numbers vary strongly between languages, e.g., US English (N = 2465), British English (N = 847), German

---

[2]http://blumenthals.com/google-lbc-categories/search.php?q=&val=hl-gl%3Den-US%28PfB%29%26ottype%3D1.

(DE) (N = 997), French (FR) (N = 1183), or Spanish (ES) (N = 2365). The place types returned in Google Maps (browser map) depend on the Google domain used (i.e., google.com, google.fr, google.it, etc.). Similarly, if users suggest a new place to be added to Google Maps, the available categories with their languages depend on the chosen Google domain. However, when downloading Google places through the Places API Web Service, only English place categories are returned, independent of the language setting. The language setting does also not affect the number of features returned. Returned place features often contain a list of place categories, where the first category appears to be the most specific one. Examples for POI categories include {doctor, health, point of interest, establishment} or {car dealer, store, point of interest, establishment}.

OSM offers a free tagging system for nodes, ways, and relations, which allows a user to add an unlimited number of attributes to features. However, the OSM community agrees to certain key-value combinations for commonly used features. Features are divided into 23 primary feature categories[3] (amenity, building, highway, etc.) which represent the key of a feature. Each key can take many different values to further specify the sub-type of the mapped feature. Key-value examples include highway = motorway, or amenity = restaurant.

Yelp places are structured in a four-tiered hierarchy which is customized for 32 countries.[4] POI classification schemes contain between about 900 and 1200 categories. A single venue can be assigned to several categories even from different hierarchical levels, e.g. Gyms (L3), Sports Clubs (L2), and Day Spas (L2). Setting a country determines which Yelp place categories can be added to Yelp. This is because many POI categories are associated with a whitelist (which specifies in which countries a category can be added) and/or a blacklist (a list of countries where the category cannot be added). For example, category "Bird Shops" has the following whitelist: NO, NL, DE, IT, SG, BE, ES, US, DK, SE. Thus with Sweden (SE) as chosen country bird shops are recognized by autocomplete during manual data entry (Fig. 1a), whereas, for example, in the UK they are not (Fig. 1b).

POI categories returned through the Yelp API depend on the local setting of the app/request as well. Yelp uses an "alias" for each category, similar to a unique category ID, but the name of the category returned by the API depends on the country setting. For example, the "landmarks" alias returns a "Landmarks & Historical Buildings" feature category for the US, and a category "Sehenswürdigkeiten" for Germany.

The list of 785 Facebook place topics can be extracted from the Facebook Graph API Explorer, which is not hierarchically structured. This list is more comprehensive than what is offered when interactively creating a Facebook page for a new business, brand, product etc. on the Facebook Web site.

All 920 Foursquare place categories are organized in a five-tiered hierarchical structure, where the top hierarchy contains 10 entries including Arts &

---

[3]http://wiki.openstreetmap.org/wiki/Map_Features.

[4]https://www.yelp.com/developers/documentation/v2/category_list.

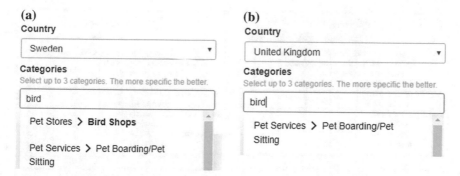

Fig. 1 Trying to add a place for a country that is listed (**a**) and not listed (**b**) on the white list of category "Bird Shops" in Yelp

Entertainment or Travel & Transport. Some venues are restricted to certain countries, as is specified in the Foursquare category documentation. For example, category "Anhui Restaurant" is available only in China and some other nearby countries.

3.2 Observed Distribution of POI Categories

Figure 2 shows the 10 most frequently used categories of POIs that were downloaded in Albuquerque, Gainesville, and London from the five social media platforms that provide POI categories. The analysis was limited to these three cities since they are the only ones that provide POI information (not necessarily category information though) for all seven platforms, and were therefore also used as common geographic areas for other data quality metrics. The top row in Fig. 2 shows that the most prominent OSM primary key is building, followed by amenity and shop, and that restaurant and hotel are the most prominent sub-types which were identified through querying OSM key-value pairs. Since shops in OSM are distributed across 73 potential shop types according to the wiki Map Features site, no shop type makes it to the top ten in that second chart. Figure 2 as a whole reveals that the most frequently used POI categories vary between analyzed platforms, comprising tourism facilities (hotels, restaurants, cafes in Yelp, OSM, and Foursquare), health infrastructure (doctors, hospitals, medical in Google, Facebook), and university buildings (Foursquare).

These differences indicate that the different platforms have strengths in different category groups and may thus complement each other in a meaningful way. It should be noted that charted category frequencies are based on only three cities in the US and Europe. They will differ from category frequencies found in other cities, especially those located in other parts of the world. Nairobi, for example, shows

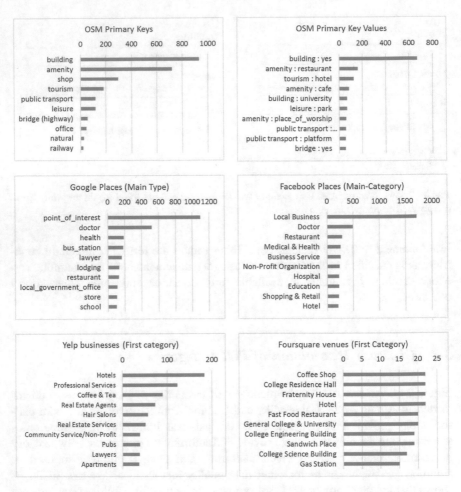

Fig. 2 Most frequent POI categories found in Albuquerque, Gainesville, and London

"sport" on 6th place in OSM primary keys, includes finance and banks in the top five Google categories, maps only a single doctor's office in Facebook, and misses fraternity housing in Foursquare altogether. These differences reflect local variations in the presence of venues between cities.

Twitter place types do not contain thematic categories but are organized in a spatial hierarchy instead. The coarsest place level is admin polygons which vary in size, density, and coverage between and within countries (Fig. 3a).

A more refined place level is cities which are found primarily in Europe, Canada, Mexico, Brazil, Japan, New Zealand, and a few other countries. Neighborhood places represent the sub-city level with a few clusters around the world, including the Netherlands, Australia, and New Zealand. Although metadata for city,

(a)

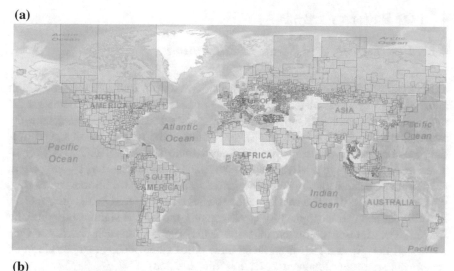

(b)

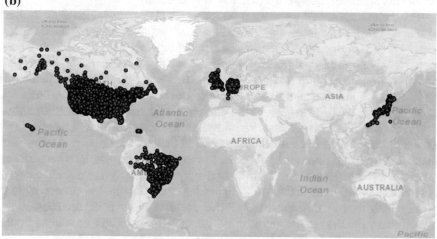

Fig. 3 Worldwide coverage of Twitter admin place type (Alaska and ocean polygons were removed from this map for clarity) (**a**), and place type Point of interest (POI) (**b**)

neighborhood, and administrative contain only a minimum bounding box as a feature geometry, Twitter uses accurate boundary polygons internally to determine which city, neighborhood, etc. a tweet needs to be assigned to during the geo-tagging process. The POI place type is the only one with point geometry and found in several countries, including Canada, the US, Brazil, Great Britain, Germany, and Japan (Fig. 3b). The POI level will be analyzed in more detail for selected cities.

4 POI Pattern Analysis

4.1 POI Density and Spatial Distribution

Table 2 summarizes descriptive statistics of POIs for the analyzed data sources in cities where they are available. Values in the upper row of each data source list the number of downloaded POIs per km^2, and values in the lower row denote the average nearest neighbor index (NNi), computed as the ratio of the observed over the mean nearest-neighbor distance (O'Sullivan and Unwin 2010). The NNi characterizes a point pattern relative to complete spatial randomness, i.e., a point pattern created by a homogenous Poisson process. An NNi < 1 indicates clustering, whereas an NNi > 1 indicates a tendency toward evenly spaced points (dispersion). A statistical test can be applied to check whether the NNi is significantly different from 1. The right-most column in Table 2 (M) reports for each data source the mean POI density and NNi from those three cities where POIs are available in all seven data sources (Albuquerque, Gainesville, London).

The highest mean POI density can be found for Instagram, which allowed users to add arbitrary place labels until 2015. As a consequence many POIs are incorrectly labeled, mislocated, duplicates, or stacked together due to positioning inaccuracies (Cvetojevic et al. 2016). Duplicate locations appeared to be at least partially cleaned out since October 2015, which is when Instagram was purchased by Facebook. Since then new Instagram places can be added through Facebook. We checked which of the Instagram places that we downloaded in October 2015 were still available in 2017, and stored these results in a newly added "available" POI attribute. Facebook POIs exhibit the second highest mean density. Also on that platform a frequent occurrence of stacked place labels at single locations poses a problem. Google shows the most consistent place density among all cities at approximately 100–200 POIs/km^2, suggesting that Google has access to high quality base data in different parts of the world. Twitter POIs demonstrate the lowest mean density, suggesting that the list of POIs is strictly controlled by the company. These POIs are only suitable for approximate geo-tagging of posted tweets, but less so for mapping and navigation purposes, which would require a more dense pattern of POIs. Foursquare venues reveal the second lowest place density, since they are limited to businesses (hotels, bars, bakeries), public buildings (city halls, university campuses and buildings, train stations), and public locations (parks, plazas) in the analyzed cities.

Yelp offers its service in five of the analyzed cities with a high variation of POI densities between cities, suggesting that different base datasets are used for this platform depending on the region.

The OSM POI density is higher in the two European than the three US cities. Compared to all other cities POI densities are lowest in Qingdao for the four data sources offered in that city, possibly due to a lower prominence of analyzed VGI and social media apps in that city. The analyzed area in London, which is a mixed business and residential district located between Hyde Park and Paddington

Table 2 Density and nearest neighbor index for POIs in analyzed cities

		Albuquerque	Cairns	Gainesville	London	Nairobi	Qingdao	Salzburg	$M_{A.G.L}$
Facebook	Dens.	244.5	530.8	247.9	1679.0	115.6	4.8	321.7	723.8
	NNi	0.45	0.57	0.51	0.58	0.44	0.66	0.53	0.51
Foursquare	Dens.	11.2	10.9	29.3	48.0	4.1	0.0	11.8	29.5
	NNi	0.71	0.68	0.64	0.54	0.72	N/A	0.62	0.63
Google	Dens.	117.6	121.2	114.7	165.5	136.6	99.0	185.8	132.6
	NNi	0.64	0.74	0.64	0.73	0.66	0.51	0.68	0.67
Instagram	Dens.	178.3	417.2	315.7	2753.6	260.1	8.1	431.6	1082.5
	NNi	0.62	0.59	0.60	0.72	0.59	0.57	0.57	0.65
OSM	Dens.	35.2	29.0	48.0	572.9	39.0	4.6	294.6	218.7
	NNi	0.70	0.68	0.74	0.75	0.55	0.68	0.72	0.73
Twitter	Dens.	5.3	0.0	7.1	25.4	0.0	0.0	0.0	12.6
	NNi	0.80	N/A	0.69	0.61	N/A	N/A	N/A	0.70
Yelp	Dens.	45.7	343.4	42.8	1160.4	0.0	0.0	708.8	416.3
	NNi	0.64	0.35	0.55	0.42	N/A	N/A	0.29	0.54

Railway station, shows the highest POI density in all data sources except for Google, which may have to do with frequent visitors in the area with its railway station and shopping streets.

NNi values < 1 in Table 2 indicate that POIs are clustered for all platforms and cities (where data is available). All clusters are significant at $p < 0.001$. Clustering can be expected since complete spatial randomness, though mathematically elegant, is often unrealistic in the physical world (O'Sullivan and Unwin 2010). Instead, point patterns typically display spatial dependence. It can either be modeled as first order effect (variation in the intensity of the process across space) or as second order effect (interaction of some kind between events). In the context of this work, a first order effect could be the tendency of restaurants and shops to be opened along selected roads of the built environment. Second order effects can result in clustering or dispersion. An example in the context of this study is the frequent co-occurrence of railway stations and restaurants (clustering), or the repulsion between public schools (dispersion).

The degree of clustering varies strongly between the datasets. Figure 4 maps for the analyzed area in Gainesville (enclosing polygon) the POIs for the most clustered (Facebook, NNi = 0.51) and the least clustered (OSM, NNi = 0.74) point patterns. The primary POI categories mapped in each platform (Fig. 2) can help to explain some of the differences in NNi values. Facebook POIs depict primarily business data, including shops, restaurants, retail and fitness clubs. A high density of such venues can be found along University Avenue (running East-West to the north), and on Community Plaza to the east. As opposed to this the OSM POI pattern shows heavy mapping activities on the UF campus (south-west portion of the map). With university buildings being further apart than businesses on a plaza or a shopping street, the clustering is less pronounced than in Facebook. Though the magnitude of the NNi for a specific data source itself is not a quality criterion, it can reveal differences in POI distributions between compared data sources.

In the OSM POI dataset, when averaging (unweighted mean) the geometry proportions across the seven analyzed cities, the share of POIs with node

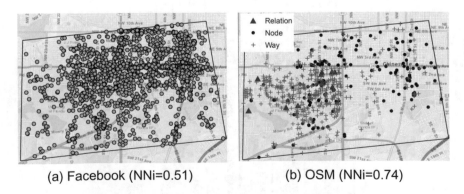

(a) Facebook (NNi=0.51)   (b) OSM (NNi=0.74)

Fig. 4 POIs in Facebook (**a**), and OSM (**b**) for the Gainesville study area

Table 3 Number and proportion of OSM node features, way features, and relations

	Albuquerque	Cairns	Gainesville	London	Nairobi	Qingdao	Salzburg	Mean
Nodes	343	197	161	500	251	135	1743	
%	49.3	77.3	27.2	49.3	53.4	49.8	72.7	54.1
Ways	346	57	422	505	217	136	582	
%	49.7	22.4	71.2	49.8	46.2	50.2	24.3	44.8
Relations	7	1	10	9	2	0	74	
%	1.0	0.4	1.7	0.9	0.4	0.0	3.1	1.1

geometries is the highest (54.1%), followed by way features (44.8%) and relations (1.1%). However, considerable inter-urban variation in the use of geometries can be observed (Table 3). For example, Gainesville exhibits a high percentage of way features (71.2%) due to many UF campus buildings being mapped as closed polyline features (Fig. 4b). A high concentration of OSM way features on university campuses can also be observed for other cities, for example, around the University of New Mexico in Albuquerque.

Relation objects are generally more sparsely found in OSM maps. Salzburg stands out as an area with a relatively high proportion (3.1%) of relation objects. This can be attributed to the detailed mapping of plazas and historic buildings with court yards, which are mapped using inner and outer polygons as part of a relation (Fig. 5). In summary it can be stated that for OSM POI analysis both point and way objects should be considered since the ratio between those two geometry types varies significantly across cities.

Fig. 5 OSM nodes, way centroids and relation centroids in the Salzburg study area

4.2   Relative Clustering of Point Patterns

While so far the level of clustering of POIs was discussed for individual platforms, statistical methods can be applied to analyze if and how POI locations from different platforms cluster relative to each other. For this purpose, a bivariate generalization of Ripley's K function, known as the Cross-K function (Dixon 2002), can be applied. The Cross-K function can be formulated as

$$K_{ij}(r) = \lambda_j^{-1} E[f(r)] \tag{1}$$

where $E[f(r)]$ is the expected number of type j events within a distance r of a randomly chosen type i event, and λ_j is the density of j events per areal unit. If the two point patterns i and j are identical, the Cross-K function collapses to the self-K function $K(r)$ which considers only locations of events but ignores information about the type of event.

Under random labeling, that is, assigning the n1 points from type 1 and n2 points from type 2 randomly to type 1 and type 2 events (keeping their original proportions), all four bivariate Cross-K functions should equal the K function, giving $K_{ii}(r) = K_{ij}(r) = K_{ji}(r) = K_{jj}(r) = K(r)$. Using place data from Gainesville as an example, this study analyzes the spatial relationship between all possible combinations of platform pairs, using events from different platforms as event types i and j in Eq. 1. Statistical inference of the difference between the observed Cross-K function and a Cross-K function generated by random labeling can be achieved through Monte Carlo simulation. Within each of the 99 completed permutations of the Monte Carlo simulation, the combined set of locations and the number of events of each type are held fixed. The labels (of the two platforms involved in the test) are randomly assigned to locations, which is followed by the computation of the Cross-K function. This establishes an upper and lower simulation envelope for random labeling at a 99% confidence level. If the observed Cross-K function falls within the simulation envelope, POIs from both platforms are similarly clustered around each other.

As an example for this analysis, Fig. 6 shows for all platform combinations in Gainesville the observed Cross-K function (black), the simulation mean from the Monte Carlo simulation (dashed red), and the 99% confidence envelope (gray area), for distances between 0 and 2000 meters. No significant attraction between two point patterns can be observed. While most pairwise platform point patterns are independent of each other, there are some platform combinations where the observed Cross-K function falls below the lower simulation envelope. This is clearly the case for some platform combinations that involve Foursquare and OSM. Whereas OSM is primarily contributed around the UF campus, business related contributions to Yelp and Facebook cluster around shopping strips and plazas (compare Fig. 4). Hence events from the platform pairs Foursquare-Yelp, OSM-Yelp, OSM-Facebook, and Foursquare-Facebook are spatially segregated (see Fig. 6). Google places are evenly distributed across the study area with no

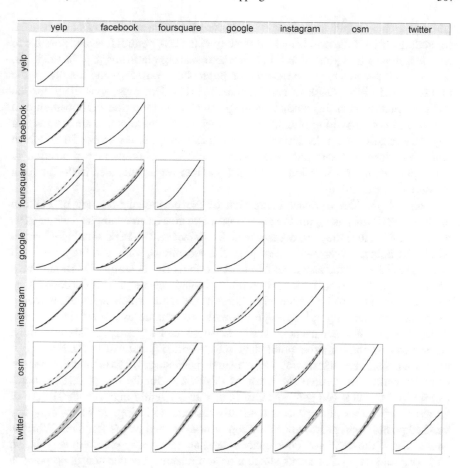

Fig. 6 Cross-K functions for Gainesville with 99% confidence envelopes

apparent clusters on the UF campus or in shopping areas. This makes it spatially segregated from Facebook (businesses) and Instagram POI locations (clustered around event places like student centers, campus food courts, hospitals, restaurant areas, market plazas). Overall, these findings suggest that the urban structure of the analyzed area is reflected in Cross-K functional patterns, and that a conflation of POIs from different sources can lead to improved data coverage for that city.

5 Positional Accuracy

All platforms analyzed in this study provide map interfaces and/or address search functions for manually adding new POIs. Provided that a user possesses basic map reading capabilities, such an approach supports accurate mapping of POIs. In addition to this, mobile apps show a user's current position, which can also be used for adding a POI if the user is located directly at the new venue. It is also possible to add coordinates to a picture in different photo sharing platforms, e.g. in Flickr. In such cases it is encouraged to map the photographer's position, and not that of the photographed object (Zielstra and Hochmair 2013). However, guidelines for the analyzed platforms in this study encourage users to add the true POI position of a venue, and not that of the photographer's position (if any pictures are involved at all). An exception are older Instagram place locations (before October 2015) which could be added by users and were often placed at a photographer's position or biased by the physical location from which the image was uploaded to the platform (Cvetojevic et al. 2016).

Figure 7 provides a visual impression of the positional accuracy of POIs in analyzed platforms, using the Salzburg downtown area as an example. Twitter POIs are not shown since they are not available for Salzburg. No POIs should be located within the Salzach river, with a few potential exceptions, such as a ferry service, a river place label, or the city label. None of the POIs from Google, OSM, and Yelp are located in the river, indicating good positional accuracy. The Facebook map reveals a few POIs to be incorrectly placed in the river, which include a barber, a shopping strip, and a graphic design business. Foursquare uses a review of added locations through super users to verify locations and to increase data reliability. A Boolean attribute in the point data set ("verified") indicates whether a POI underwent such a check or not. Only a small percentage of Foursquare venues is actually marked as verified. In Salzburg this is true for 96 out of 2012 features (4.8%), and for all seven analyzed cities this rate is slightly higher with 910 out of 13040 (7.0%). This filter process is clearly discernible in Fig. 7, where only few verified points (light green) appear on top of non-verified points (orange). Whereas some POIs of the unfiltered Foursquare dataset are incorrectly placed in the Salzach river (e.g. old city hall, a snack stand, a person's name, a coffee shop), no point in the filtered dataset is. This indicates that the revision through super users has a positive effect on the positional accuracy of Foursquare POIs.

The Instagram POI file used for this study was downloaded in October 2015. Next, using the Instagram API it was verified if a POI was still available in 2017. This information was then coded in a Boolean "available" attribute. POIs not available any more were possibly cleaned due to inspection after Instagram has been purchased through Facebook. Using this attribute, yellow dots in the Instagram map indicate possibly reviewed (and retained) POIs, whereas the red dots show the locations of the remaining unverified POIs from the original dataset. The map suggests that a significant percentage of Instagram locations was removed since 2015, namely 27.3% of POI for Salzburg and 27.5% for all seven analyzed cities. Several POIs that are incorrectly placed in the river disappear when considering only "available" Instagram features, including a bus stop, the old city hall, a pub, and a road. However, remaining POIs in the river (yellow dots) include bars, restaurants, or shops, still revealing POI accuracy problems.

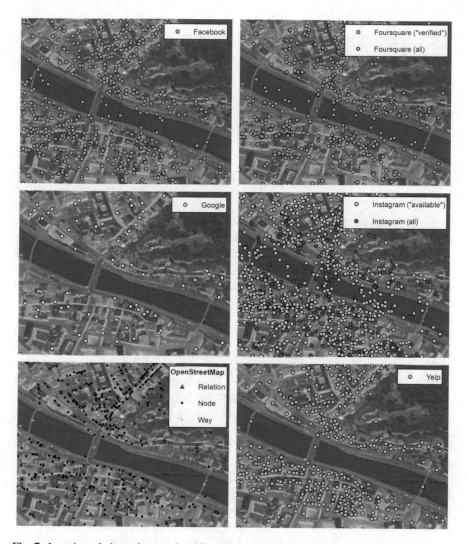

Fig. 7 Location of places features for different platforms in the Salzburg downtown area

To quantify the distance offsets between mapped POIs (based on coordinates provided on the platforms) and their true location, a sample of place points was selected for each of the available six data sources from the Salzburg downtown area. Using the name tag of a selected POI and the authors' local knowledge of the area, the corresponding true location was identified (if possible) and the offset computed. If a place was mapped on the street directly in front of the correct building, it was counted as correct as well and assigned an offset distance of 0. Seasonal POIs (e.g. a Christmas market) were also taken into account for the

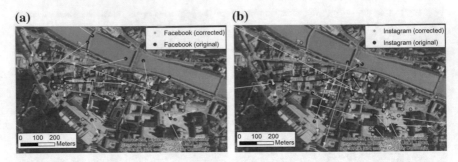

Fig. 8 Offset vectors for POIs in Salzburg for Facebook (**a**) and Instagram (**b**)

analysis. As an example, Fig. 8 shows for the samples of Facebook POIs (a) and Instagram POIs (b) their original positions (red) and corrected positions (green).

Yellow arrows denote the offset vectors from the published to the corrected position. For Facebook 24 POIs (43.6%) had to be corrected with offsets ranging up to 25,514 m. For Instagram this was the case for 26 POIs (40.6%) with a maximum offset of 715.4 km. Instagram has four outliers with an offset >10 km, whereas Facebook has only one.

Figure 9 plots the histograms of offset distances (in meters) for the six evaluated platforms in downtown Salzburg, using a logarithmic scale on the x-axis. The median distance offset is zero for all platforms, which means that at least half of all POIs in each platform is correctly placed. Google and OSM sample POIs do not reveal any positional errors, closely followed by Yelp which has moderate offsets in 12.7% of the cases. This finding suggests that, at least for the chosen test site, mapping platforms (Google, OSM) and business platforms with strong quality control (Yelp) provide most reliable POI positions. This finding is in-line with the distribution of point patterns observed in Fig. 7. Foursquare (verified POIs only) has the next smallest error rate (32.5%), followed by Instagram (40.6%) and Facebook (43.6%). Besides higher error rates, the latter two platforms are also the only ones with positional errors of over 10 km. Hence these two social media platforms together with an unverified (complete) Foursquare POI dataset achieve a lower POI reliability than other analyzed platforms. This indicates that social media platforms with little to no quality control through the governing company perform poorly in terms of positional accuracy when compared to mapping platforms or platforms that implemented stricter quality control measures (i.e. approval by moderators). To be able to provide more generalizable conclusions, however, offset measures would have to be expanded to other cities as well.

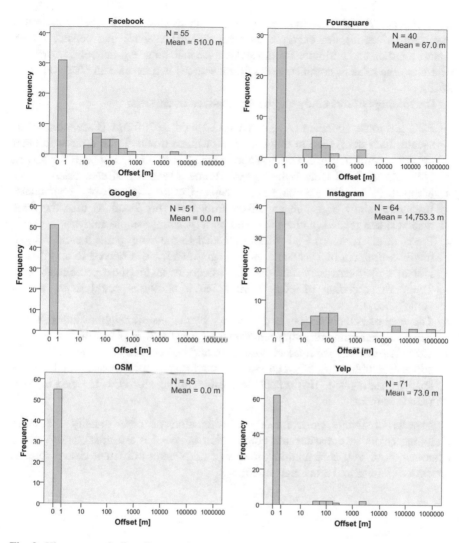

Fig. 9 Histograms of offset distances for evaluated features in downtown Salzburg

6 Discussion and Conclusions

The study examined various aspects of data quality and POI clustering for seven mapping and social media platforms in seven study sites across the world. The findings provide information to help determine the suitability of a given POI source for an intended geo-application, such as an LBS. Various quality metrics (e.g. nearest neighbor index, density, bulk uploads, spatial offsets) were compared between different platforms in the absence of ground truth data. Even with a correct a POI reference dataset available, POIs from the different platforms would first have to be matched to reference features in order to determine certain quality measures (e.g. categorization). An example for such detailed analysis that considers both feature location and attributes for manual feature matching (e.g. schools) in order to determine the relative completeness of data sources is presented in (Jackson et al. 2013).

The findings of this study can be summarized as follows:

- POIs are more abundant in selected social media platforms (Facebook, Foursquare, Instagram) than in mapping and business oriented platforms with strict quality control (Google, OSM, Yelp). Twitter is an exception with the lowest POIs density in all areas (where present), and a lack of POI categories.
- Mapped POIs of the three social media platforms (Facebook, Foursquare, Instagram) show higher mean offsets from their true locations than the three map/business related platforms, based on a Salzburg sample analysis.
- Presence of erroneous POI stacks uploaded to the same point location is primarily a problem of Facebook and Instagram and was observed in all cities.
- Different platforms map different POI categories as the most prominent ones. Therefore conflation of POIs from different platforms could improve POI completeness.
- The level of POI clustering, as determined by the nearest neighbor index, differs between platforms, reflecting a different topical focus of platforms.
- Cross-K functions for marked point patterns showed that point patterns cluster sometimes differently between pairs of platforms, which means that POIs are spatially segregated. This reflects also different types of POIs mapped in compared platforms.

Aspects of future work include consideration of other quality measures, including errors of omission and commission in selected test areas, and a closer examination of POI contribution patterns of users across different crowd-sourced platforms (Juhász and Hochmair 2016).

References

Barron C, Neis P, Zipf A (2014) A comprehensive framework for intrinsic OpenStreetMap quality analysis. Trans GIS 18(6):877–895

Cvetojevic S, Juhász L, Hochmair HH (2016) Positional accuracy of twitter and instagram images in urban environments. GI_Forum 1:191–203

Dixon PM (2002) Ripley's K function. In: El-Shaarawi AH, Piegorsch WW (eds) Encyclopedia of environmetrics. Wiley, Chichester

Duckham M, Winter S, Robinson M (2010) Including landmarks in routing instructions. J Location Based Serv 4(1):28–52

Fan H, Zipf A, Fu Q, Neis P (2014) Quality assessment for building footprints data on OpenStreetMap. Int J Geogr Inf Sci 28(14):700–719

Gröchenig S, Brunauer R, Rehrl K (2014) Estimating completeness of VGI datasets by analyzing community activity over time periods. In: Huerta J, Schade S, Granell C (eds) Connecting a digital Europe through location and place. Lecture notes in geoinformation and cartography. Springer, Berlin, pp 3–18

Hastings JT (2008) Automated conflation of digital gazetteer data. Int J Geogr Inf Sci 22 (10):1109–1127

Heipke C (2010) Crowdsourcing geospatial data. ISPRS J Photogramm Remote Sens 65:550–557

Hochmair HH, Zielstra D (2013) Development and completeness of points of interest in free and proprietary data sets: a Florida case study. In: Jekel T, Car A, Strobl J, Griesebner G (eds), GI_Forum 2013. Creating the GISociety. Wichmann, Berlin, pp 39–48

Jackson SP, Mullen W, Agouris P, Crooks A, Croitoru A, Stefanidis A (2013) Assessing completeness and spatial error of features in volunteered geographic information. ISPRS Int J Geo-Inf 2:507–530

Juhász L, Hochmair HH (2016) Cross-linkage between Mapillary street level photos and OSM edits. In: Sarjakoski T, Santos MY, Sarjakoski T (eds) Geospatial data in a changing world: selected papers of the 19th AGILE conference on geographic information science. Lecture notes in geoinformation and cartography. Springer, Berlin, pp 141–156

Juhász L, Hochmair HH (2017) Where to catch 'em all?'—a geographic analysis of Pokémon Go locations. Geo-spat Inf Sci 20(3):241–251

Juhász L, Rousell A, Arsanjani JJ (2016) Technical guidelines to extract and analyze VGI from different platforms. Data 1(3):15

Li L, Xing X, Xia H, Huang X (2016) Entropy-weighted instance matching between different sourcing points of interest. Entropy 18(2):45

Lim KH, Chan J, Leckie C, Karunasekera S (2015) Personalized tour recommendation based on user interests and points of interest visit durations. In: 24th international joint conference on artificial intelligence (IJCAI 2015), Buenos Aires, Brazil

McKenzie G, Janowicz K, Adams B (2014) A weighted multi-attribute method for matching user-generated points of interest. Cartogr Geogr Inf Sci 41(2):125–137

Mülligann C, Janowicz K, Ye M, Lee W-C (2011) Analyzing the spatial-semantic interaction of points of interest in volunteered geographic information. In: Egenhofer MJ, Giudice NA, Moratz R, Worboys MF (eds) Conference on spatial information theory (COSIT 2011). LNCS 6899. Springer, Berlin, pp 350–370

Nothegger C, Winter S, Raubal M (2004) Computation of the salience of features. Spat Cogn Comput 4:113–136

O'Sullivan D, Unwin DJ (2010) Geographic information analysis, 2nd edn. Wiley, Hoboken, New Jersey

Rösler R, Liebig T (2013) Using data from location based social networks for urban activity clustering. In: Vandenbroucke D, Bucher B, Crompvoets J (eds) Geographic information science at the heart of Europe. Lecture notes in geoinformation and cartography. Springer, Berlin

Senaratne H, Mobasheri A, Ali AL, Capineri C, Haklay M (2017) A review of volunteered geographic information quality assessment methods. Int J Geogr Inf Sci 31(1):138–167

Zhao B, Sui DZ (2017) True lies in geospatial big data: detecting location spoofing in social media. Ann GIS 23(1):1–14

Zielstra D, Hochmair HH (2013) Positional accuracy analysis of Flickr and Panoramio images for selected world regions. J Spat Sci 58(2):251–273

Zielstra D, Hochmair HH, Neis P (2013) Assessing the effect of data imports on the completeness of OpenStreetMap—A United States case study. Trans GIS 17(3):315–334

Mapping Spatiotemporal Tourist Behaviors and Hotspots Through Location-Based Photo-Sharing Service (Flickr) Data

Joey Ying Lee and Ming-Hsiang Tsou

Abstract Social media services and location-based photo-sharing applications, such as Flickr, Twitter, and Instagram, provide a promising opportunity for studying tourist behaviors and activities. Researchers can use public accessible geo-tagged photos to map and analyze hotspots and tourist activities in various tourist attractions. This research studies geo-tagged Flickr photos collected from the Grand Canyon area within 12 months (2014/12/01–2015/11/30) using kernel density estimate (KDE) mapping, Exif (Exchangeable image file format) data, and dynamic time warping (DTW) methods. Different spatiotemporal movement patterns of tourists and popular points of interests (POIs) in the Grand Canyon area are identified and visualized in GIS maps. The frequency of Flickr's monthly photos is similar (but not identical) to the actual tourist total numbers in the Grand Canyon. We found that winter tourists in the Grand Canyon explore fewer POIs comparing to summer tourists based on their Flickr data. Tourists using high-end cameras are more active and explore more POIs than tourists using smart phones photos. Weekend tourists are more likely to stay around the lodge area comparing to weekday tourists who have visited more remote areas in the park, such as the north of Pima Point. These tourist activities and spatiotemporal patterns can be used for the improvement of national park facility management, regional tourism, and local transportation plans.

Keywords Spatiotemporal · Hotspot analysis · Geo-tagged photos
Tourism · Flickr · Grand Canyon

J. Y. Lee · M.-H. Tsou (✉)
The Center for Human Dynamics in the Mobile Age, Department of Geography,
San Diego State University, San Diego, CA, USA
e-mail: mtsou@mail.sdsu.edu

J. Y. Lee
e-mail: nuo5218@gmail.com

© Springer International Publishing AG 2018
P. Kiefer et al. (eds.), *Progress in Location Based Services 2018*, Lecture Notes
in Geoinformation and Cartography, https://doi.org/10.1007/978-3-319-71470-7_16

1 Introduction

The tourism industry plays a key role in the economic development of many countries. Statistics from the UN's World Tourism Organization indicate that the tourism industry contributes up to 40% of the gross domestic products (GDPs) of developing countries (Ashley et al. 2007). To boost the tourism industry, the further development of existing tourism locations and identification of new tourism attractions are both recognized as crucial approaches in many countries. Tourism geography research that provides a spatial view of attractions can greatly help tourism industry development.

Over the past several years, many approaches toward mapping tourist behaviors and hotspots using photo-sharing service data have emerged (García-Palomares et al. 2015; Hawelka et al. 2014; Sun and Fan 2014; Vu et al. 2015). Conventional tourism management involves a mixed method approach using quantitative research and qualitative research based on questionnaire surveys, focus groups, and interviews. Traditional methodologies, such as questionnaires and interviews have limited capability in data collection (Chen and Chen 2012; Sun and Budruk 2015). The methodologies of tourism geography research continue to evolve as technology advances. Recently, more and more photos were taken by smartphones and GPS-equipped cameras in popular tourist attractions with geo-tagged information (Tsou 2015). Many geo-tagged photos were uploaded to photo-sharing websites such as Flickr, Instagram, and Panoramio allowing public access. Researchers can use the public application programming interfaces (APIs) to download and analyze these public accessible geo-tagged photos and analyze their spatiotemporal patterns.

This research studies geo-tagged Flickr photos collected from the Grand Canyon area within 12 months (2014/12/01–2015/11/30) using kernel density estimate (KDE) mapping, Exif (Exchangeable image file format) data, and dynamic time warping (DTW) methods. The Grand Canyon is one of the most popular tourist attractions in the U.S. and it is located across two states: Arizona and Utah (277 miles long, and up to 18 miles wide). There are over five million tourists visited the Grand Canyon during the last ten years. Natural resource management and transportation plans became important issue for the National Park Service (NPS) Agency. One key question in tourism management is to identify when and where exactly tourists are. Geo-tagging photos can indicate where and when photos have been taken by tourists, and thus can be used for tourism management.

This study utilized the space-time analysis framework for analyzing tourist behaviors and hotspots. Space-time analysis in geography was developed in the 1970s (Taaffe 1974; Palm and Pred 1974; Cullen 1972; Sauer 1974). Space-time geography can provide a comprehensive analysis framework for many research topics, such as criminology, public health, and tourism management. Space-time geography research focuses on some unique time analysis methods, such as duration, accessibility, and trajectory by using both spatial and temporal variables. Social media data can be a great data source for conducting space-time analysis since the data include both space and time variables (Yuan and Nara 2015; Issa et al. 2017). The spatio-temporal patterns of tourists' behaviors are bound by the spatial distributions of different destinations, and they are easily affected by spatio-temporal constraints. Therefore, the patterns within the analytical construct of the space-time prism can be explored using time geography (Chen and Kwan 2012).

2　Relate Work

2.1　Analyzing Travel Behaviors Using Geo-tagged Photos

Researchers have developed various methods for acquiring tourist behavior data, including surveys, GPS tracking, and interviews. Since GPS devices have become inexpensive and affordable after 2000, many studies have combined GPS data with questionnaires to analyze tourist behaviors (McKercher et al. 2012). Gao et al. (2013) used the check-in social media data to perform traffic forecasting, disaster relief, and advertising services. Girardin et al. (2008a, b), Popescu and Grefenstette (2011), and Majid et al. (2013) explored the spatio-temporal pattern through Flickr photos. Girardin et al. (2008) used Flickr photos to explore the tourist behaviors. Popescu and Grefenstette (2011) used historical photos from certain Flickr users to build a personal tourist recommendation system. Majid et al. (2013) used geo-tagged Flickr photos to predict users' tourist destination preferences. Some studies use tag frequency of social media images to acquire tourist behavior patterns (Sun and Fan 2014). Therefore, analyzing social media pictures with their geo-tagged information and time stamps can be a promising method to improve tourist management and identify regional hotspots of POIs. García-Palomares et al. (2015) research identified tourist hotspots by analyzing social media data, as well as revealed the spatio-temporal patterns of the identified tourist hotspots in European cities. Furthermore, their study highlighted the difference between residents' and tourists' daily attractions and travel routes. García-Palomares et al. (2015) study relies on using spatial statistical methods (hexagons with cluster analysis) to determine tourist hotspots and using geo-tagged photos to identify tourist attractions in Barcelona. Kádár's (2014) analysis of geographically positioned photography retrieved from Flickr with tourist arrivals and registered hotel bed nights (from TourMIS website) for 16 European cities. There are high correlations between bed nights and geo-tagged Flickr images. Birenboim (2016) utilized Ecological Momentary Assessment (EMA) to conduct surveys for tourist experiences in a high resolution spatiotemporal scale. Önder et al. (2016) traced Austria tourists' travel routes using their digital footprints. In their research, they collected photos with the geo-tagged "Austria" from 2007 to 2011 using the Flickr API. To differentiate tourist photos from non-tourist ones, they used a "time span" concept where a user who uploads two different photos in two different places within a certain period is identified as a tourist. This study also used multi-level scales to evaluate tourist footprints. The result showed that Flickr data could better represent tourism information on a city level rather than a regional level. Their research suggested that although Flickr could be used in tracking tourists' digital footprints, the accuracy of the user tracking may vary in different locations, depending on whether it is in a region level or a city level.

Aggregated geo-tagged social media can also reveal groups' semantic meanings and group activities. Kisilevich et al. (2010) used P-DBSCAN (Density-based spatial clustering of applications with noise) method to detect attractive destination from aggregated geo-tagged photos. Kennedy and Naaman (2008) used text mining to explore place semantics from Flickr tag data, while Cranshaw et al. (2012) used Foursquare data to investigate socially dynamic neighborhoods using clustered groups based on their social similarities. These research studies identified clustered groups using Flickr photos, which can reveal human mobility and to explore the patterns of human mobility.

Vu et al. (2015) used geo-tagged photos to explore travel behaviors in Hong Kong. They built a Hong Kong inbound tourist Flickr photo dataset and used a Markov chain model for travel pattern mining. Their research demonstrated that the Markov chain model could be applied to predict the probability of tourist routes between two tourist spots, and the result could be used by the government to improve transportation services. The Markov chain model could also be applied to model the tourist flows (Vu et al. 2015).

All these previous studies mentioned above did not utilize kernel density kernel density estimate (KDE) mapping nor dynamic time warping (DTW) methods for the analysis of tourist activtivies and hotspots, which are the major methodolgical contributions in this paper for tourism geography.

3   Data Collection

The data downloaded from Flickr within the Grand Canyon area in December 2014 to November 2015, which is from winter, spring, summer, and autumn, included 38,127 photos. The collection boundary of the Grand Canyon area had illustrated in Fig. 1. In this study, three types of data are acquired from Flickr APIs: time, location, and context data (Exif). The time data can be collected through timestamps of photos. The location data can be retrieved from users' instant locations via the mobile devices' coordinates or the check-in places they send along with photos (geotagged). When Flickr users upload their photos, they can choose whether they want to keep the Exif (Exchangeable image file format) info and coordinates or not. In this study, we will only collect the photo information containing coordinates and then used the photo id to retrieve their Exif information if available. The Exif data include detail information about the camera devices, such as "Manufacturer", "Model", "Date and Time

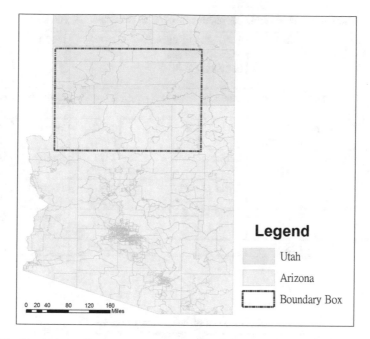

Fig. 1 The Flickr data collection boundary (the red box) across Utah and Arizona

(original)". We can use the Exif data to identify photos taken by smart phones (iPhones, Andriod phones, etc.) and cameras (Canon, Nikon, etc.).

One limitation of this study is the uncertainty of timestamps and locations. The timestamps in Flickr photos could be the time of taking photos or uploading photos. The geo-tagged locations can be modified or changed by users. Different types of spatiotemporal analysis (such as seasonal or weekend/weekday comparison) could be affected by the uncertainty of these data collection.

In this study, we used python program to collect Flickr photo information via its APIs and then stored in MongoDB database framework. The basic statistics of these photo data we collect are shown in Table 1.

As Table 1 shows, this study collected 38127 photo information in 2014/12/01– 2015/11/30. In which 25395 were collected with coordinates (geo-tagged) in 2015. Among these geo-tagged photo information collected in 2015, 7471 (29.4%) of photos were taken in weekends, 17924 (70.6%) of photos were taken in weekdays. For the monthly change, May is the highest month for Flickr photo uploaded count, the winter months, which is from December to February, are the lowest month.

As Fig. 2, among these photos taken by the camera, Canon and Nikon were the most popular camera devices. 353 photos were taken by *Canon EOS 6D*, 207 photos were taken by *Nikon D600*, 123 photos were taken by *Canon EOS 7D*, and 113 photos were *Nikon D7100*. These cameras are all digital single-lens reflex camera (DSLR), which means the users are likely professional users. As Fig. 3, among those photos taken by phones, iPhone was the most popular device.

Table 1 Descriptive analysis table for photo collection in the Grand Canyon area

Total photo collected in 2015 (2014/12/01– 2015/11/30)	38127		Total photo collected with coordinates (geo-tagged) in 2015	25395	(66%)	
	Count by different days		Count by devices		Count by seasons	
	Weekend	7471 (29.4%)	Non-specific	23660 (93%)	Winter	2625 (10.3%)
	Weekday	17924 (70.6%)	Cameras	1377 (5.4%)	Spring	10056 (39.6%)
			Smart phones	358 (1.6%)	Summer	7105 (28.0%)
					Autumn	5609 (22.1%)
Total		25395		25395		25395

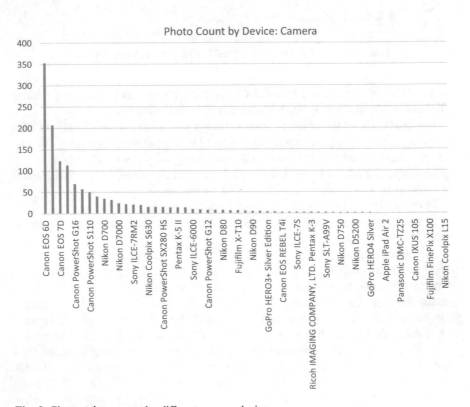

Fig. 2 Photos taken counts by different camera devices

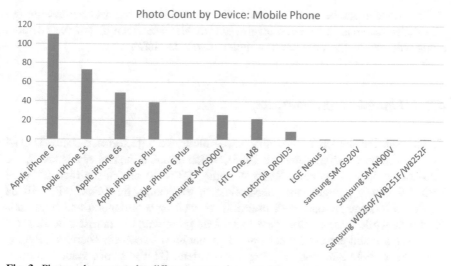

Fig. 3 Photos taken counts by different smart phone models

110 photos were taken by iPhone 6 and 73 photos were taken by iPhone 5s. 49 photos were taken by iPhone 6s and 39 photos were iPhone 6s Plus. 26 photos were taken by iPhone 6 plus.

4 Research Method

4.1 Kernel Density Estimation Mapping

To analyze the statistical outcome and identify hotspots of tourist behaviors, Kernel Density Estimation (KDE) mapping has been implemented in this research. KDE mapping is able to identify the hotspots visually from large datasets (Okabe et al. 2009; Tsou et al. 2013a). Using KDE has been widely used to create raster files to explore hotspots in social media research. Han et al. (2015) used KDE to identify hotspots using Twitter data and by exploring Twitter activity. Han also established the differential maps to compare the changes in activity by using the raster-based "map algebra tool" developed by ESRI after KDE hotspot maps were created. Following the method used by Tsou et al. (2013b), the below formula was applied to the raster formatted maps for all case studies.

Differential Map = (Each Cell Value of Map A/Maximum Cell Value of Map A) − (Each Cell Value of Map B/Maximum Cell Value of Map B)

One important variable in the KDE method is the kernel radius. Adopting different sizes of kernel radius will generate different scale of hotspot analysis. This study utilized two spatial scales of KDE for tourist activity analysis. The first level is 50 km which can be used to identify the general (large regions) hotspots in the Grand Canyon area. The second level is 200 m which can identify smaller hotspots along with roads and trails (with a higher spatial resolution).

4.2 Dynamic Time Warping

To compare the similarities and difference among the trajectories, distance is used as the common variable (Tan et al. 2005), which means if the distance is lower, the similarity is higher. Dynamic Time Warping (DTW) is one of the methods used with distance to compare the trajectories. In this study, the Flickr API will be employed to generate data with user IDs, as well as coordination and time data. However, while these photos were located as point data the tourist trajectory for each user will still present a crucial problem that must be solved. Therefore, Python was used to write a module to solve this problem. DTW distance value is a comparison value between two users. In this research, the DTW distance value of two users was calculated by Python. When the distance is lower, the similarity is higher.

5 Major Findings

Two types of tourist activity analysis were conducted by using the 2015 geo-tagged Flickr data in the Grand Canyon area: spatiotemporal hotspot analysis (with two case studies), and tourist trajectory analysis. Figure 4 illustrated the spatial distribution of geotagged photos (top) in the Grand Canyon area and the activity hotspot map (bottom) using 50 km radius KDE method. The hotspot map illustrated two popular tourist locations within the study area: Grand Canyon Village area (Visitor Center) and Emerald Pools. Therefore, we selected the two sub-regions as our case studies.

5.1 Spatiotemporal Hotspot Analysis

The first case study area is the Grand Canyon (GC) Village (Visitor Center) area, which is not far from the south entrance of the park. GC Village would usually be the first stop for visitors. It provides all kinds of facilities like hotels, visitor center, restaurants, and gift shops (Fig. 5). Figure 5 illustrated some hotspots of geo-tagged Flickr photos using 200 m radius KDE method. The hotspots (red color) are near Grant Canyon Village, Hopi House and some scenery locations.

The second case study area, Emerald Pools (Fig. 6), in Zion was in the heart of Zion Canyon, near Zion Lodge. It has a variety of accommodations and a dining

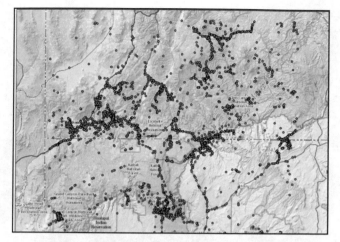

Activety Hotspots in the Grand Canyon Area

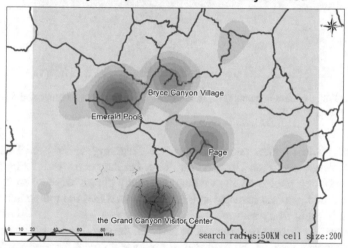

Fig. 4 The Flickr geotagged photos (top) and the activity hotspot map (bottom) using 50 km radius KDE method

room for visitors. To the west of Zion Lodge locates Lady Mountain, which is one of Zion's landmarks.

To explore how these hotspots changed pattern, differential map has applied on the two case studies. There are three major colors used throughout the two differential maps: (1) blue, (2) green, and (3) red. The blue areas show decreased photo activity density, the green areas show constant density, and the red areas show increased photo activity density.

In the Grand Canyon visitor center area, the photos taken in summer are dispersed off the trails and broader regions comparing to winter (Fig. 7). In case study 2 (Fig. 8), Emerald Pools, the photos taken in summer also shows similar patterns

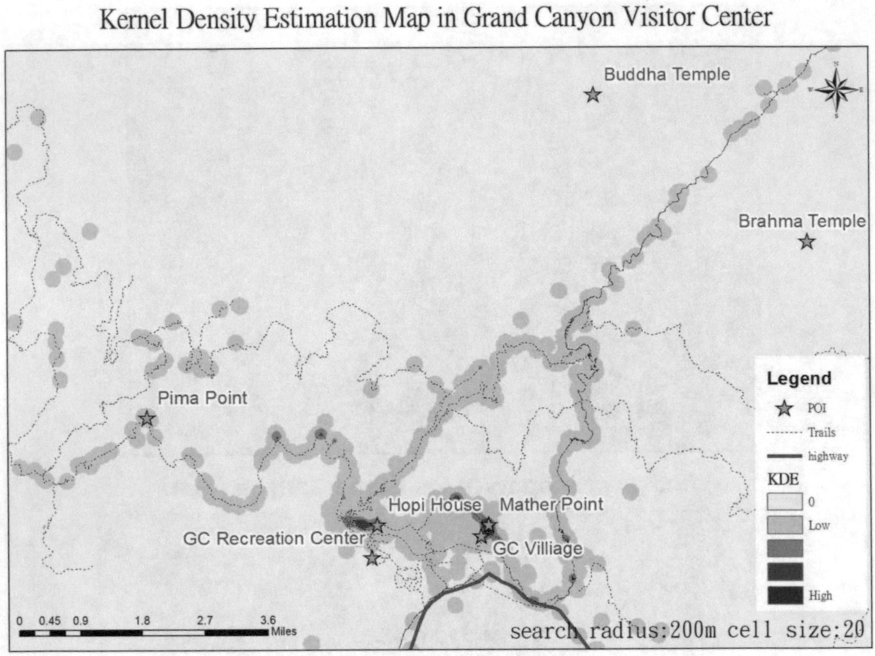

Kernel Density Estimation Map in Grand Canyon Visitor Center

Fig. 5 The Flickr photos hotspots (red) in the Grand Canyon Visitor Center and Village Areas using 200 m radius KDE

(more trails). Most photos taken in winter are clustering within Zion Lodge, Echo Canyon Passage, and Zion Observation Point. On the northwest of Fig. 8 map, it revealed only two photos taken in winter, but many photos were taken in summer. For the seasonal patterns revealed in Fig. 8, it showed that the tourist activity areas have been influenced by season.

Not only the seasonal patterns can be revealed through Flickr data, weekday and weekend patterns also can be explored. For exploring weekday and weekend patterns in the Grand Canyon area, Fig. 9 used differential mapping to identify the differences between weekdays and weekends. Photos taken in weekdays are more disperse in the park, such as the north of Pima Point comparing photos taken in weekends. Not only the spatial difference of their disperse but also the average travel times spent in the Grand Canyon area are different between weekdays and weekends (Table 2). The weekday travel duration is 68 h, and the weekend travel duration is only 45 h. The other case study also showed similar patterns (Fig. 10).

In this research, the Exif information of each photo had been collected, per the statistics, most of the camera users use DSLR to take the photos, and most of the phone users used iPhone. In different devices, such as camera and phone, tourist behavior may be different. Although the three case study areas are not in the same area, the similar spatio-temporal patterns still can be identified. The camera users can travel apart from main tourist area, even choose the unpaved trail. The smart

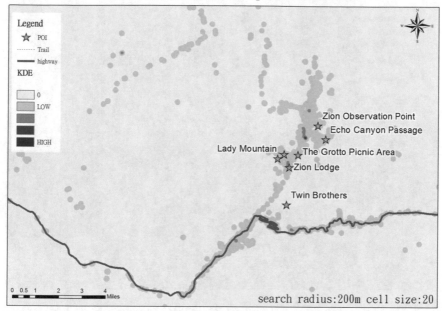

Fig. 6 The Flickr photos hotspots in the Emerald Pools Areas

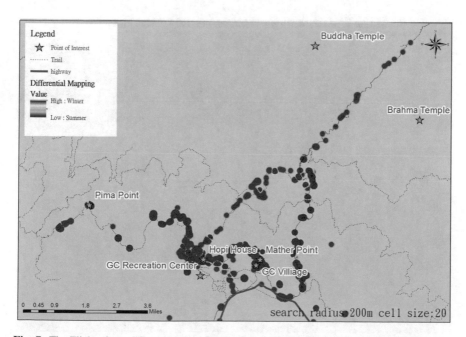

Fig. 7 The Flickr photo differential map for the Grand Canyon Visitor Center case study: using seasonal difference (Blue: Summer, Red: Winter)

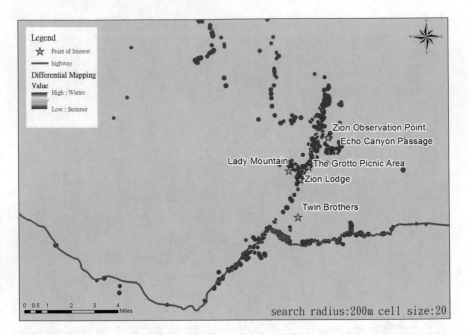

Fig. 8 The Flickr photo differential map for the Emerald pools case study: using seasonal difference (Blue: Summer, Red: Winter)

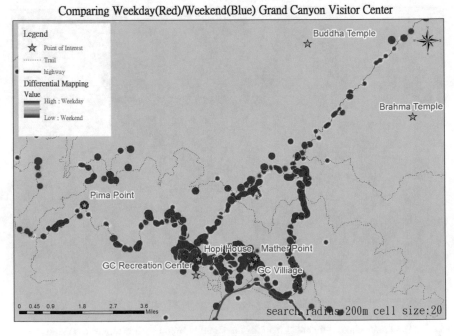

Fig. 9 The Flickr photo differential map for the Grand Canyon Visitor Center case study: Comparing weekday (blue) and weekend (red) difference

Table 2 The average travel times on weekdays and weekends

	Average travel duration (h)
Weekday travels	68
Weekend travels	45
All	63

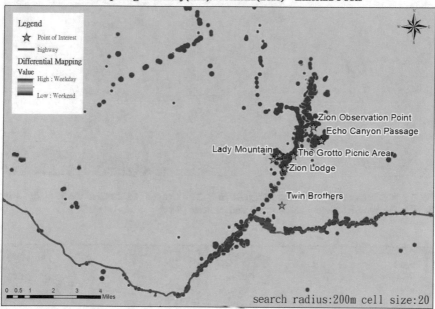

Fig. 10 The Flickr photo differential map for the Emerald Pools case study: Comparing weekday (blue) and weekend (red) difference

phone users usually travel in the recommended locations suggests by National Park Service. In the Grand Canyon Village, the camera users' footprint could be found at the "unpaved trail" in the west of Pima Point, which is different with smart phone users (Fig. 11). In Emerald Pools, it can be found the photos taken by phone are more likely cluster near Zion observation point and The Grotto Picnic area, which is the tourist destination suggested by national park service (Fig. 12).

5.2 Travel Time and Dynamic Time Warping for Trajectory Analysis

Geo-tagged photographs on Flickr platform showed many photos taken by tourists or local residents. The criterion for determining whether the photographs were

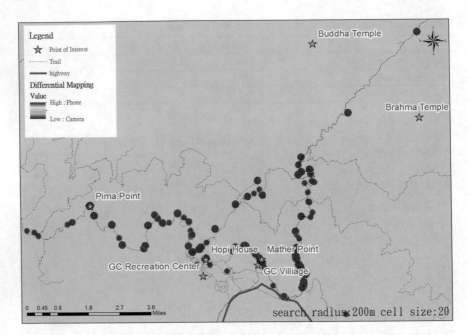

Fig. 11 The Flickr photo differential map for the Grand Canyon Visitor Center case study: using the different devices: smart phones (Red) and cameras (Blue)

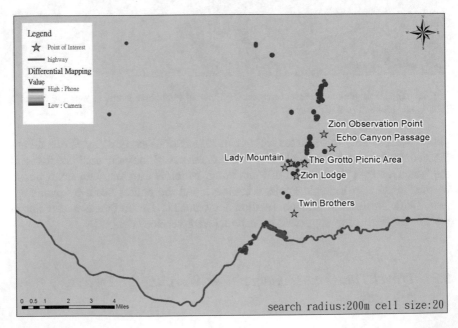

Fig. 12 The Flickr photo differential map for the Emerald pools case study: using the different device: phone (red) and camera (blue)

Table 3 The average travel time for visitor in different season

Total Flickr user visitors (estimated)	735	Total visitor average travel duration	63.53 h
winter	108	Average travel duration	49.03 h
spring	272	Average travel duration	68.81 h
summer	221	Average travel duration	59.00 h
autumn	134	Average travel duration	70.56 h

taken by visitors or local residents in this study is the time period during which each user had taken pictures: if this period exceeded one month, then the photographs were classified as taken by residents; if the period was less than one month (720 h), then the users were classified as tourists. Table 3 illustrated the estimated temporal patterns of visitors in Grant Canyon area and their average travel duration.

Selected Camera Users

To analyze visitors' trajectory patterns, we selected the top three photo-uploaded users who took the highest numbers of photos by cameras as our case studies. Although these top users can not represent the average visitor's movement patterns in the Grant Canyon area, we used these cases to demonstrate the feasibility of DTW for trajectory analysis. To protect users' privacy, we used "Camera user A", "Camera user B", and "Camera user C" to label these top users. "Camera User A" joined Flickr in April 2013, and the user indicated that "*Not a regular user, just wanted to share my photos to give back to the community for some of the great photos I've seen here.*" "Camera User B" joined Flickr in September 2009. The user B uploaded 11714 photos on Flickr platform, and 10700 photos are geo-tagged. "Camera user C" is similar to user B, he joined Flickr in August 2007, and the user has 7600 geo-tagged photos of total 9177 photos uploaded on Flickr platform (Fig. 13). To compare with the similarity of each user's trajectory, we used DTW analysis to measure their similarity in the distance. The DTW value of Camera users A, B, and C are on Table 4, per their DTW distance, camera user A is 194, which is higher than average. For camera user B and C, they have similar DTW distance value and similar routing per their trajectories.

Selected Smart Phone Users

For the smart phone user group, we selected the top three photo-uploaded users who took the highest numbers of photos by Smart Phones. Users who used the phone to take the photos will be defined as "leisure tourist." To protect their privacy, we used "Phone user D", "Phone user E," and "Phone user F" to label these users. "Phone user D" joined Flickr in September 2013. The user D uploaded 241 geo-tagged photos of total 356 photos. "Phone user E" joined Flickr in May 2008, the user uploaded 539 geo-tagged photos of total 4059 photos. User E identified himself/herself as a Montana resident, like hiking, backpacking, car camping, road and trail running, cross-country skiing, snowshoeing and travel.

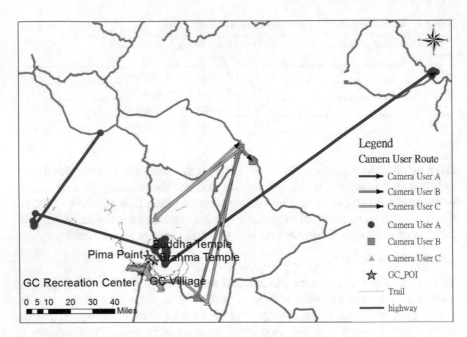

Fig. 13 Selected three camera users' routes (A, B, and C)

Table 4 Camera users' DTW distance value (User B and C are more similar)

	DTW distance average value (compared to other 734 routes)	Camera User A route	Camera User B route	Camera User C route
Camera User A route	194 (less representative)	–	160	99
Camera User B route	121	160	–	56
Camera User C route	77 (more representative)	99	56	–

"Phone user F" joined Flickr in April 2010. User-F had uploaded 901 geo-tagged photos, and 768 of them are public. To compare with the similarity of each user's trajectory, the DTW distance value of phone users D, E and F had been shown in Table 5. Phone users are more likely travel within the area suggest by National Park Service, and their travel destination is more similar. Such as phone user D and F, their DTW value is only 13, and their trajectory has overlay within Emerald pools area (Fig. 14).

Table 5 Selected phone users DTW value (User D route and F route are very similar)

	DTW distance average value (compared to other 734 routes)	Phone User D route	Phone User E route	Phone User F route
Phone User D route	61	–	69	13 (very similar routes)
Phone User E route	60	69	–	199
Phone User F route	41	13 (very similar routes)	199	–

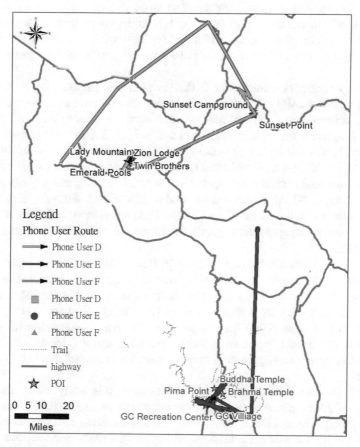

Fig. 14 Selected three smart phone users' routes (D, E, F)

   This study only selected top three contributors of camera and smart phone users
to demonstrate the feasibility of using DTW to compare the similarity of visitors'
trajectories in the Grand Canyon area. By calculating the DTW distance value, we
can find out which user may have more similar trajectory movements comparing to
all other users. The strength of DTW analysis is to provide a quantatative value to
compare the similarity of visitors' trajectory movement. The weakness is the
missing of spatial factors and location-based analysis in the DTW analysis.

6   Conclusion

For tourism study, social media data is a new world to explore. In the past, data
collection was expensive, monopolized. With social media data, researchers could
collect high-resolution spatiotemporal data from public social media APIs and
analyze tourist activities and behaviors. In this study, we collected and cleaned the
geo-tagged Flickr photo data, and then applied two spatio-temporal analysis
methods (KDE and DTW) to explore the tourists' spatial and temporal activity
patterns.

   The major scientific contribution in this research is to demonstrate the feasibility
of using kernel density estimate (KDE) mapping for tourism hotspot analysis and
dynamic time warping (DTW) methods for visitor's trajectory analysis. Previous
tourism geography research works mentioned in Sect. 2 (literature review) did not
utilize any kernel density methods nor DTW methods. This research also illustrated
that adopting different sizes of kernel radius will generate different scale of hotspot
analysis. This study utilized two spatial scales of KDE for tourist activity analysis.
The first level is 50 km which can be used to identify the general (large regions)
tourism hotspots in the Grand Canyon area. The second level is 200 m which can
identify smaller hotspots along with roads and trails (with a higher spatial
resolution).

   This research identified unique activity patterns between different types of users
on Flickr: camera users are exploring remote areas beyond traditional tourist
attractions. Smart phone users are more likely clustered within the lodge area and
viewpoints suggested by the tour guides. For temporal pattern analysis, this
research identified weekday tourists are more "activate" comparing to weekend
visitors. In the Grand Canyon areas, Flickr photo data can also identify the seasonal
pattern: the winter photo amount is the lowest and the increased trend for spring and
summer.

   There are several limitations and challenges in this study. First, the demographics of Flickr photo users might be biased comparing to the general visitor
profiles in the Grand Canyon area. User privacy concerns and restriction are an
important issue for using Flickr data. Although this study only collected public
accessible Flickr photos, the detailed trajectory analysis might reveal some personal
information regarding specific users. Finally, the user's travel trajectories may not

reflect the most reality tourists' trajectories since we only collect the top 3 most active users.

Two future research directions could be explored in Flickr-based tourism research: computer image processing and text analysis. Computer image process technology using machine learning tools and deep learning methods could be used to identify the content of photos in the Grand Canyon area to explore the activity type in each photo. Text analysis, such as topic modeling or latent dirichlet allocation (LDA) methods, can be used to aggregate the texts and tags associated with each photos and provide additional information for various analysis, such as emotional analysis, social network analysis, and user profile analysis.

Acknowledgements This material is based upon work supported by the National Science Foundation, under Grant No. 1416509 and Grant No. 163464. Any opinions, findings, and conclusions or recommendations expressed in this material are those of the author and do not necessarily reflect the views of the National Science Foundation.

References

Ashley C, De Brine P, Lehr A, Wilde H (2007) The role of the tourism sector in expanding economic opportunity. John F. Kennedy School of Government, Harvard University, Cambridge

Birenboim A (2016) New approaches to the study of tourist experiences in time and space. Tour Geogr 18(1):9–17

Chen CF, Chen PC (2012) Research note: exploring tourists' stated preferences for heritage tourism services—the case of Tainan city, Taiwan. Tour Econ 18(2):457–464

Chen X, Kwan MP (2012) Choice set formation with multiple flexible activities under space–time constraints. Int J Geogr Inf Sci 26(5):941–961

Cranshaw J, Schwartz R, Hong JI, Sadeh N (2012) The livehoods project: utilizing social media to understand the dynamics of a city

Cullen IG (1972) Space, time and the disruption of behaviour in cities. Environ Plann A 4(4):459–470

Gao H et al (2013) Exploring temporal effects for location recommendation on location-based social networks. In: Proceedings of the 7th ACM conference on recommender systems

García-Palomares JC, Gutiérrez J, Mínguez C (2015) Identification of tourist hot spots based on social networks: a comparative analysis of European metropolises using photo-sharing services and GIS. Appl Geogr 63:408–417

Girardin F, Calabrese F, Dal Fiore F, Ratti C, Blat J (2008a) Digital footprinting: uncovering tourists with user-generated content. IEEE Pervasive Comput 7(4)

Girardin F et al (2008b) Digital footprinting: uncovering tourists with user-generated content. IEEE Pervasive Comput 7(4):36–43

Han SY, Tsou MH, Clarke KC (2015) Do global cities enable global views? Using Twitter to quantify the level of geographical awareness of U.S. cities. PLoS ONE 10(7):e0132464. https://doi.org/10.1371/journal.pone.0132464

Hawelka B, Sitko I, Beinat E, Sobolevsky S, Kazakopoulos P, Ratti C (2014) Geo-located Twitter as proxy for global mobility patterns. Cartogr Geogr Inf Sci 41(3):260–271

Issa E, Tsou MH, Nara A, Spitzberg B (2017) Understanding the spatio-temporal characteristics of Twitter data with geotagged and nongeotagged content: two case studies with the topic of flu and Ted (movie). Ann GIS 23:219–235

Kádár B (2014) Measuring tourist activities in cities using geotagged photography. Tour Geogr 16 (1):88–104

Kennedy LS, Naaman M (2008) Generating diverse and representative image search results for landmarks. In: Proceedings of the 17th international conference on world wide web. ACM, pp 297–306

Kisilevich S, Mansmann F, Keim D (2010) P-DBSCAN: a density based clustering algorithm for exploration and analysis of attractive areas using collections of geo-tagged photos. In: Proceedings of the 1st international conference and exhibition on computing for geospatial research and application. ACM, p 38

Majid A et al (2013) A context-aware personalized travel recommendation system based on geotagged social media data mining. Int J Geogr Inf Sci 27(4):662–684

McKercher B, Shoval N, Ng E, Birenboim A (2012) First and repeat visitor behaviour: GPS tracking and GIS analysis in Hong Kong. Tour Geogr 14(1):147–161

Okabe A, Satoh T, Sugihara K (2009) A kernel density estimation method for networks, its computational method and a GIS-based tool. Int J Geogr Inf Sci 23(1):7–32

Önder I, Koerbitz W, Hubmann-Haidvogel A (2016) Tracing tourists by their digital footprints: the case of Austria. J Travel Res 55(5):566–573

Palm R, Pred AR (1974) A time-geographic perspective on problems of inequality for women (No. 236). Institute of Urban & Regional Development, University of California

Popescu A, Grefenstette G (2011) Mining social media to create personalized recommendations for tourist visits. In: Proceedings of the 2nd international conference on computing for geospatial research and applications

Sauer CO (1974) The fourth dimension of geography. Ann Assoc Am Geogr 64(2):189–192

Sun YY, Budruk M (2015) The moderating effect of nationality on crowding perception, its antecedents, and coping behaviours: a study of an urban heritage site in Taiwan. Curr Issues Tour 1–19

Sun Y, Fan H (2014) Event identification from georeferenced images. In: Connecting a digital Europe through location and place. Springer International Publishing, pp 73–88

Taaffe EJ (1974) The spatial view in context. Ann Assoc Am Geogr 64(1):1–16

Tan PN, Steinbach M, Kumar V (2005). Introduction to data mining, 1st edn

Tsou MH (2015) Research challenges and opportunities in mapping social media and big data. Cartogr Geogr Inf Sci 42(sup1):70–74

Tsou MH, Kim IH, Wandersee S, Lusher D, An L, Spitzberg B, Gupta D, Gawron JM, Smith J, Yang JA, Han SY (2013a) Mapping ideas from cyberspace to realspace: visualizing the spatial context of keywords from web page search results. Int J Digit Earth 7:4. https://doi.org/10.1080/17538947.2013.781240

Tsou MH, Yang JA, Lusher D, Han SY, Spitzberg B, Gawron JM, Gupta D, An L (2013b) Mapping social activities and concepts with social media (Twitter) and web search engines (Yahoo and Bing): a case study in 2012 US Presidential Election. Cartogr Geogr Inf Sci 40 (4):337–348. https://doi.org/10.1080/15230406.2013.799738

Vu HQ, Li G, Law R, Ye BH (2015) Exploring the travel behaviors of inbound tourists to Hong Kong using geotagged photos. Tour Manag 46:222–232

Yuan M, Nara A (2015) Space-time analytics of tracks for the understanding of patterns of life. In: Space-time integration in geography and GIScience. Springer Netherlands, pp 373–398

Printed in the United States
By Bookmasters